Antenna Engineering and Radio Wave Propagation with MATLAB®

For a listing of recent titles in the
Artech House Antennas and Propagation Library,
turn to the back of this book.

Antenna Engineering and Radio Wave Propagation with MATLAB®

Osama W. Ata

ARTECH HOUSE
BOSTON | LONDON
artechhouse.com

Library of Congress Cataloging-in-Publication Data
A catalog record for this book is available from the U.S. Library of Congress.

British Library Cataloguing in Publication Data
A catalog record for this book is available from the British Library.

ISBN 13: 978-1-68569-099-1

Cover design by Joi Garron

685 Canton St.
Norwood, MA

10 9 8 7 6 5 4 3 2 1

To my late father and mother,
Wadie and Asma

To my late wife, Sirin
who are celebrating in Heaven

To my wife, Salpi, my son Ramsey, and my daughter, Razan

To all youths studying communications engineering and researching antennas

I dedicate this work
with rejoicing and eternal love

Contents

Preface

> "The basic laws of the universe are simple, but because our senses are limited, we can't grasp them. There is a pattern in creation. If we look at this tree outside, whose roots search beneath the pavement for water, or a flower which sends its sweet smell to the pollinating bees, or even our own selves and the inner forces that drive us to act, we can see that we all dance to a mysterious tune and the piper who plays this melody from an inscrutable distance—whatever name we give this—Creative Force, or God—escapes all book knowledge ..."
>
> —*Albert Einstein, philosopher-scientist*

Antenna Engineering and Radio Wave Propagation with MATLAB® is a comprehensive textbook for graduate and postgraduate students, researchers, and engineers in antenna engineering and radio wave propagation. It bridges fundamental concepts and advanced techniques, covering wire antennas, dipoles, arrays, and specialized horn and aperture antennas, with methods such as ray theory and numerical approaches. Key topics include low sidelobe suppression and phase-corrected apertures, exemplified by hyperbolic secant (HYSEC) horn antennas.

Chapters 1 to 4 describe antenna fundamentals including the short current filament and the half-wave dipole antenna.

Chapters 5 to 8 focus on core antenna designs. Chapter 5 covers linear arrays and radiation from multiple elements. Chapter 6 explores log-periodic dipole arrays (LPDA) and their design procedure. Chapter 7 details rectangular and circular aperture antennas, and Chapter 8 addresses reflector antennas, aperture distribution, and far-field calculations.

Chapters 9 to 12 tackle advanced antenna designs and experimental evaluations. Chapter 9 addresses sidelobe suppression in E-sectoral horn antennas, computational design methods, and radiation pattern calculations. Chapter 10 applies ray theory to phase-corrected horn antennas, focusing on trajectory equations and aperture amplitude distributions. Chapter 11 presents experimental evaluation methods for HYSEC horn far-field and aperture field measurements. Chapter 12 discusses a novel phase-corrected compact antenna design.

Chapter 13 delves into radio wave propagation through the atmosphere, exploring tropospheric refraction, ionospheric propagation, and the effects of the Earth's magnetic field.

Chapters 14 and 15 introduce fifth generation (5G) technology, deployable satellite antennas, reconfigurable intelligent surfaces (RIS), and advanced antenna applications for biomedical uses, including wave propagation in lossy media. The book includes 50 MATLAB codes, offering readers practical tools for hands-on learning and problem-solving in modern wireless communication systems.

Acknowledgments

I deeply acknowledge all engineering and architecture students, graduates, and postgraduates I taught over the years, who indeed inspired me to write this book as a culmination of my taught lectures and research, and project proposals they submitted; some of which were very interesting and simply creative.

I also acknowledge colleagues and friends who supported me during this journey. Special thanks go to Professor T. M. Benson and the late memories of Professors F. Benson and A. Marincic. Finally, I am deeply thankful for the patience and support of my wife and children throughout the long months of writing this work.

CHAPTER 1

Antenna Wave Electromagnetics

1.1 Introduction

For centuries, humans relied on visual signals such as smoke, flags, and beacon fires for long-distance communication. The discovery and utilization of electromagnetic waves revolutionized this process, enabling wireless transmission of information. Antennas play a crucial role in this advancement by converting electrical signals into electromagnetic waves for radiation and reception. They act as a bridge between guided wave transmission lines and free-space propagation, facilitating communication without physical connections. The electromagnetic spectrum used for such transmission spans from 3 kHz to 300 GHz, covering various applications from low-frequency broadcasting to high-frequency satellite communications.

The theoretical foundation of antennas originates from Maxwell's equations, formulated in the nineteenth century, which describe the relationship between electric and magnetic fields and predict the existence of electromagnetic waves. In 1887, Heinrich Hertz experimentally validated Maxwell's predictions by generating and detecting radio waves, demonstrating that electromagnetic energy could propagate through free space. This laid the groundwork for practical wireless communication. Guglielmo Marconi, an Italian inventor, further expanded on Hertz's work by pioneering long-distance radio transmission, including the first transatlantic wireless signal in 1901. These early developments led to the rapid evolution of radio technology, shaping modern wireless communication.

Antennas are now indispensable in various applications, including broadcasting, radar, satellite communication, remote sensing, and mobile networks. Their design significantly influences system performance, with key parameters such as gain, directivity, radiation pattern, polarization, and impedance matching playing vital roles. While transmission lines remain practical for short distances and lower frequencies, high-frequency applications demand antennas due to increased losses in guided media. The advent of fiber optics has provided an alternative for certain applications, but antennas remain the primary solution for wireless systems, particularly in mobile, aerospace, and military applications.

A significant advancement in antenna technology is the development of microstrip antennas, which offer advantages such as low profile, lightweight construction, and ease of integration with electronic circuits. These antennas have become widely used in modern wireless communication systems, including satellite navigation, radio frequency identification (RFID), and fifth generation (5G) networks. However, challenges such as bandwidth limitations, efficiency, and miniaturization require innovative design techniques. Researchers continue to explore advanced feeding mechanisms, polarization diversity, and novel materials to enhance microstrip antenna performance.

With increasing complexity in antenna design, computational electromagnetic tools have become essential. CST Microwave Studio (CST-MWS) is a leading software used for electromagnetic simulation, employing numerical methods such as the Finite Integration Technique (FIT) to analyze and optimize antenna structures. This tool enables engineers to evaluate radiation patterns, impedance characteristics, and electromagnetic interactions in complex antenna systems, significantly reducing development time and cost.

As wireless technology continues to evolve, the demand for high-performance antennas capable of operating across diverse environments will grow. This book provides a comprehensive study of antenna engineering, from fundamental principles to advanced design techniques. It is designed to serve students, researchers, and practicing engineers, equipping them with the knowledge and tools necessary to develop innovative antenna solutions for contemporary and future wireless communication systems.

1.2 Maxwell's Equations

The work of James Clerk Maxwell in 1873 that summarized the state of electromagnetic science at the time led to the discovery of electromagnetic wave propagation by Hertz and Marconi. Maxwell's work was based on a large body of empirical and theoretical knowledge developed by Gauss, Ampere, Faraday, and others.

$$\nabla \times \boldsymbol{E} = -\frac{\partial \boldsymbol{B}}{\partial t} \tag{1.1}$$

$$\nabla \times \boldsymbol{H} = \boldsymbol{J} + \frac{\partial \boldsymbol{D}}{\partial t} \tag{1.2}$$

$$\nabla \cdot \boldsymbol{D} = \rho \tag{1.3}$$

$$\nabla \cdot \boldsymbol{B} = 0 \tag{1.4}$$

where $\boldsymbol{E}$ is the electric field strength (V/m), $\boldsymbol{B}$ is the magnetic flux density (Wb/m^2 or Tesla), $\boldsymbol{H}$ is the magnetic field strength (A/m), $\boldsymbol{J}$ is the current density (A/m^2), $\boldsymbol{D}$ is the electric flux density (C/m^2), and ρ is the volume charge density (C/m^3).

Equation (1.1) represents the differential form of Faraday's law, (1.2) represents that of Ampere's law, (1.3) represents that of Gauss's law, and (1.4) demonstrates that the magnetic field is formed in continuous closed loops.

Most notably, if the divergence of both sides of (1.2) is applied, and (1.3) is substituted, then:

$$\nabla \cdot \nabla \times \boldsymbol{H} = \frac{\partial \rho}{\partial t} + \nabla \cdot \boldsymbol{J} = 0 \tag{1.5}$$

Equation (1.5) demonstrates that the outflow of the current field is equal to the reduction in the time rate of the volume charge density or that the charge, as an energy source, is conserved. This would not have been realized without Maxwell's addition of $\partial D/\partial t$ to the originally electrostatic form of (1.2).

One can retrieve the integral form of (1.3) by applying the divergence theorem that the closed surface integral of a vector equals the volume integral of the divergence of that vector:

$$\int \nabla \cdot D\,dv = \oint D \cdot ds = \int \rho\,dv = Q \tag{1.6}$$

where Q is the total charge included inside the closed surface.

Applying Stokes's theorem to (1.1), which demonstrates that the closed line integral to a vector equals the surface integral of the curl of that vector, would lead to:

$$\int \nabla \times E \cdot ds = \oint E \cdot dl = -\frac{\partial}{\partial t}\int B \cdot ds = -\frac{\partial \Phi_B}{\partial t} \tag{1.7}$$

where Φ_B is the magnetic flux.

Similarly, applying Stokes's theorem to (1.2) yields:

$$\int \nabla \times H \cdot ds = \oint H \cdot dl = \int J \cdot ds + \frac{\partial}{\partial t}\int D \cdot ds = I + \frac{\partial \Phi_E}{\partial t} \tag{1.8}$$

where I is the total current and Φ_E is the electric flux.

With the fields being harmonic time-dependent, the electric field m can be expressed as:

$$E(x,y,z,t) = E(x,y,z)\cos(\omega t + \phi) \tag{1.9}$$

In the phasor term, the electric field can be conveniently expressed as:

$$E(x,y,z) = E(x,y,z)e^{j\phi} \tag{1.10}$$

Assuming cosine-based phasors so that conversion from phasor quantities to real-time quantities is accomplished:

$$E(x,y,z,t) = \mathrm{Re}\left[E(x,y,z)e^{j\omega t}\right] \tag{1.11}$$

Hence, assuming $e^{j\omega t}$ time dependence, Maxwell's curl equations become:

$$\nabla \times E = -j\omega B \tag{1.12}$$

$$\nabla \times H = J + j\omega D \tag{1.13}$$

and in a source-free environment, distant from a radiating source, $J = 0$ and $\rho = 0$.

1.3 Boundary Conditions

It is often necessary to find solutions to Maxwell's equations in nonhomogenous regions, in which there are boundary media with different constitutive parameters.

Maxwell's equations in a differential form require known boundary values for a complete and unique solution. A solution of Maxwell's equations will be found, after resolving boundary conditions.

Consider the closed surface, S, in Figure 1.1 where (1.6) holds. The boundary is the surface, cutting through the closed surface cylinder, which is associated with a current charge density ρ_s.

In the limit $h \to 0$, the area of the curved (side) surface would be negligible for any field flux to flow, leaving the normal components, D_{n1} and D_{n2}, to flow through the bottom and top surfaces of area ΔS. Hence, (1.6) would reduce to:

$$\Delta S D_{n2} - \Delta S D_{n1} = \Delta S \rho_s \tag{1.14}$$

which, in vector form, may be written as:

$$\hat{\boldsymbol{n}} \cdot (\boldsymbol{D}_2 - \boldsymbol{D}_1) = \rho_s \tag{1.15}$$

where $\hat{\boldsymbol{n}}$ is a unit vector, normal to the boundary.

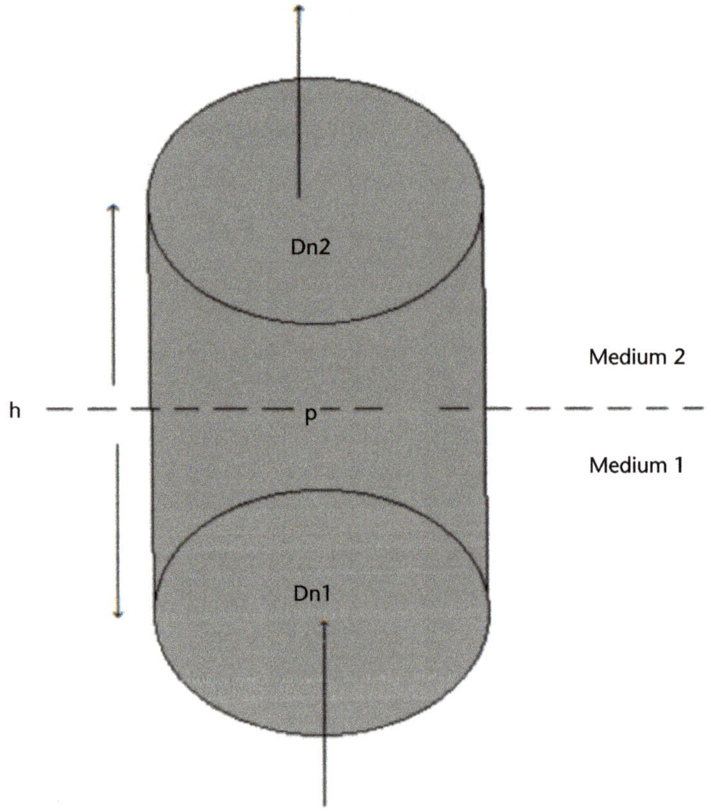

Figure 1.1 A closed surface with a penetrating charged boundary that separates two different media in electromagnetic properties.

A similar argument for the magnetic flux leads to:

$$\hat{n} \cdot (B_2 - B_1) = 0 \tag{1.16}$$

which demonstrates that the net flux of the magnetic flux density through a closed surface is always 0. This is because magnetic charges can never be isolated and hence will form closed magnetic field loops, in a similar way to a magnet with two opposite poles.

Consider the closed contour, C, in Figure 1.2 where (1.7) holds. The boundary is the surface, cutting through the closed rectangular contour of Δl length and negligible width h.

In the limit $h \rightarrow 0$, the area of the rectangular surface would be negligible for any field flux to change, in it, with time. Hence, the right side of (1.7) is 0. E_{t1} and E_{t2} flow along the closed contour. Hence, (1.7) would reduce to:

$$\Delta l E_{t1} - \Delta l E_{t2} = 0 \tag{1.17}$$

which, in vector form, may be written as:

$$\hat{n} \times (E_2 - E_1) = 0 \tag{1.18}$$

The normal electric field components along the negligible width, h, would reduce to 0.

A similar argument for the magnetic field strength leads to:

$$\hat{n} \times (H_2 - H_1) = J_s \tag{1.19}$$

A special case is considered if one of the media is a perfect conductor, where all field components must vanish. The boundary conditions would reduce to:

$$\hat{n} \cdot D = \rho_s \tag{1.20}$$

$$\hat{n} \cdot B = 0 \tag{1.21}$$

$$\hat{n} \times E = 0 \tag{1.22}$$

$$\hat{n} \times H = J_s \tag{1.23}$$

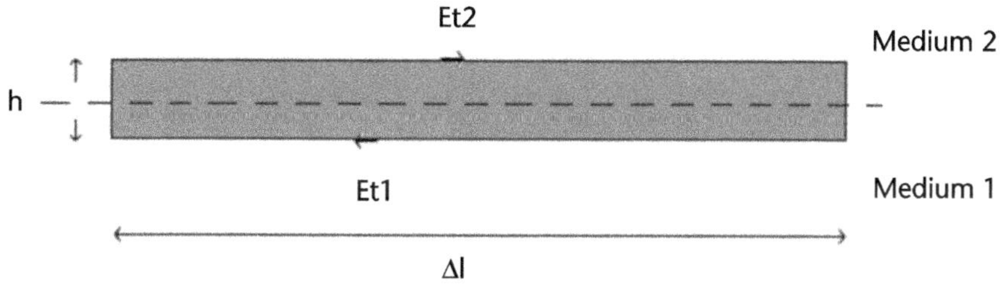

Figure 1.2 A closed rectangular loop with a penetrating boundary that separates different media in electromagnetic properties.

1.4 The Wave Equation

So far, the electric and magnetic fields were shown to be mutually related, in a time-varying situation, by Maxwell's equations and the boundary conditions imposed by the electric and magnetic fields.

In a source-free, linear, isotropic, homogenous region, Maxwell's curl equations, in phasor forms, are:

$$\nabla \times E = -j\omega\mu H \tag{1.24}$$

$$\nabla \times H = j\omega\varepsilon E \tag{1.25}$$

These give rise to the following simultaneous first-order, two-dimensional (2D) partial differential equations:

$$\frac{\partial E_z}{\partial E_y} - \frac{\partial E_y}{\partial E_z} = -j\omega\mu H_x \tag{1.26}$$

$$\frac{\partial E_x}{\partial E_z} - \frac{\partial E_z}{\partial E_x} = -j\omega\mu H_y \tag{1.27}$$

$$\frac{\partial E_y}{\partial E_x} - \frac{\partial E_x}{\partial E_y} = -j\omega\mu H_z \tag{1.28}$$

$$\frac{\partial H_z}{\partial H_y} - \frac{\partial H_y}{\partial H_z} = j\omega\mu E_x \tag{1.29}$$

$$\frac{\partial H_x}{\partial H_z} - \frac{\partial H_z}{\partial H_x} = j\omega\mu E_y \tag{1.30}$$

$$\frac{\partial H_y}{\partial H_x} - \frac{\partial H_x}{\partial H_y} = j\omega\mu E_z \tag{1.31}$$

Taking the curl of (1.24) and (1.25) yields:

$$\nabla \times \nabla \times E = -j\omega\mu\nabla \times H = -j\omega\mu\left(jw\varepsilon E\right) = \omega^2\mu\varepsilon E \tag{1.32}$$

But

$$\nabla \times \nabla \times E = \nabla(\nabla \cdot E) - \nabla^2 E \tag{1.33}$$

and $\nabla \cdot E = 0$, in a source-free region.

Hence,

$$\nabla^2 E + \omega^2\mu\varepsilon E = 0 \tag{1.34}$$

Similarly,

$$\nabla^2 H + \omega^2 \mu \varepsilon H = 0 \tag{1.35}$$

where $k = \omega\sqrt{\mu\varepsilon}$ is defined as the wavenumber or propagation constant of the medium.

1.5 Plane Waves in a Lossless Medium

In a lossless medium, μ and ε are real numbers, so k is real. A basic plane wave solution can be found by considering an electric field with only an x-component, which is uniform in the x and y directions (i.e., no field variations in the x and y directions).

It follows that $\partial/\partial x = \partial/\partial y = 0$ and (1.34) reduces to:

$$\frac{\partial^2 E_x}{\partial z^2} + k^2 E_x = 0 \tag{1.36}$$

The general solution of (1.36) follows:

$$E_x(z) = E^+ e^{-jkz} + E^- e^{jkz} \tag{1.37}$$

In the time domain, the above solution, at frequency ω, can be rewritten as:

$$E_x(z,t) = E^+ \cos(\omega t - kz) + E^- \cos(\omega t + kz) \tag{1.38}$$

The first term of the right side of the equation is a wave traveling in the $+z$ direction, whereas the second term is a wave traveling in the $-z$ direction.

Choose a forward-traveling wave, in the $+z$ direction, and maintain a fixed point:

$$\omega t - kz = C\text{, where C is a constant} \tag{1.39}$$

Differentiating yields:

$$\frac{dz}{dx} = \frac{\omega}{k} = v_p \tag{1.40}$$

where in free space, $v_p = \sqrt{1/\mu_0\varepsilon_0} = c = 3 \times 10^8$ m/s.

A complete specification of the electromagnetic field must include the magnetic field. From (1.27), $\partial E_z/\partial x = 0$ and hence:

$$\frac{\partial E_x}{\partial z} = -j\omega\mu H_y \tag{1.41}$$

Substituting (1.37) into (1.41):

$$-jk\left(E^+ e^{-jkz} - E^- e^{jkz}\right) = -j\omega\mu H_y \tag{1.42}$$

and

$$H_y = \frac{1}{\eta}\left(E^{+}e^{-jkz} - E^{-}e^{jkz}\right) \tag{1.43}$$

where $\eta = \omega\mu/k = \sqrt{\mu/\varepsilon}$ is the wave impedance of the plane wave. In free space, $\sqrt{\mu_0/\varepsilon_0} = 377\Omega$.

Note, as shown in Figure 1.3, that E and H are orthogonal to each other and both vectors are orthogonal to the direction of propagation.

1.6 Plane Waves in a General Lossy Medium

Now consider the effect of a lossy medium. If the medium is conductive with a conductivity σ, Maxwell's curl equations can be written as:

$$\nabla \times E = -j\omega B = -j\omega\mu H \tag{1.44}$$

$$\nabla \times H = j\omega\varepsilon E + \mathrm{J} = j\omega\varepsilon E + \sigma E \tag{1.45}$$

$$\nabla \times \nabla \times E = -j\omega\mu\left(j\omega\varepsilon E + \sigma E\right) = \omega^2\mu\varepsilon\left(1 - \mathrm{j}\frac{\sigma}{\omega\varepsilon}\right)\mathrm{E} \tag{1.46}$$

In a free source region, $\nabla \cdot E = 0$ and the left side of (1.46) reduces to $-\nabla^2 E$. Hence, the equation becomes:

$$\nabla^2 \mathrm{E} + \omega^2\mu\varepsilon\left(1 - \mathrm{j}\frac{\sigma}{\omega\varepsilon}\right)\mathrm{E} = 0 \tag{1.47}$$

We can define a complex propagation constant, $\gamma = \alpha + j\beta$, where:

$$\gamma^2 = -\omega^2\mu\varepsilon\left(1 - \mathrm{j}\frac{\sigma}{\omega\varepsilon}\right) \tag{1.48}$$

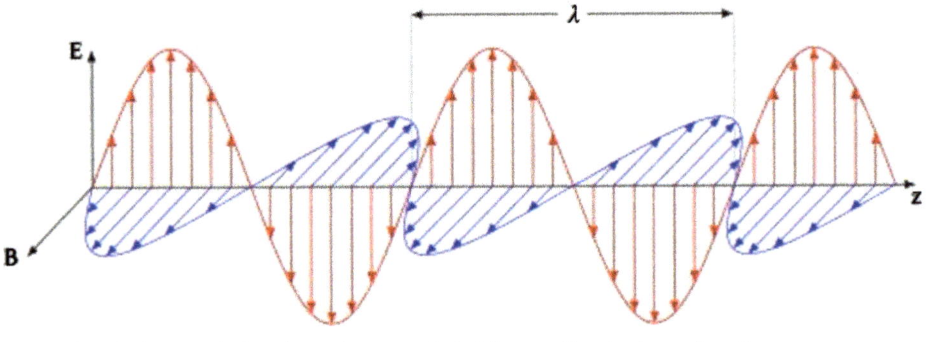

Figure 1.3 Orthogonal electric and magnetic fields traveling in the *z* direction.

and

$$\gamma = j\omega\sqrt{\mu\varepsilon}\sqrt{\left(1 - \mathrm{j}\frac{\sigma}{\omega\varepsilon}\right)} \tag{1.49}$$

Substituting for γ in (1.48), and taking the real and imaginary parts, the following two equations emerge:

$$\alpha^2 - \beta^2 = -\omega^2\mu\varepsilon \tag{1.50}$$

$$2\alpha\beta = \omega\mu\varepsilon \tag{1.51}$$

Solving the above two equations simultaneously gives:

$$\alpha^2 = \frac{\omega^2\mu\varepsilon}{2}\left(\sqrt{1 + \frac{\sigma^2}{\omega^2\varepsilon^2}} - 1\right) \tag{1.52}$$

$$\beta^2 = \frac{\omega^2\mu\varepsilon}{2}\left(\sqrt{1 + \frac{\sigma^2}{\omega^2\varepsilon^2}} + 1\right) \tag{1.53}$$

Now (1.47) reduces to:

$$\frac{\partial E_x}{\partial z} - \gamma^2 E_x = 0 \tag{1.54}$$

which has a solution:

$$E_x(z) = E^+e^{-\gamma z} + E^-e^{\gamma z} \tag{1.55}$$

The positive traveling wave has a propagation factor of the form:

$$e^{-\gamma z} = e^{-\alpha z}e^{-j\beta z} \tag{1.56}$$

where $e^{-j\alpha z}$ is a damping factor and $e^{-j\beta z}$ represents a wave traveling in the $+z$ direction.

For a good lossy (conducting) medium where $\sigma >> \omega\varepsilon$, (1.49) can be further reduced to become:

$$\gamma \approx j\omega\sqrt{\mu\varepsilon}\sqrt{\left(\frac{\sigma}{j\omega\varepsilon}\right)} = \sqrt{j}\sqrt{\omega\mu\sigma} = (1 + j)\sqrt{\frac{\omega\mu\sigma}{2}} \tag{1.57}$$

Hence,

$$\alpha = \beta = \sqrt{\frac{\omega\mu\sigma}{2}} \tag{1.58}$$

and the skin depth of the lossy medium, defined as, δ, which is the distance over which the magnitude of the electric or magnetic field is reduced by a factor of $1/e \cong 0.368$, and is expressed as:

$$\delta = \frac{1}{\alpha} = \sqrt{\frac{2}{\omega\mu\sigma}} \text{ m} \tag{1.59}$$

The wave impedance for a lossy medium, η, can be derived by substituting a wave in the $+z$ direction from (1.55) into (1.41) to give:

$$-\gamma E_x = -j\omega\mu H_y \tag{1.60}$$

By substituting for, γ, from (1.57) into (1.61):

$$\eta = \frac{E_x}{H_y} = \frac{j\omega\mu}{\gamma} = (1+j)\sqrt{\frac{\omega\mu}{2\sigma}} \ \Omega \tag{1.61}$$

It is interesting to know that, in a good lossy medium and upon dividing (1.57) by (1.61), one gets:

$$\frac{\gamma}{\eta} = \sigma \text{ S/m} \tag{1.62}$$

Example 1.1 A sand-drying process is proposed by passing a layer of wet sand on a horizontally moving belt through a microwave beam of frequency 10 GHz. The beam propagates vertically and can be regarded as a plane wave. If the wet sand has a conductivity $\sigma = 50$ S/m and a relative permittivity of 2.5, what is an appropriate value for the thickness of the sand layer?

Solution 1.1 The sand can be considered to be a good conductor if $\sigma/\omega\varepsilon >> 1$. Hence,

$$\frac{\sigma}{\omega\varepsilon} = \frac{50}{2\pi \times 10^{10} \times 8.855 \times 10^{-12} \times 2.5} = 35.9$$

and the sand can be regarded as a good conductor. Now, since the fields decrease in the sand due to the attenuation, the sand layer cannot be much thicker than the skin depth, δ, where:

$$\delta = \sqrt{\frac{2}{\omega\mu\sigma}} = \sqrt{\frac{2}{2\pi \times 10^{10} \times 4\pi \times 10^{-7} \times 50}} \text{ m} = 0.71 \text{ mm}$$

Since the sand layer cannot be that thin, the process is not efficient.

Problems for Students

1.1 Maxwell's equations form the foundation for understanding electromagnetic wave propagation.

(a) Write down Maxwell's equations in an integral form and explain each term.
(b) Derive the differential form of Gauss's law for electric fields and interpret the result in the context of a vacuum.
(c) Explain the significance of each of Maxwell's equations in the context of the antenna theory.

1.2 Given the boundary conditions at the interface between two dielectric media, where the first medium has permittivity ε_1 and permeability μ_1 and the second medium has permittivity ε_2 and permeability μ_2:
(a) Derive the boundary conditions for the electric and magnetic fields at the interface.
(b) Discuss the physical meaning of these boundary conditions and how they apply to wave propagation in antennas.
(c) Using the boundary conditions, calculate the reflection and transmission coefficients for an electromagnetic wave incident on a dielectric interface at normal incidence.

1.3 Consider a time-harmonic electromagnetic wave in a nonconductive medium.
(a) Starting from Maxwell's equations, derive the wave equation for the electric field.
(b) Solve the wave equation for a plane wave in a vacuum and express the solution in terms of the wave vector $\vec{k}$ and angular frequency ω.
(c) Discuss how the solution represents the behavior of the electric field in free space.

1.4 Consider a plane electromagnetic wave traveling in a lossless medium with permittivity ε and permeability μ.
(a) Derive the expressions for the electric field $\vec{E}$ and magnetic field $\vec{H}$ in the medium.
(b) Calculate the intrinsic impedance of the medium and discuss its significance for wave propagation in free space.
(c) For a wave traveling in the $+z$ direction, write the time-dependent expressions for $\vec{E}$ and $\vec{H}$.

1.5 In a general lossy medium, the permittivity is complex, $\varepsilon = \varepsilon' - j\varepsilon''$, and similarly for the permeability.
(a) Starting from the wave equation, derive the expression for the propagation constant $\gamma = \alpha + j\beta$ in a lossy medium.
(b) Explain the physical meaning of α and β and how they relate to attenuation and phase shift of the wave.
(c) Given that the conductivity σ is related to the imaginary part of the permittivity, calculate the attenuation constant α for a wave propagating in a conducting medium.

1.6 For a plane wave in free space:
(a) Show the relationship between the electric and magnetic fields (i.e., $\vec{E}$ and $\vec{H}$) for a linearly polarized wave.
(b) Using Maxwell's equations, demonstrate that the directions of $\vec{E}$, $\vec{H}$, and the wave vector $\vec{k}$ are mutually perpendicular.
(c) Calculate the Poynting vector $\vec{S}$ for a plane wave and explain its physical interpretation.

1.7 Consider the superposition of two plane waves with different frequencies and directions of propagation.
   (a) Using the principle of superposition, express the electric field of the resulting wave.
   (b) Determine the resulting interference pattern in both the near field and the far field.
   (c) Explain how the interference phenomenon can be used in antenna design to improve radiation patterns.

1.8 Consider a wave propagating through a lossy dielectric medium with $\varepsilon = \varepsilon_r\varepsilon_0$, $\mu = \mu_0$, and conductivity σ.
   (a) Derive the complex propagation constant γ, and express both the attenuation constant α and phase constant β.
   (b) If the conductivity of the medium increases, how does this affect the propagation constant and wave attenuation?
   (c) Calculate the depth of penetration ($1/\alpha$) for a wave propagating in such a medium at a frequency of 2 GHz, given that the conductivity is 5 S/m and $\varepsilon_r = 2.2$.

Problems for Instructors

1.1 Derivation of Maxwell's Equations from Integral to Differential Form: Derive the differential forms of Maxwell's equations from their integral counterparts, using the divergence and Stokes's theorems. Extend the problem by considering nonstandard boundary conditions, such as those involving complex materials such as metamaterials.

1.2 Boundary Conditions at Interfaces: Consider an interface between two media with contrasting permittivity and permeability (e.g., air and dielectric material). Derive the boundary conditions for both the electric and magnetic fields at the interface. Investigate how the boundary conditions change when the interface is between a lossy medium and a perfect electric conductor (PEC).

1.3 Wave Equation in Complex Media: Derive the wave equation for a medium with both conductivity and permittivity, leading to a complex refractive index. Explore how this wave equation changes when the material exhibits anisotropy.

1.4 Energy Considerations for Plane Waves in Lossless and Lossy Media: Calculate the energy carried by a plane wave in both lossless and lossy media. Derive expressions for power loss due to absorption in the lossy medium, considering different levels of conductivity. Discuss the impact of lossy media on antenna performance.

1.5 Reflection and Transmission at Oblique Incidence: Analyze the reflection and transmission of a plane wave at oblique incidence on a dielectric boundary. Extend the problem to consider the case where the boundary is between a lossless and a lossy medium, and derive the reflection and transmission coefficients for both TE and TM polarizations.

1.6 Wave Propagation in Conductive Media (Skin Depth): Derive the skin depth for electromagnetic wave propagation in a conductive medium. Discuss the significance of skin depth in the design of antennas for different frequency ranges. How does skin depth affect radiation efficiency and material selection for antennas operating in gigahertz frequencies?

1.7 Practical Application of Plane Wave Solutions in Antenna Design: Given the solution to the wave equation for a plane wave, explore how this theoretical knowledge applies to the design of a half-wave dipole antenna. Use the wave equation to predict the near-field and far-field radiation patterns of the antenna.

Selected Bibliography

Balanis, C. A., *Antenna Theory: Analysis and Design*, Second Edition, New York: John Wiley & Sons, 1997.

Benson, F. A., and T. M. Benson, *Fields, Waves, and Transmission Lines*, London: Chapman and Hall, 1991.

Stutzman, W. L., and G. A. Thiele, *Antenna Theory and Design*, New York: John Wiley & Sons, 2012.

Ulaby, F. T., E. Michielssen, and U. Ravaioli, *Fundamentals of Applied Electromagnetics*, Sixth Edition, Boston, MA: Pearson, 2010.

CHAPTER 2

Basic Antenna Parameters

2.1 Introduction

An antenna acts to convert guided waves on a transmission structure into free space waves. Most antennas are reciprocal devices and behave the same on transmit as on receive. Antennas are treated as transmitting or receiving as appropriate for the particular situation. In the receiving mode, antennas act to collect incoming waves and direct them to a common feed point where a transmission line is attached. In some cases, antennas focus radio waves just as lenses focus optical waves. In all cases, antennas have directional characteristics; that is, electromagnetic power density is radiated from a transmitting antenna with an intensity that varies with the angle around the antenna.

Figure 2.1 shows a typical communication system that includes the transmitting and receiving antennas. It also shows the transmitter, the receiver, and the free space region between the antennas.

The transmitting antenna transmits electromagnetic energy, from a single input port, into free space and the receiving antenna collects the impinging electromagnetic energy into its output port. The system gain, which is not the same as the antenna gain, is described as:

$$G_s = \frac{P_{\text{out, receiver}}}{P_{\text{in, transmitter}}} \tag{2.1}$$

System gain is influenced by several key factors:

1. *Antenna properties:* The gain of the transmitting ($G_t(\theta, \phi)$) and receiving ($G_r(\theta, \phi)$), antennas determines their efficiency in converting input power to radiated power and vice versa.

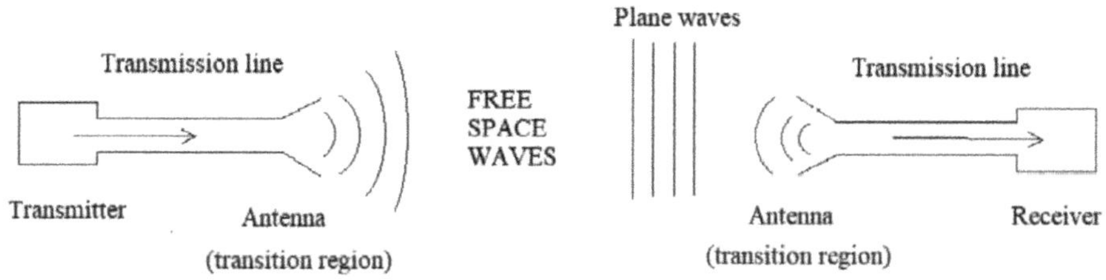

Figure 2.1 A typical antenna-to-antenna communication system.

2. *System configuration:* Proper antenna alignment ensures that the maximum gain direction of each antenna is oriented toward the other. Path loss is given by

$$P_L = \left(\frac{4\pi r}{\lambda}\right)^2 \tag{2.2}$$

3. *Transmission medium:* Lossy media attenuate signal amplitude due to the attenuation constant (α), expressed as $e^{-\alpha z}$. Environmental obstructions (e.g., buildings, trees) cause signal loss and multipath distortion. The curvature of the Earth also impacts the maximum transmission range.
4. *Antenna compatibility:* Polarization mismatches (e.g., linear versus circular, vertical versus horizontal) lead to polarization losses, modeled as:

$$PLF = \cos^2 \varphi \tag{2.3}$$

for linearly polarized antennas with an offset angle ϕ. Additionally, both antennas must operate within the same frequency bandwidth to ensure system compatibility.

2.2 Antenna Parameters

Before embarking on the description of the basic antenna parameters, it is worth mentioning that the function of an antenna can be classified into three main categories:

- To match energy from the transmission line to free space, in the case of a transmitting antenna, or to match incoming electromagnetic energy from free space into the transmission line, in the case of the receiving antenna.
- To direct the electromagnetic energy into a required direction, since an antenna has directional characteristics.
- To perform specialized functions such as radar detection or radar scanning.

The basic antenna parameters are the following:

- Input impedance, Z_A, of an antenna is the ratio of the voltage to the current at the antenna terminals. The usual goal is to match the antenna input impedance to the characteristic impedance of the connecting transmission line. This is to practically reduce any reflection to a very small level that would allow the desired energy to be completely radiated or received by the antenna. The antenna input impedance is a complex one and can be described as:

$$Z_A = \frac{V_A}{I_A} = R_A + jX_A \tag{2.4}$$

where R_A is the antenna resistance and X_A is the antenna reactance.

When the stored magnetic and electric energy values are equal, a condition of resonance occurs and the reactive part of Z_A vanishes. For a thin dipole antenna, this occurs when the antenna length is close to a multiple of a half-wavelength.

- Mutual impedance, Z_{AB}, is considered as the impedance generated between two dipole antennas, in proximity to each other, for example, a distance $d < \lambda$ apart. If current flows in antenna A, its radiated energy induces an electromotive force (emf) equals E_B where the voltage V_B across the port would be such that $V_B = -E_B$. The negative sign with induced emf is because the changing magnetic flux produces an electric field, which opposes the change in flux.

The major directional characteristics of an antenna can be described by the following parameters:

- Radiation pattern, $F(\theta, \varphi)$, gives the angular variation of radiation at a fixed distance from an antenna when the antenna is transmitting. Radiation is quantified by the value of power density S at a fixed distance r from the antenna. When receiving, the antenna responds to an incoming wave from a given direction according to the pattern value in that direction. The typical pattern in Figure 2.2 shows the pattern's main beam and sidelobes. The directive antenna, with a single narrow main beam, is used in point-to-point communications. In some applications, the shape of the main beam is important. However, an omnidirectional antenna with constant radiation in one plane is used in broadcast situations. It is often desired to scan the main beam of an antenna over a region of space. This can be accomplished by moving an antenna, by electronic scanning, or by a combination of mechanical and electronic scanning.

  It is common practice to show planar sections of the radiation pattern instead of the complete three-dimensional (3D) radiation pattern. The two most important views of an antenna radiation pattern are those of the principal

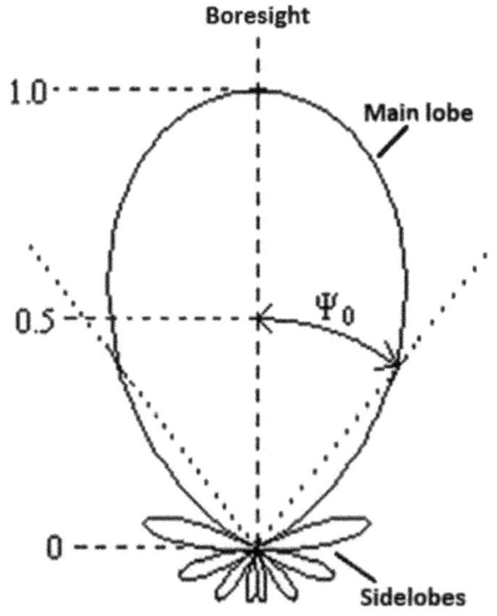

Figure 2.2 Antenna radiation pattern.

E-plane and H-plane patterns, obtained from sections containing the maximum radiated E-field and H-field, respectively.

- Beamwidth, BW, is the half-power beamwidth of the radiation pattern, which defines the angular width between two points on the main beam, at which the radiated power/unit area is one-half of the maximum. BW may also define the angular width between the first nulls (zero radiation points), on both sides of the main lobe.
- Sidelobes, as shown in Figure 2.2, can be sidelobes and a back lobe. Typically, there could be second, third, and more sidelobes, in a typical radiation pattern. These are generally undesired but unavoidable since they cause interference in radar detection applications and consume part of the overall radiation energy desired for the main beam. The sidelobes and back lobe have radiation levels and techniques for reducing the sidelobe level (SLL) that are not within this scope. SLL can be calculated if the amplitudes of the main lobe, A, and side lobe, S, are known, respectively:

$$SLL = 10\log_{10}\left(\frac{A}{S}\right) \tag{2.5}$$

- Directivity function, $D(\theta, \varphi)$, expresses how much greater the peak radiated power density is for an antenna than it would be if all the radiated power were distributed uniformly around the antenna. Figure 2.2 shows the radiation pattern of a real antenna compared to an isotropic spatial distribution. The spatial enhancement that can be achieved by an antenna is evident. Gain G is directivity reduced by the losses on the antenna.
- The power density, P_d, at a distance, r, from a spherically isotropic radiator (s.i.r.), which is a hypothetical source that radiates uniformly in all directions, can be formulated as:

$$P_d = \frac{P_T}{4\pi r^2} \tag{2.6}$$

where P_T is total radiated power and $4\pi r^2$ is the surface area of the radiated power, at range r.

- The maximum directivity, D, expresses the power density in a maximum direction for a given antenna, compared to the power density from an s.i.r. with the same transmitting power. In decibels, it can be expressed as:

$$D_{\text{dB}} = 10\log_{10} D \tag{2.7}$$

The gain of the antenna is almost similar to the maximum directivity, where the radiated power density of the antenna is compared to the input power, part of which is lost as it converts to radiated power.

This is expressed as:

$$G = \eta D \tag{2.8}$$

Here G expresses the antenna gain and η is the radiation efficiency where $\eta < 1$. The radiation efficiency is the actual radiated power of the antenna, divided by its input power. To calculate the radiation efficiency, the gain and the directivity must be measured first. Factors such as impedance mismatch and polarization mismatch have to be minimized during these measurements. The radiation efficiency is then calculated using (2.8).

- Frequency bandwidth, FBW, is the range of frequencies with acceptable antenna performance as measured by one or more of the parameters. The FBW of broadband antennas is expressed as the ratio of the upper to the lower frequencies, where the antenna performance is acceptable:

$$FBW = \frac{f_{\max}}{f_{\min}} \tag{2.9}$$

Broadband antennas with FBW as large as 40:1 have been designed. Such antennas are referred to as frequency-independent antennas. For narrowband antennas, the FBW is expressed as a percentage of the frequency difference over the center frequency:

$$FBW = \frac{f_{\max} - f_{\min}}{f_0} \times 100\% \tag{2.10}$$

where the center frequency is calculated as follows:

$$f_0 = \frac{f_{\max} + f_{\min}}{2} \tag{2.11}$$

- Polarization describes the vector nature of electric fields radiated by an antenna. The figure traced out with time by the tip of the instantaneous electric field vector determines the polarization of the wave. A straight-wire antenna radiates a wave with linear polarization parallel to the wire. Another popular polarization is circular. In general, polarization is elliptical. Dual polarized antennas enable the doubling of communication capacity carrying separate information on orthogonal polarizations over the same physical link at the same frequency.
- Effective area, A_e (i.e., effective antenna aperture), is the ratio of the available power at the terminals of the antenna to the power flux density of a plane wave incident upon the antenna, which is matched to the antenna in terms of polarization. If no direction is specified, the direction of maximum radiation is implied. A_e is expressed as:

$$A_e = \frac{P_r}{P_d} \tag{2.12}$$

where P_r is the power received by the antenna in (W) and P_d is the power flux density of the incident wave in (W/m^2).

The relationship between G and A_e is given as:

$$G = \frac{4\pi}{\lambda^2} A_e \tag{2.13}$$

where λ is the wavelength of the received signal.

For aperture-type antennas, the effective area is smaller than the physical aperture area. Aperture antennas with constant amplitude and phase distribution across the aperture have the maximum effective area, which is practically equal to the geometrical area. The effective aperture of wire antennas is much larger than the surface of the wire itself. Sometimes, the aperture efficiency of an antenna is estimated as the ratio of the effective antenna aperture and its physical area:

$$\varepsilon_{ap} = \frac{A_e}{A_p} \tag{2.14}$$

- Radiation resistance, R_r, is defined as the resistance that would dissipate the same power as that transmitted for the same current. It therefore relates the transmitted power, P_t, to the voltage (or current, I) at the antenna terminal as:

$$R_r = \frac{2P_t}{|I|^2} \tag{2.15}$$

If the antenna loss is very small, then R_r would be the real part of the complex input impedance.

2.3 Basic Antenna Types

Antennas can be divided into four basic types by their performance as a function of frequency. These are electrically small, resonant, broadband, and aperture antennas. Electrically small antennas are used at very high frequency (VHF) and below. Resonant antennas are mainly used from high frequency to low gigahertz frequencies. Broadband antennas are mainly used from VHF to middle gigahertz frequencies. Aperture antennas are mainly used at ultrahigh frequency (UHF) and above.

- *Electrically small antennas:* These are much less than a wavelength in extent. From electromagnetics, the wavelength λ is related to frequency f through the speed of light c as $\lambda = c/f$. Electrically small antennas are simple in structure and their properties are not sensitive to construction details. The vertical monopole used for amplitude modulation (AM) reception on cars is a good example. It is about 0.003λ long and has a pattern that is nearly omnidirectional in the horizontal plane. This is often a desirable property, but its low input resistance and high input reactance are serious disadvantages. Also, small antennas are inefficient because of ohmic losses on the structure.

- *Resonant antennas:* These are popular when a simple structure with good input impedance over a narrow band of frequencies is needed. It has a broad main beam and low or moderate (a few decibels) gain. The half-wavelength-long dipole is a prominent example. Figure 2.3 shows a Yagi antenna composed of half-wave dipoles and a reflector.
- *Broadband antennas:* Many applications require an antenna that operates over a wide frequency range. A broadband antenna has acceptable performance as measured with one or more parameters (pattern, gain, and/or impedance) over a 2:1 bandwidth ratio of upper to lower operating frequency. A broadband antenna is characterized by an active region. Propagating waves originate at the feed point and travel without radiation to the active region where most of the power is radiated. A broadband antenna with circular geometry has an active region where the circumference is one wavelength and produces circular polarization. An example is the spiral antenna illustrated in Figure 2.4 which can have a 20:1 bandwidth. A broadband antenna made up of linear elements or straight edges has an active region where these elements are about a half-wavelength in extent and produces linearly polarized radiation parallel to the linear elements. Since only a portion of a broadband antenna is responsible for radiation at a given frequency, the gain is low. However, it may be an advantage to have gain that is nearly constant with frequency, although low. Also, the traveling wave nature of a broadband antenna means that it has a real-valued input impedance.
- *Aperture antennas:* These have an opening through which propagating electromagnetic waves flow. A horn antenna is a good example; it acts as a funnel, directing the waves into the connecting waveguide. The aperture is usually several wavelengths long in one or more dimensions. The pattern usually has a narrow main beam, leading to high gain. The main beam narrows with increasing frequency for a fixed physical aperture size. The bandwidth is moderate as much as 2:1. Figure 2.5 demonstrates two examples on aperture antennas: a reflector TV antenna and a pyramidal horn antenna.

As technology continues to evolve, antennas play a pivotal role in enabling advanced applications across various industries. The following five antenna types and concepts represent the forefront of innovation, shaping the future of wireless

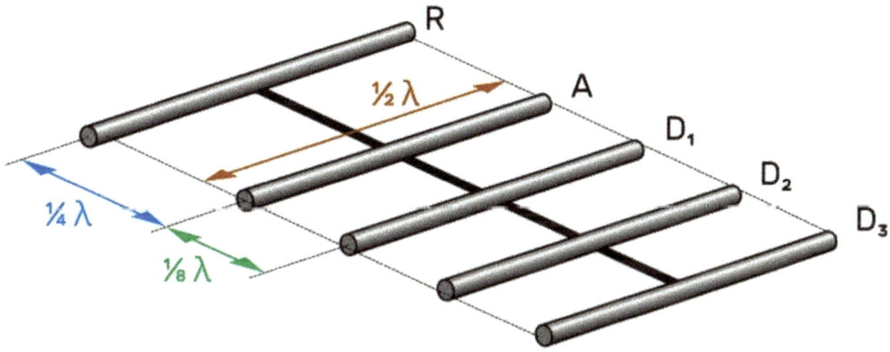

Figure 2.3 Yagi antenna composed of half-wave dipoles and a reflector.

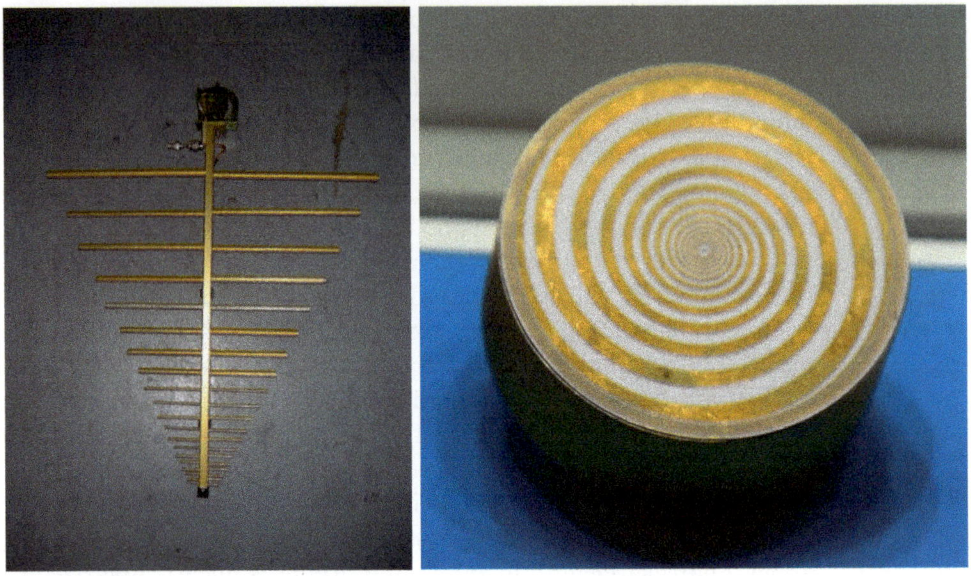

(a) (b)

Figure 2.4 (a) Log-periodic dipole antenna. (Created by Dante Alighieri, licensed under CC BY-SA 3.0. Source: Wikimedia Commons.) (b) Spiral antenna. (Created by Bin im Garten. Source: Wikimedia Commons.)

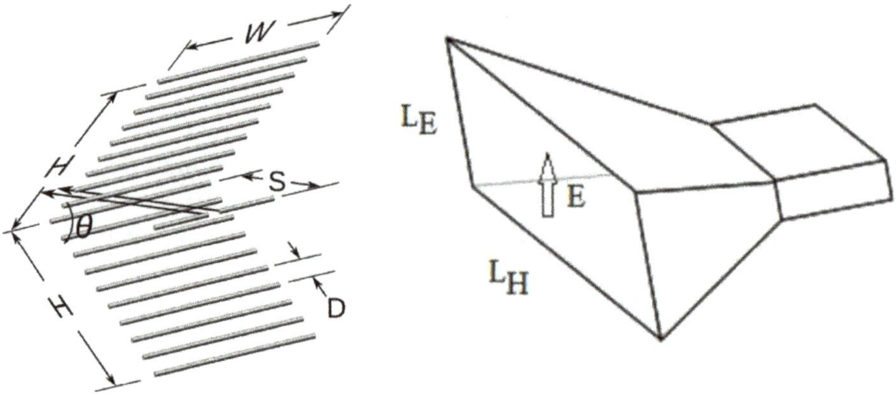

(a) (b)

Figure 2.5 (a) Reflector TV antenna. (Created by an unknown author, licensed under CC BY PD US no notice. Source: Wikimedia.) (b) Pyramidal horn antenna. (Created by Avpu. Source: Wikimedia Commons.)

communications, energy transfer, and beyond. From artificial intelligence (AI)-driven optimization to groundbreaking metamaterials, these antennas are not only transforming traditional systems but also opening doors to new, efficient solutions in fields such as 5G/6G, medical technology, and space exploration. By embracing these cutting-edge concepts, we can anticipate a significant shift in how we connect, communicate, and power devices in the coming decades. Some of the most innovative antenna concepts driving the future of communications, energy transfer, and beyond are:

- *Metamaterial antennas:* These antennas utilize engineered materials with unique structures, enabling properties not found in natural materials, such as miniaturization and enhanced performance. Metamaterial antennas offer improved gain, directivity, and compact designs, providing solutions where traditional materials fall short. Applications include advanced radar, wireless communications, and stealth technologies. Figure 2.6 demonstrates a metamaterial antenna.
- *AI-driven smart antennas:* By incorporating AI, these antennas adapt their radiation patterns in real time to optimize signal reception and transmission. AI algorithms allow for dynamic beamforming and interference management, improving performance in complex environments. Applications include 5G/6G networks, autonomous vehicles, and Internet of Things (IoT) systems.
- *Terahertz and millimeter-wave antennas:* Operating at high frequencies, these antennas are crucial for next-generation wireless communication systems and imaging technologies. With their ability to transmit large amounts of data over short distances, they enable ultrafast wireless communications. Applications include 5G/6G communications, medical imaging, and security systems. Figure 2.7 demonstrates a 15-THz broadband spiral antenna for astronomy.
- *Reconfigurable antennas:* These antennas can dynamically alter their frequency, polarization, or radiation pattern to adapt to changing operational requirements. This flexibility makes them suitable for diverse environments where conditions frequently change. Applications include aerospace, defense, mobile communications, and IoT. Figure 2.8 demonstrates a reconfigurable pixel antenna.
- *Wireless power transfer (WPT) antennas:* Specialized antennas designed to transmit power wirelessly over distances, enabling contactless charging and other energy transfer applications. These antennas use resonant inductive coupling or electromagnetic fields for energy transmission. Applications include electric vehicle charging, healthcare (e.g., powering medical devices), and consumer electronics. Figure 2.9 shows a 15-W wireless charging stand.

Figure 2.6 Metamaterial antenna. (Created by C. Holloway/NIST. Source: Wikimedia Commons.)

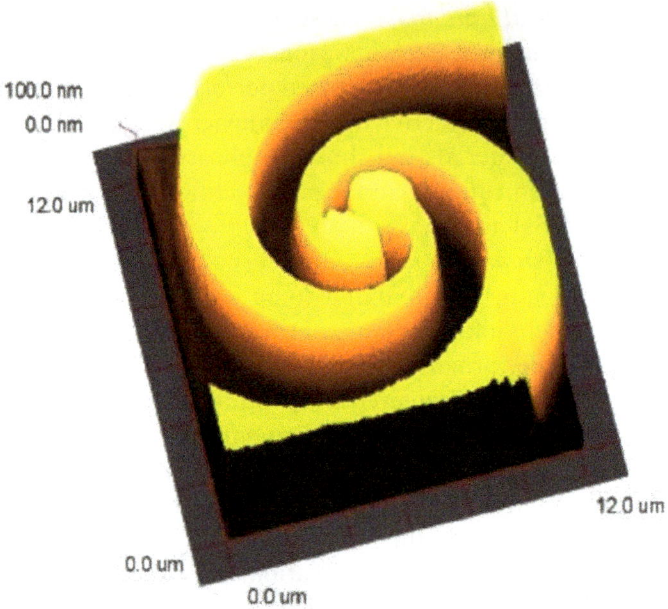

Figure 2.7 A 1.5-THz broadband spiral antenna for astronomy. (Created by von Richter, Heiko Helmut. Source: Wikimedia Commons.)

Figure 2.8 Reconfigurable pixel antenna. (Created by Eagle5eagle. Source: Wikimedia Commons.)

Figure 2.9 A 15-W wireless charging stand. (Created by KKPCW. Source: Wikimedia Commons.)

Problems for Students

2.1 Consider a half-wave dipole antenna operating at a frequency of 3 GHz in free space.
   (a) Calculate the resonant impedance of the antenna.
   (b) Explain the significance of the impedance matching between the antenna and the transmission line.
   (c) If the antenna is fed by a coaxial cable with a characteristic impedance of 50Ω, calculate the reflection coefficient at the feed point.

2.2 Given a parabolic reflector antenna with a diameter of 2m operating at a frequency of 10 GHz:
   (a) Calculate the directivity of the antenna using the formula for parabolic reflector antennas.
   (b) Derive the relationship between directivity and gain, and estimate the antenna's gain assuming a 50% efficiency.
   (c) Explain how directivity and gain are related but not the same, and discuss the impact of efficiency on the gain of an antenna.

2.3 For a linear antenna, the radiation pattern is given by $f(\theta) = \cos(\theta)$, where θ is the angle from the axis of the antenna.
   (a) Plot the radiation pattern for θ between 0° and 180°.

(b) Calculate the beamwidth of the antenna, which is defined as the angle between the half-power points (3 dB points) of the main lobe.
(c) Explain the significance of beamwidth in the context of antenna performance.

2.4 A dipole antenna has a total radiated power of 50W and a resistance of 75Ω.
(a) Calculate the efficiency of the antenna if the resistive loss is 5W.
(b) If the antenna operates at 3 GHz, calculate the input impedance at the resonant frequency and explain how it relates to the antenna's efficiency.
(c) Discuss the impact of material and design factors on antenna efficiency.

2.5 An antenna radiates linearly polarized waves with the electric field vector oriented at 45° to the ground.
(a) Explain what is meant by the polarization of an antenna.
(b) If the antenna is receiving a wave polarized at 90°, calculate the polarization mismatch loss.
(c) What is the effect of polarization mismatch on antenna performance, and why is polarization alignment important in wireless communication?

2.6 Consider a horn antenna with an aperture area of 1 m^2 and an efficiency of 60%.
(a) Calculate the effective aperture of the antenna.
(b) If the antenna operates at a frequency of 15 GHz, calculate the directivity of the antenna.
(c) Explain how aperture efficiency affects the performance of the antenna and its directivity.

2.7 For a dipole antenna operating at 5 GHz, the impedance bandwidth is measured to be 200 MHz.
(a) Calculate the fractional bandwidth of the antenna.
(b) Discuss the factors that influence the bandwidth of an antenna and explain how bandwidth relates to the antenna's ability to operate over a range of frequencies.
(c) If the antenna is used in a system operating at multiple frequencies, how can the bandwidth be optimized for the application?

2.8 For a half-wave dipole antenna, the radiation resistance is known to be 73Ω.
(a) Calculate the efficiency of the antenna if the antenna resistance is 100Ω and the loss resistance is 10Ω.
(b) Discuss the relationship between radiation resistance and antenna efficiency, and explain how it affects the overall performance.
(c) How would the efficiency change if the antenna's resistance increased due to poor material quality?

Problems for Instructors

2.1 Comprehensive Gain and Directivity Analysis: Given an antenna with a known radiation intensity pattern, derive the total radiated power and calculate the antenna's directivity and gain for specific frequencies. Discuss the influence of frequency on these parameters.

2.2 Radiation Pattern Interpretation: An antenna's far-field radiation pattern is provided in both 2D and 3D. Instruct students to:

(a) Analyze the radiation pattern to extract the half-power beamwidth (HPBW) and sidelobe levels.
(b) Calculate the directivity using both graphical and analytical methods.
(c) Explain how the pattern impacts the antenna's practical applications.

2.3 Polarization and Impedance Matching: A dipole antenna operates at 300 MHz with a given polarization. Define the polarization types and evaluate the importance of polarization matching with a receiving antenna in a communication system. Additionally, calculate the impedance matching required for optimal power transfer if the input impedance is not 50Ω.

2.4 Antenna Efficiency Evaluation: For a given antenna with specified resistive and reactive losses, calculate the antenna efficiency. Then compare this efficiency with the theoretical maximum and discuss the practical limits of antenna efficiency in real-world applications.

2.5 Field Regions and Measurement Setup: Outline the procedure to determine the near-field, transition zone, and far-field regions of a parabolic reflector antenna. Then design an experimental setup to measure the antenna's radiation pattern, gain, and directivity in the far-field region.

2.6 Reciprocity Theorem Application: Using the reciprocity theorem, explain why the radiation pattern of a transmitting antenna is identical to that of the same antenna in the receiving mode. Apply this to calculate the receiving antenna's effective aperture for a known gain and operating wavelength.

2.7 Multi-Antenna System Design: Propose a multi-antenna system (array) based on the given specifications of individual antenna elements. Focus on achieving a specific directivity, sidelobe suppression, and beam-steering capabilities. Instruct students to:
(a) Design the geometry and spacing of the array.
(b) Analyze the resulting radiation pattern and directivity.
(c) Discuss practical design challenges in real-world implementations.

Selected Bibliography

Balanis, C. A., *Antenna Theory: Analysis and Design*, Second Edition, New York: John Wiley & Sons, 1997.

Stutzman, W. L., and G. A. Thiele, *Antenna Theory and Design*, New York: John Wiley & Sons, 2012.

CHAPTER 3

Short Current Filament: Fundamental Building Block of Antennas

3.1 Introduction

The short current filament in Figure 3.1 is an antenna element whose length and radial dimensions are much less than a wavelength. It constitutes the most basic and realistic radiating antenna, after the fictitious radiating point source, which assumes the radiation of spherical electromagnetic waves. In this chapter, we derive the necessary formulas of the basic antenna by utilizing a developed parameter, the magnetic vector potential, which will prove very useful, if not essential, in the analysis of half-wave and full-wave dipoles, and linear arrays.

The magnetic vector potential, symbolized as **A**, is a vector quantity defined such that its curl is equal to the magnetic field:

$$\mathbf{B} = \nabla \times \mathbf{A} \tag{3.1}$$

Together with the electric scalar potential φ, the magnetic vector potential can be used to specify the electric field **E** as well. Therefore, many electromagnetic equations can be written either in terms of the fields **E** and **B** or equivalently in terms of the potentials φ and **A**.

Consider (1.44) and (1.45). Taking the curl of (1.44) and substituting (1.45) yield:

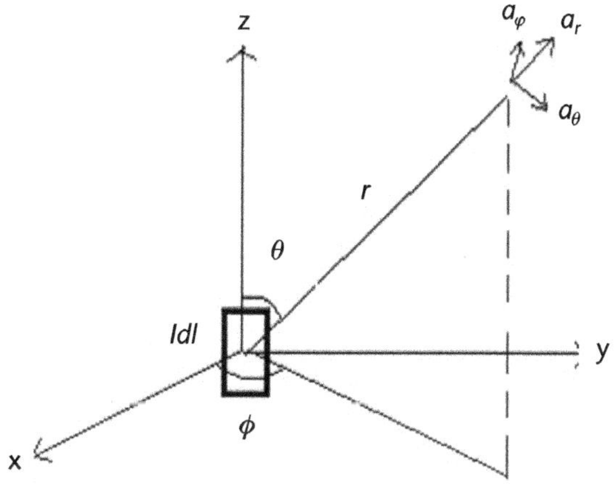

Figure 3.1 Short current filament, *Idl*, in a spherical coordinate system.

$$\nabla \times \nabla \times \mathrm{E} = -j\omega\mu(j\omega\varepsilon\mathrm{E} + \mathrm{J}) = \omega^2\mu\varepsilon E - j\omega\mu J$$
$$= k^2E - j\omega\mu J \tag{3.2}$$

The above equation is the one to be solved to find ***E*** in terms of ***J***. In practice, however, a simple equation to solve is obtained by introducing the vector potential **A** and scalar potential φ.

The vector potential in (3.1) is appropriately defined, since upon substituting it in the fourth Maxwell's equation (1.4), one gets the null equation $\nabla \times \nabla \times \mathbf{A} = 0$, which conforms with (1.4).

Further substituting (3.1) into (1.44) gives:

$$\nabla \times E = -j\omega(\nabla \times A) \tag{3.3}$$

Rearranging,

$$\nabla \times (E + j\omega A) = 0 \tag{3.4}$$

We can define a scalar potential by applying the second null equation $\nabla x(\nabla\Phi) = 0$; hence:

$$E + j\omega A = -\nabla\Phi \tag{3.5}$$

The negative sign is introduced since a reduction in scalar potential contributes to an enhancement in the electric field.

Consider now (1.45) and multiply both sides with μ:

$$\nabla \times \mu H = \nabla \times \nabla \times \mathrm{A} = j\omega\mu\varepsilon E + \mu\mathrm{J} \tag{3.6}$$

Substitute (3.5) into (3.6) yields:

$$\nabla \times \nabla \times A = j\omega\mu\varepsilon(-j\omega A - \nabla\Phi) + \mu J \tag{3.7}$$

Since $\nabla \times \nabla \times A = \nabla\nabla \cdot A - \nabla^2 A$, then substituting and rearranging (3.6) gives:

$$\nabla^2 A + k^2 A = -\mu J + \nabla(\nabla \cdot A + j\omega\mu\varepsilon\Phi) \tag{3.8}$$

To simplify the above equation, the term between the brackets is reduced to what is known as the Lorentz condition, which equals 0. Hence,

$$\nabla^2 A + k^2 A = -\mu J \tag{3.9}$$

The above equation is known as the inhomogeneous Helmholtz equation.

A parallel equation to (3.9) for the scalar potential in terms of the charge density, ρ, can be derived, if (3.5) and the Lorentz condition are substituted in the third Maxwell's equation (1.3).

$$\begin{aligned}\nabla \cdot \varepsilon(-j\omega A - \nabla\Phi) &= -j\omega\varepsilon\nabla \cdot A - \varepsilon\nabla^2\Phi \\ &= -j\omega\varepsilon(-j\omega\mu\varepsilon\Phi) - \varepsilon\nabla^2\Phi = \rho\end{aligned} \tag{3.10}$$

Consequently,

$$\nabla^2\Phi + k^2\Phi = -\frac{\rho}{\varepsilon} \tag{3.11}$$

To find E in terms of A, we only need to substitute for Φ, from the Lorentz condition into (3.5). This yields:

$$E = -j\omega A + \frac{\nabla\nabla \cdot A}{j\omega\mu\varepsilon} \tag{3.12}$$

3.2 Radiation from the Short Current Filament

Consider a short filament of current located at the origin and oriented along the z-axis. For this source, the vector potential has only a z-component. Hence,

$$\left(\nabla^2 + k^2\right)A_z = -\mu J_z \tag{3.13}$$

where $I = J_z \cdot \mathbf{dS}$ and $\mathbf{dS}$ is the cross-sectional area of the current element of length dl.

In spherical coordinates and dropping derivatives of θ and φ,

$$\nabla^2 A_z = \frac{1}{r^2}\frac{\partial}{\partial r} r^2 \frac{\partial A_z}{\partial r} \tag{3.14}$$

Let $A_Z = \psi(r)/r$; then:

$$\frac{dA_z}{dr} = \frac{1}{r}\frac{d\psi}{dr} - \frac{1}{r^2}\psi \tag{3.15}$$

Substituting (3.15) into the right side of (3.14) yields:

$$\frac{1}{r^2}\frac{\partial}{\partial r}\left(r\frac{\partial\psi}{\partial r} - \psi\right) = \frac{1}{r^2}\left(r\frac{\partial^2\psi}{\partial r^2} + \frac{\partial\psi}{\partial x} - \frac{\partial\psi}{\partial x}\right) \tag{3.16}$$

Now, substituting (3.16) into the left side of (3.13) and equating it to 0 gives:

$$\frac{\partial^2\psi}{\partial r^2} + k^2\psi = 0 \tag{3.17}$$

This is a simple harmonic motion equation with the following general solution:

$$\psi(r,t) = C_1 e^{-jkr+j\omega t} = C_1 e^{j\omega\left(t-\frac{r}{c}\right)} \tag{3.18}$$

The solution of A_z is hence of the form $c_1 e^{-jkr}/r$ where the phase is retarded by kr.

We now integrate (3.13) over a small spherical volume of radius r_0, in order to find C_1.

$$\iiint \nabla^2 A_z \, dv = -k^2 \iiint A_z \, dv - \mu \iiint J_z \, dv \tag{3.19}$$

Taking the left side:

$$\iiint \nabla^2 A_z \, dv = \iiint \nabla(\nabla \cdot A_z) dv - \iiint \nabla \times \nabla \times A_z \, dv \tag{3.20}$$

The second product on the right side of (3.20) equals 0, and upon applying the divergence theorem:

$$\iiint \nabla^2 A_z \, dv = \oint \nabla (A_z \cdot dS) = \oint \nabla \left(A_z \cdot a_r r_0^2 \sin\theta d\theta d\varphi\right) \tag{3.21}$$

Now

$$\nabla(A_z \cdot a_r) = \frac{\partial A_z}{\partial r} = -(1 + jkr) C_1 \frac{e^{-jkr}}{r^2} \tag{3.22}$$

Substituting (3.22) into (3.21) gives:

$$\oint \nabla (A_z \cdot dS) = \lim_{r_0 \to 0} - \int_0^{2\pi} \int_0^{\pi} (1 + jkr_0) \frac{C_1 e^{-jkr_0}}{r_0^2} r_0^2 \sin\theta \, d\theta \, d\varphi = -4\pi C_1 \tag{3.23}$$

Regarding the right side of (3.19):

$$-k^2 \lim_{r_0 \to 0} \iiint A_z \, dv = 0 \tag{3.24}$$

This is because $A_Z \propto 1/r_0$ and $dv \propto$; hence, $A_Z \, dv$ is vanishingly small and the integration becomes equal 0.

Also, the second term on the right side of (3.19) reduces to:

$$-\mu \iiint J_z \, dv = -\mu (J_z \cdot dS) dl = -\mu I dl \tag{3.25}$$

Equating (3.23) to (3.25) yields:

$$C_1 = \frac{\mu I dl}{4\pi} \tag{3.26}$$

and

$$A_z = \frac{C_1 e^{-jkr}}{r} a_z = \frac{\mu I dl e^{-jkr}}{4\pi r} a_z \tag{3.27}$$

where

$$a_z = a_r \cos\theta - a_\theta \sin\theta$$

Since A is completely derived, it is possible to find the magnetic field, H from (3.1). It is only necessary to compute the following matrix:

$$\nabla \times A = \frac{C_1}{r^2 \sin\theta} \begin{vmatrix} a_r & r a_\theta & r\sin\theta a_\varphi \\ \frac{\partial}{\partial r} & \frac{\partial}{\partial \theta} & \frac{\partial}{\partial \varphi} \\ \frac{e^{-jkr}}{r}\cos\theta & -e^{-jkr}\sin\theta & 0 \end{vmatrix} \tag{3.28}$$

Consequently,

$$H = \frac{1}{\mu} \nabla \times A = \frac{Idl}{4\pi} \sin\theta \left(\frac{jk}{r} + \frac{1}{r^2} \right) e^{-jkr} a_\varphi \tag{3.29}$$

Similarly, it is possible to derive the electric field, E by substituting for the vector potential, A from (3.27) into (3.12).

Hence,

$$E = -jZ_0 \frac{Idl}{4\pi k} \sin\theta \left(-\frac{k^2}{r} + \frac{jk}{r^2} + \frac{1}{r^3} \right) e^{-jkr} a_\theta \tag{3.30}$$

where $Z_0 = \sqrt{\mu/\varepsilon}$ is known as the intrinsic impedance.

Now when r is large compared to a wavelength, λ, the only important terms in E and H, besides θ, is the term k^2/r.

Consequently, the following far fields follow:

$$H = j\frac{kIdl}{4\pi r} \sin\theta e^{-jkr} a_\varphi \tag{3.31}$$

$$E = jZ_0 \frac{kIdl}{4\pi r} \sin\theta e^{-jkr} a_\theta \tag{3.32}$$

and

$$\frac{|E|}{|H|} = Z_0 \tag{3.33}$$

The complex Poynting vector, P_y, which is the power radiated per unit area (i.e., power density) can be computed, by utilizing the conjugate of H in (3.31) and E in (3.32):

$$P_y = \frac{1}{2} E \times H^* = \frac{1}{32\pi^2 r^2} II^* Z_0 dl^2 k^2 \sin^2 \theta a_r \tag{3.34}$$

Figures 3.2 to 3.4 demonstrate the electric field, magnetic field, and power density radiated from a short current filament as expressed by (3.32), (3.31), and (3.34), respectively.

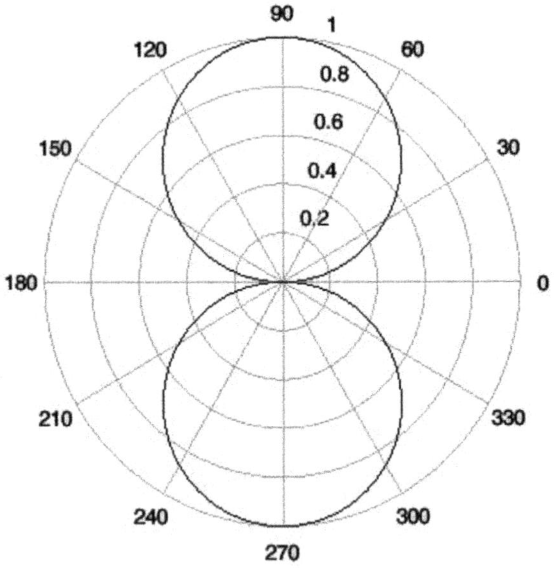

Figure 3.2 Electric field radiated from a short current filament in the *XZ* plane.

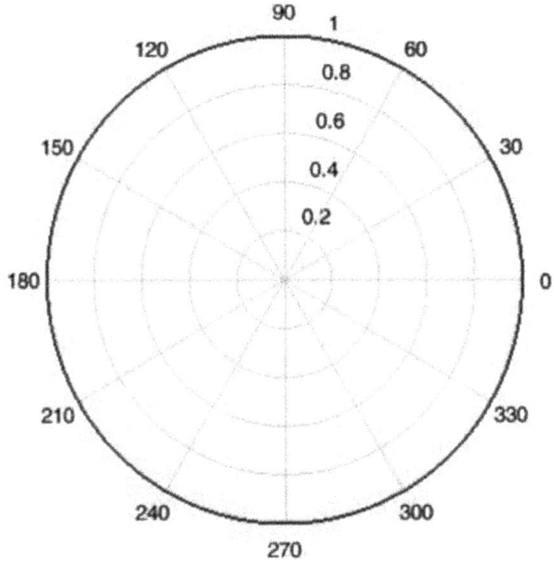

Figure 3.3 Magnetic field radiated from a short current filament in the *XY* plane.

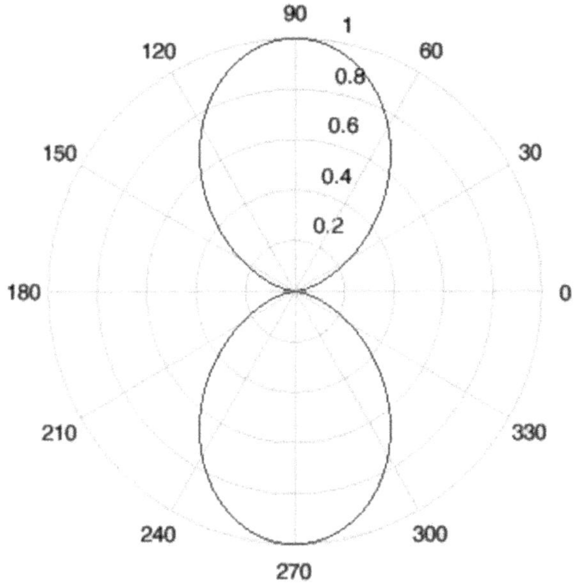

Figure 3.4 Power intensity radiated from a short current filament in the *XZ* plane.

3.3 Directivity and Gain Derivations

An antenna does not radiate uniformly in all directions. The variation of the intensity of radiation with angular direction in space is described by the directivity function, $D(\theta, \varphi)$ for the antenna.

The intensity of radiation is the power radiated per unit solid angle, Ω, and this is obtained by multiplying the Poynting vector flux density by r^2.

Hence, since $d\Omega = dS/r^2$, then:

$$\frac{dP_r}{d\Omega} = \frac{1}{2} r^2 \,\mathrm{Re}\, \mathbf{E} \times \mathbf{H}^* \mathbf{a}_r = \frac{1}{32\pi^2} II^* Z_0 dl^2 k^2 \sin^2\theta \tag{3.35}$$

The directivity function is defined as:

$$\mathrm{D}(\theta,\varphi) = 4\pi \frac{\text{Power radiated per unit solid angle}}{\text{Total radiated power}} = 4\pi \frac{(dP_r/d\Omega)}{P_r} \tag{3.36}$$

Once the total radiated power, P_r is computed, it can be substituted in (3.36):

$$P_r = \frac{1}{32\pi^2} II^* Z_0 dl^2 k^2 \int_0^{2\pi}\int_0^{\pi} \sin^2\theta (\sin\theta d\theta d\varphi) \tag{3.37}$$

where $d\Omega = \sin\theta d\theta d\varphi$and the solution to the double integral become $8\pi/3$. Hence,

$$P_r = \frac{1}{12\pi} II^* Z_0 dl^2 k^2 \tag{3.38}$$

Now substituting (3.35) and (3.38) back into (3.36) yields:

$$D(\theta,\varphi) = 1.5\sin^2\theta \tag{3.39}$$

Hence, the maximum directivity (or simply, directivity) equals 1.5 at $\theta = \pi/2$. It means that the directivity is 1.5 times the power intensity of what an isotropic radiator would produce at a constant range in space. An isotropic radiator is a fictitious point source that radiates uniformly in all directions.

MATLAB Program 1

```
% Matlab program to compute and plot in 3D the directivity
% of a short current filament.
theta = linspace(0, 2* pi, 1000);
phi = linspace(0, pi, 500);
[phi, theta] = meshgrid(phi, theta);
Dir= 1.5.*(sin(theta)).^2;
[x,y,z] = sph2cart(phi, theta, Dir);
z([z]< -1.5) = NaN;
mesh(x,y,z)
```

Figure 3.5 demonstrates a 3D plot of the directivity for a short current filament.

The gain of the antenna is defined like that for the directivity, except that the total input power to the antenna, rather than the total radiated power, is used as the reference. The difference is a measure of the efficiency of the antenna, η, so that:

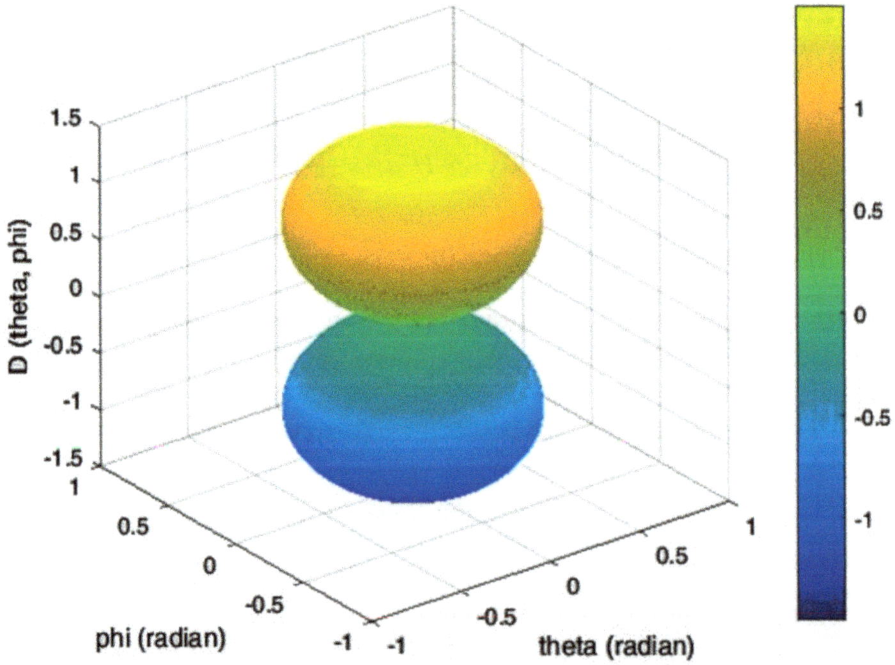

Figure 3.5 Directivity function $D(\theta, \varphi)$ for a short current filament.

$$P_r = \eta P_{in} \quad (3.40)$$

where P_r is the total radiated power and P_{in} is the total input power to the antenna. The gain function is described as:

$$G(\theta,\varphi) = 4\pi \frac{dP_r / d\Omega}{P_{in}} \quad (3.41)$$

The gain of the antenna is often incorporated into a parameter called the effective isotropic radiated power (EIRP), where:

$$EIRP = P_{in} \times G_{max} \quad (3.42)$$

3.4 Radiation Resistance Derivation

The radiation resistance of an antenna, R_r, is that equivalent resistance that would dissipate the same amount of power as the antenna radiates when the current in that resistance equals the input current at the antenna terminals. R_r is found from:

$$P_r = \frac{1}{2}|I^2|R_a \quad (3.43)$$

It is possible to derive for R_a, by substituting for P_r, from (3.38) and noting that $Z_0 = 120\pi$ and $k = 2\pi/\lambda$. It follows that:

$$R_r = \frac{Z_0}{6\pi} k^2 dl^2 = 80\pi^2 \left(\frac{dl}{\lambda}\right)^2 \quad (3.44)$$

So for a short current filament, where $dl = 0.1\lambda$, then $R_r = 7.9\Omega$.

3.5 Effective Aperture of a Short Current Filament

The power received by an antenna, such as a short current filament, can be associated with a collecting area. The collecting area for any antenna is called the effective aperture, symbolized as A_e (m^2).

Accordingly if P_d (watts/m^2) is the power density at the antennas and P_r (watts) is the received power, then

$$P_r = P_d A_e \quad (3.45)$$

Suppose that V is the induced voltage in a doublet (i.e., short current filament) of length dl, radiation resistance R_r, then

$$V = Edl \quad (3.46)$$

where E is the electric field.

From circuit equivalence of induced voltage and maximum power received by the radiation resistance, the radiation power works out as

$$P_r = \frac{V^2}{4R_r} = \frac{E^2 dl^2}{4R_r} \tag{3.47}$$

Now the maximum power density

$$P_d = |E \times H| = \frac{E^2}{\eta_0} \tag{3.48}$$

where $\eta_0 = 120\ \pi$ is the intrinsic impedance of free space.

Substituting (3.48) into (3.47) gives

$$P_r = P_d A_e = \frac{V^2}{4R_r} = \frac{E^2}{\eta_0} A_e \tag{3.49}$$

It follows that

$$A_e = \frac{120\pi V^2}{4E^2 R_r} = \frac{120\pi E^2 dl^2}{4E^2 R_r} = \frac{30\pi dl^2}{R_r} \tag{3.50}$$

Now substituting (3.44) for R_r into (3.50) yields

$$A_e = \frac{3\lambda^2}{8\pi} \tag{3.51}$$

At this point, it is possible to work out the effective aperture of any antenna. This is enabled by comparing the gains and effective apertures of an isotropic radiating point source and a doublet.

Let G_1 and A_1 be the gain and effective aperture area of an isotropic radiating point and G_2 and A_2 be the gain and the effective aperture area of a doublet. Since the received power would proportionally increase as the gain of an antenna increases, then we get

$$\frac{G_1}{A_1} = \frac{G_2}{A_2} \tag{3.52}$$

It is possible now to find out A_1 by substituting for $G_1 = 1$, $G_2 = 1.5$, and A_2 from (3.51). Hence,

$$A_1 = \frac{\lambda^2}{4\pi} \tag{3.53}$$

For any antenna with a gain G and an effective aperture area A, this can be compared with A_1 and G_1 in a similar way to (3.52). This gives:

$$A = \frac{G\lambda^2}{4\pi} \tag{3.54}$$

3.6 The Radar Equation

By finding the equation for the effective aperture area of an antenna, we can proceed for deriving the radar equation in its simple form.

Suppose a radar antenna has an antenna gain G_T and is transmitting at power P_T. The power density at an object located a distance r is:

$$P_d = \frac{P_T}{4\pi r^2} G_T \tag{3.55}$$

The target area of the object is defined as σ m^2 such that the power reflected back isotropically is $P_d\,\sigma$. It therefore produces a power density P_d' at the transmitter where

$$P_d' = \frac{P_T G_T \sigma}{\left(4\pi r^2\right)^2} \tag{3.56}$$

If a received antenna at the radar's site has a gain G_R and an effective area A, where $A = G_R\lambda^2/4\pi$, then

$$P_R = P_d' A = \frac{P_T G_T \sigma}{\left(4\pi r^2\right)^2}\left(\frac{G_R\lambda^2}{4\pi}\right) \tag{3.57}$$

$$\frac{P_R}{P_T} = \frac{G_T G_R \sigma \lambda^2}{(4\pi)^3 r^4} \tag{3.58}$$

This is the basic radar equation relating the transmitted and received powers. If the minimum detectable signal-to-noise ratio at the receiver is S/N, then $P_r = S$ and $N = kTB$ where the Boltzmann's constant $k = 1.380649 \times 10^{-23}$ J/K, T is the absolute temperature, and B is the bandwidth of the system. It follows after substituting for P_R that

$$r^4 = \frac{P_T G_T G_R \sigma \lambda^2}{(4\pi)^3 kTB(S/N)} \tag{3.59}$$

$$r = \left[\frac{P_T G_T G_R \sigma \lambda^2}{(4\pi)^3 kTB(S/N)}\right]^{1/4} \tag{3.60}$$

In other words, the maximum range depends on the fourth root of P_T, given that all other parameters in the equation are constant.

Solved Examples

Example 3.1 An aerial of gain 8 dB has a radiation resistance of 75Ω. The aerial current is 12A. A receiving aerial 12 km away has an effective area of 2 m^2. If the path loss is 0.5 dB/km, calculate the received signal. If the transmitter aerial is replaced by an improved model with a gain of 14 dB, what will be the new current to give the same received signal?

Solution 3.1

Transmitted Power Calculation: $P_t = (12)^2 \times 75 = 144 \times 75 = 10{,}800\text{W}$

Converting to dBW: $P_t = 40.3$ dBW

Free-Space Path Loss Calculation

The total path loss in dB: $L_p = 0.5 \times d = 0.5 \times 12 = 6$ dB

Received Power Calculation

Using the Friis transmission formula in decibel form:

$$P_r\,(\text{dBW}) = P_t\,(\text{dBW}) + G_t\,(\text{dB}) - L_p\,(\text{dB}) + 10\log10\,(A_r) - 10\log10\,(4\pi d^2)$$

Substituting the values:

$$P_r = 40.3 + 8 - 6 + 3 - 93.6 = -48.3\text{ dBW}$$

New Transmitting Current Calculation

With the improved antenna ($G_t' = 14$ dB), the received power remains the same:

$$P_r' = P_t' + 14 - 6 + 3 - 93.6$$

Since $P_r' = P_r P_r' = P_r P_r' = P_r$:

$$P_t' = -48.3 + 82.6 = 34.3\text{ dBW (i.e., } 14.8\ \mu\text{W)}$$

Converting back to watts:

$$P_t' \approx 2{,}690\text{W}$$

Hence, $I = \text{sqrt}(P_t'/75) = 5.99\text{A}$

Example 3.2 An antenna which is 1/50 wavelength long has a current of 5A. Determine at a distance of 1.2 km in the direction of maximum radiation, the radiation resistance, the magnetic and electric fields, and the power density.

Solution 3.2 Radiation Resistance Calculation

For a short dipole, the radiation resistance R_r is given by:

$$R_r = 80\pi^2(l/\lambda)$$

Substituting $l/\lambda = 1/50$:

$$R_r = 80\pi^2(1/50)^2 \approx 0.316\Omega$$

Magnetic Field H Calculation

For a short dipole, the magnetic field at distance r in the far-field region is given by:

$$H = kI_0\ l/4\pi r\ \sin\theta$$

Substituting $l = \lambda/50$:

$$H = (5 \times 2\pi/\lambda \times \lambda/50)/(4\pi \times 1{,}200) = 10/240{,}000 = 4.17 \times 10^{-5}\ \text{A/m}$$

Electric Field E Calculation

The electric field is related to the magnetic field as:

$$E = \eta \text{H} = 377 \times 4.17 \times 10^{-5} = 0.0157\ \text{V/m}$$

Power Density S Calculation

The power density is given by:

$$S = E^2/\eta \approx 0.0157^2/377 = 6.53 \times 10^{-7}\ \text{W/m}^2$$

Problems for Students

3.1 A radio-wave transmission system uses two aerials of gain 13 dB and 14.8 dB, with a separation of 6 km. The frequency is 150 MHz. If the transmitter power is 2.4 kW, calculate the received signal assuming no atmospheric attenuation ($G = 4\pi A_e/\lambda^2$), where A_e is the aerial's effective area.

3.2

(a) Express P_1/P_2 in decibels for the following:

P_1	P_2
100W	0.2W
5 mW	15W
7.5 mW	15 mW
12 mW	3 mW

(b) Find P for the following conditions:
- (i) +33 dB on 5 μW;
- (ii) +0.6 dB on 1.2W;
- (iii) −1.2 dB on 75 mW;
- (iv) −23 dB on 6W.

(c) Express the following powers: 6 dBm, –26 dBm, 33 dBm.

3.3 A line-of-sight communication system has a transmission power of 120W, a transmitter aerial gain of 24 dB, a receiving aerial gain of 18 dB, an aerial separation of 15 km, a maximum path attenuation of 15 dB, and an operating frequency of 6 GHz. Determine the received signal in watts and in dBm.

3.4 A system has a receiver of noise figure 7.5 dB and bandwidth 55 MHz. What will be the minimum S/N?

3.5 A line-of-sight communication system has the following characteristics: transmitter and receiver aerial gain of 16 dB, path length of 24 km, maximum path attenuation of 0.8 dB/km, operating wavelength of 6 cm, receiver noise figure of 9 dB, and bandwidth of 120 MHz. If the minimum allowable S/N is to be equal 20, determine the minimum allowable transmitter power.

3.6 Consider an antenna modeled as a short current filament with a length of $L = 0.25\lambda$ and a current of $I = 10$A, operating at a frequency of 3 GHz. The antenna is used for communication over a distance of 500m.

(a) Calculate the radiation resistance of the antenna using the appropriate formula for a short current filament.

(b) Compute the directivity of the antenna based on its physical characteristics, and explain the relationship between directivity and gain for such antennas.
(c) If the transmitting power is 1 kW, determine the radiated power using the radiation resistance. Assume that the efficiency of the antenna is 90%.
(d) The receiving antenna is located 500m from the transmitter and has an effective area of $A_e = 0.5\ \text{m}^2$. Using the formula $G = 4\pi A_e/\lambda^2$, calculate the received signal strength in decibels.
(e) Assuming the received signal is subject to a free-space path loss and the antenna is aligned for maximum radiation, calculate the received power at the receiver.
(f) If the path loss is 80 dB over the 500-m distance, find the received power in watts at the receiver.

Problems for Instructors

3.1 Derivation of Electric and Magnetic Fields from a Short Current Filament: Derive the time-averaged Poynting vector for the fields generated by a short current filament in free space. Extend the result to evaluate the power radiated at various distances from the filament, including the near-field and far-field regions.

3.2 Generalization of Radiation Patterns: Derive the 3D radiation pattern for a short current filament. Then extend the analysis to include oblique radiation, where the current filament is inclined at an angle with respect to the observation point.

3.3 Influence of the Reactive Near Field: Discuss the influence of the reactive near field on the overall radiation properties of the short current filament. Pose the problem of calculating how energy storage in the near field alters the apparent directivity in practical antenna design.

3.4 Effective Aperture in Complex Media: Calculate the effective aperture of a short current filament in a lossy medium (such as seawater or ionosphere). Discuss how medium properties influence the antenna performance and extend this to practical design considerations for underwater or ionospheric communication.

3.5 Radar Equation for Complex Scattering: Use the radar equation to determine the maximum range for detecting a small object with a known radar cross-section (RCS) in a nonideal atmospheric condition. Instruct students to adjust parameters such as temperature and pressure to account for practical losses in radar detection scenarios.

3.6 Radiation Resistance with Complex Loads: Analyze how the radiation resistance of a short current filament changes when placed in proximity to a ground plane or within a confined space (e.g., inside a waveguide or cavity). Develop a model for practical antenna design constraints in such environments.

3.7 Extension of Gain Calculations to Array Configurations: Extend the gain derivation for a single short current filament to an array of filaments in a uniform linear or circular array. Instruct students to calculate the total gain and beamwidth of the array, considering mutual coupling effects between filaments.

Selected Bibliography

Balanis, C. A., *Antenna Theory: Analysis and Design*, Second Edition, New York: John Wiley & Sons, 1997.

Marshall, G. G., and S. V. Skitek, *Electromagnetic Concepts and Applications*, Third Edition, Upper Saddle River, NJ: Prentice Hall, 1986.

Rao, N. N., *Elements of Engineering Electromagnetics*, Second Edition, Upper Saddle River, NJ: Prentice Hall, 1987.

Stutzman, W. L., and G. A. Thiele, *Antenna Theory and Design*, New York: John Wiley & Sons, 2012.

CHAPTER 4

Half-Wave Dipole Antenna

4.1 Introduction

A very widely used antenna is the half-wave dipole antenna. It is a linear current whose amplitude varies as one-half of a sine wave with a maximum at the center. For simplicity, we will assume this to be a filament of current. In addition, it could be imagined to flow on an infinitely thin, perfectly conducting, and half-wavelength-long wire. This is a good approximation of a wire half-wave dipole that has a diameter much smaller than its length. The advantage of a half-wave dipole is that it can be made to resonate and present a zero input reactance, thus eliminating the need for tuning to achieve a conjugate impedance match. The input impedance of dipole antennas is discussed in detail, but, for now, we point out that to obtain a resonant condition for a half-wave dipole, the physical length must be somewhat shorter than a free-space half-wavelength, and as the antenna wire thickness is increased, the length must be reduced more to achieve resonance.

The following are the advantages of a half-wave dipole antenna:

1. Input impedance is not sensitive.
2. It matches well with transmission line impedance.
3. It has reasonable length.
4. The length of the antenna matches with size and directivity.

The disadvantages of a half-wave dipole antenna are mainly that:

- It is not very effective due to a single element.
- It can work better only with a combination.

Concerning the applications of a half-wave dipole antenna:

- It is used in radio receivers.
- It is used in television receivers.
- When employed with others, it used for a wide variety of applications.

4.2 Monopole Antenna

A thin, linear dipole antenna can be effectively shortened by replacing half of its length with an infinite, perfectly conducting ground plane, as shown in Figure

4.1(a). The ground plane acts as a mirror, creating an image current that compensates for the missing half of the dipole in Figure 4.1(b). As a result, both antennas exhibit identical radiation characteristics. However, their self-impedance differs—the monopole antenna in Figure 4.1(a) has one-half the self-impedance of the balanced dipole in Figure 4.1(b). This occurs because the voltage required at the feed point of the monopole is only one-half of that required for the dipole, assuming that both are energized by the same current. Consequently, the self-impedance of a thin $\lambda/4$ monopole is approximately 37Ω, compared to 73Ω for a $\lambda/2$ dipole.

The input impedance behavior of a monopole antenna, with a total length l and a cylindrical rod diameter d, is illustrated in Figures 4.2 and 4.3. These curves, originally based on experimental measurements in [1], are provided for three different l/d ratios: 400 (thin), 100 (moderate), and 20 (thick). Since a monopole's impedance is effectively one-half that of a dipole of twice its length, its impedance variations follow a similar resonance pattern but at lower values.

For comparison, a half-wave dipole typically exhibits an input resistance of approximately 73Ω and an input reactance of about $j42.5\Omega$. Since reactance affects impedance matching, it must be compensated for maximum power transfer, often by adjusting the dipole length to $l \approx 0.48\lambda$. The input reactance of a dipole is highly frequency-sensitive, as it depends on l/d, a characteristic of narrowband resonant structures. Another resonance occurs around $l \approx 0.9\lambda$, where the radiation resistance can exceed 1,500Ω due to the formation of a current minimum at the feed point, rather than the usual maximum seen in a half-wave dipole.

Measured data from Figures 4.2 and 4.3 provide deeper insight into these impedance variations, illustrating how element thickness and electrical length influence monopole impedance characteristics in practical designs.

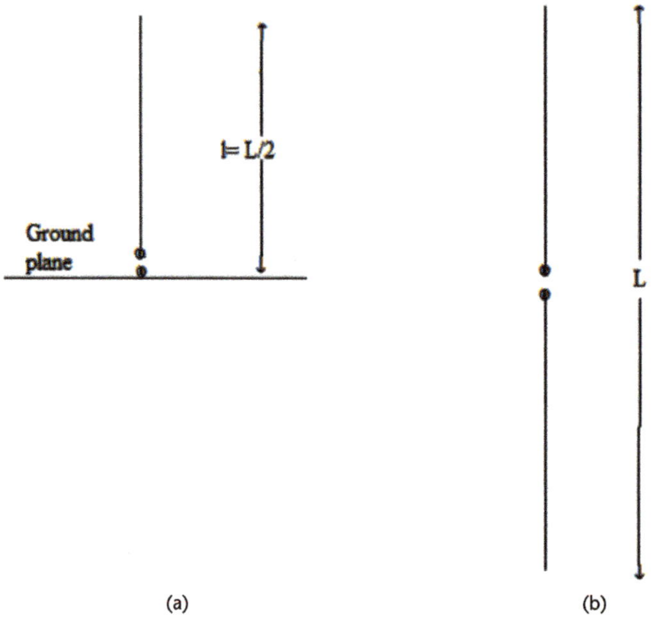

Figure 4.1 (a) Monopole antenna of length $L/2$. (b) Dipole antenna of length L.

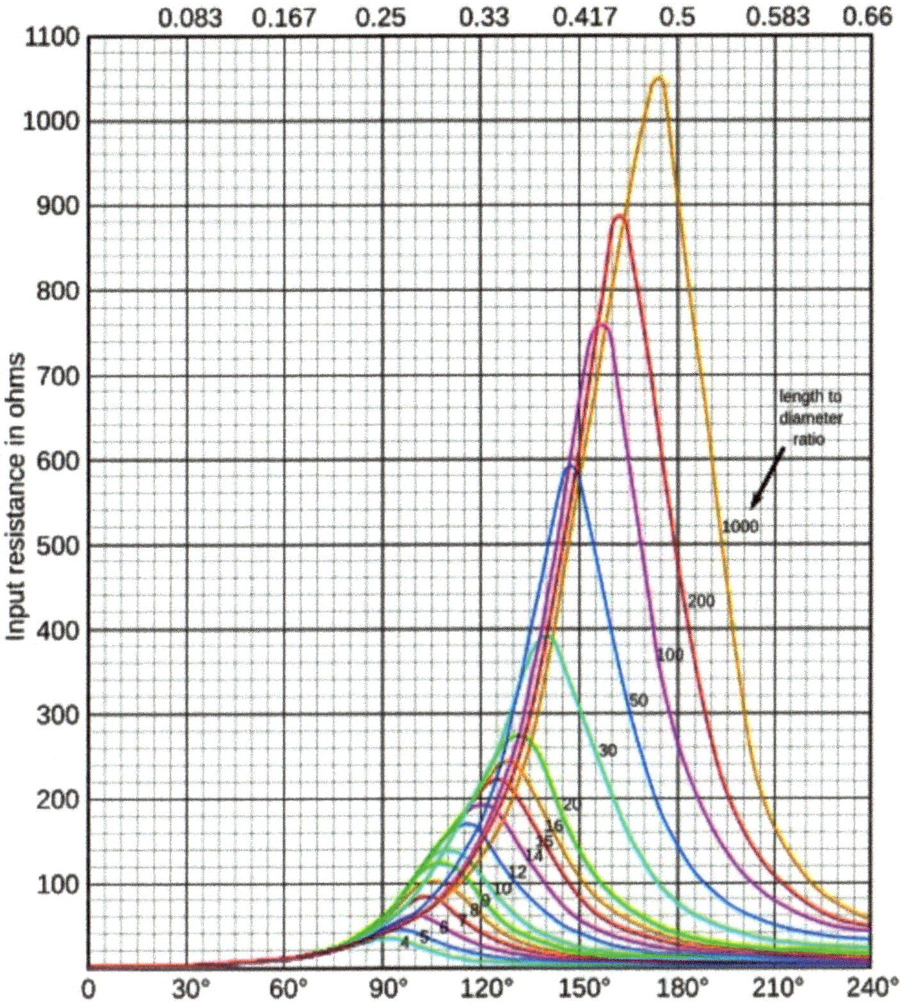

Figure 4.2 Measured input resistance of monopole antenna. Valid SVG created with Inkscape. (Source: Wikimedia Commons.)

4.3 Analysis of Half-Wave Dipole

In deriving the fields of a half-wave dipole, it is possible to consider its length as a series of short current filaments. This provides the advantage of utilizing the vector potential **A** and magnetic field **H** for the short current element, for the integration process along the range of the dipole length, shown in Figure 4.4.

Let the vector potential, contributed from element dz' at range R be as described in (4.1):

$$A = \frac{\mu I dl e^{-jkR}}{4\pi R} a_z \quad (4.1)$$

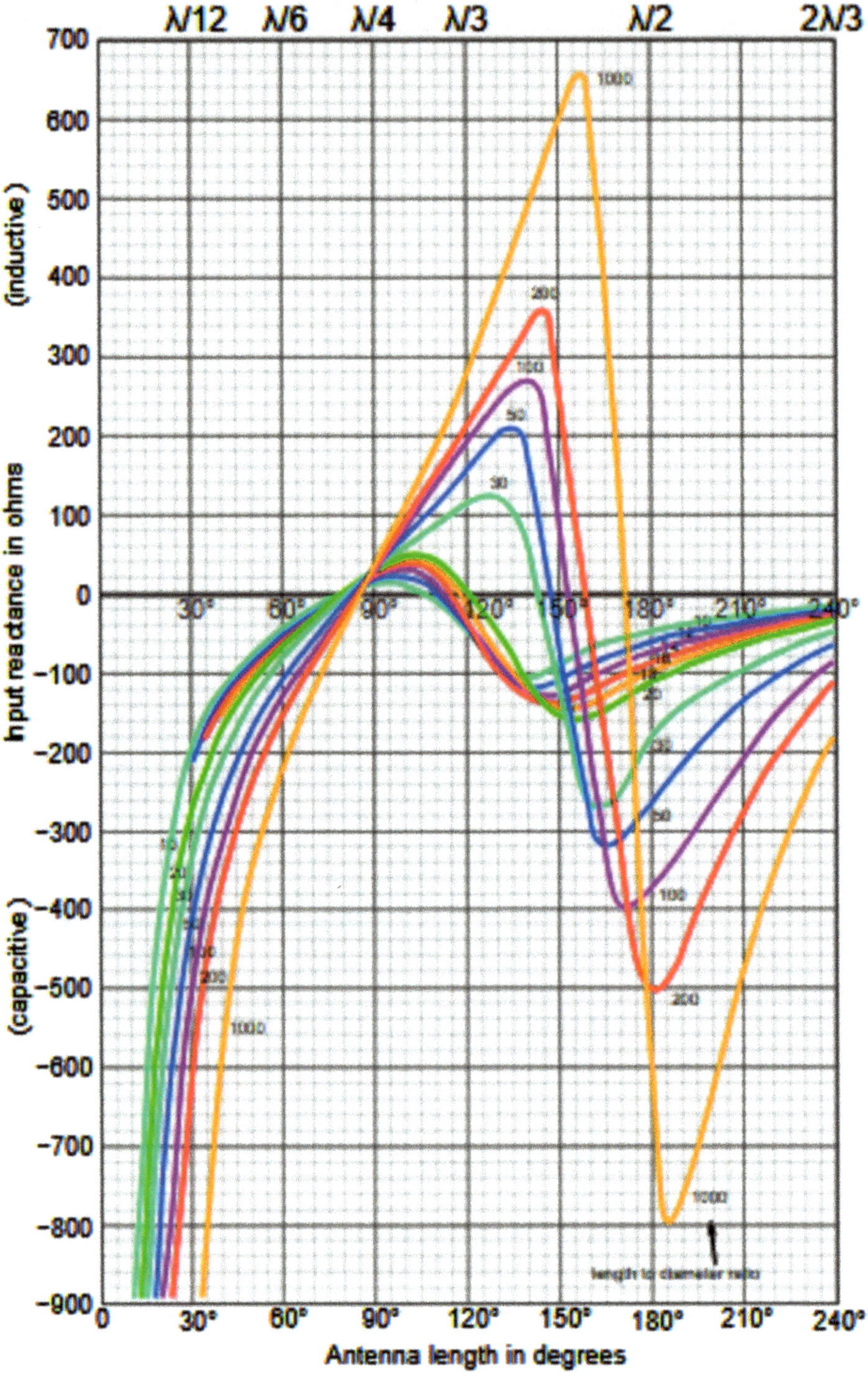

Figure 4.3 Measured input reactance of monopole antenna. Valid SVG created with Inkscape. (Source: Wikimedia Commons.)

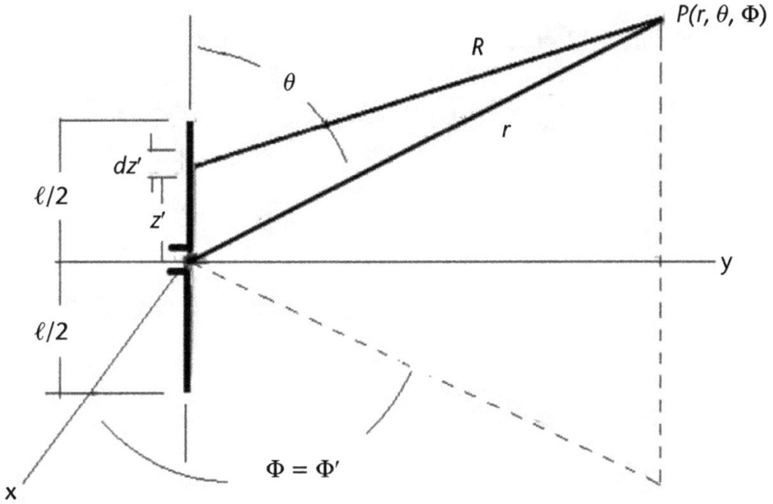

Figure 4.4 Schematic diagram of a half-wave dipole.

Then, in the far-field zone, where $R \approx r$ and $r >> \lambda$:

$$dH = \frac{1}{\mu}\nabla \times A = \frac{Idl}{4\pi}\sin\theta\left(\frac{jk}{r}\right)e^{-jkR}a_{\varphi} \tag{4.2}$$

It is noted that R is retained in the phasor, e^{-jkR}, since a phase change of 2π occurs for a distance of λ.

A half-wave dipole antenna consists of two quarter-wavelength conductors placed end to end for a total length of approximately $\ell = \lambda/2$. The current distribution is that of a standing wave, approximately sinusoidal along the length of the dipole, with a node at each end and an antinode (peak current) at the center (i.e., feed point).

Hence,

$$I(z) = I_0 \cos(kz) \tag{4.3}$$

We can now substitute for I in (4.2) and integrate:

$$H = \frac{jkI_0}{4\pi r}\sin\theta \int_{-\lambda/4}^{\lambda/4} e^{-jkR}\cos(kz)\,dza_{\varphi} \tag{4.4}$$

where $R = r - z\theta$. It follows that:

$$H = \frac{jkI_0}{4\pi r}\sin\theta e^{-jkr} \int_{-\lambda/4}^{\lambda/4} \cos(kz)e^{jkz\cos\theta}\,dza_{\varphi} \tag{4.5}$$

Substituting for cos(kz) in the integral of (4.5), in terms of its equivalence in exponential terms where $\cos(kz) = e^{jkz} + e^{-jkz}/2$:

$$I = \frac{1}{2}\int_{-\lambda/4}^{\lambda/4}\left(e^{jkz(1+\cos\theta)} + e^{-jkz(1-\cos\theta)}\right)dz = \frac{2\cos\left(\frac{\pi}{2}\cos\theta\right)}{k\sin^2\theta} \tag{4.6}$$

where (4.6) gives the solution of the integral and the magnetic field in (4.5) consequently becomes:

$$H = \frac{jI_0e^{-jkr}}{2\pi r}\left(\frac{\cos\left(\frac{\pi}{2}\cos\theta\right)}{\sin\theta}\right)a_\varphi \tag{4.7}$$

and the electric field, in the far field:

$$E = jZ_0|H|a_\theta \tag{4.8}$$

Figure 4.5 shows the electric radiated field of a half-wave dipole, compared to that of a short current filament.

A MATLAB program has been developed to show the 2D elevation and azimuth planes of the E-plane and H-plane radiation patterns, respectively, as shown in Figure 4.6, as well as the 3D radiation pattern as shown in Figure 4.7.

MATLAB Program for Plotting the 2D Elevation and Azimuth Planes of the E-Plane and H-Plane as Well as the 3D Radiation Patterns

```
% This script generates 3D radiation pattern plots for a half
wave dipole
%  including the E-plane and H-plane radiation patterns. The
script creates:
```

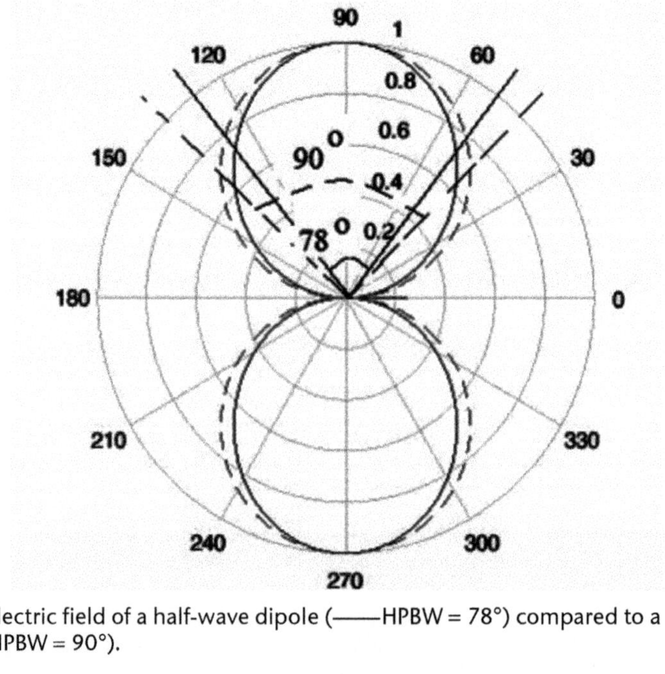

Figure 4.5 Electric field of a half-wave dipole (—— HPBW = 78°) compared to a short current filament (---- HPBW = 90°).

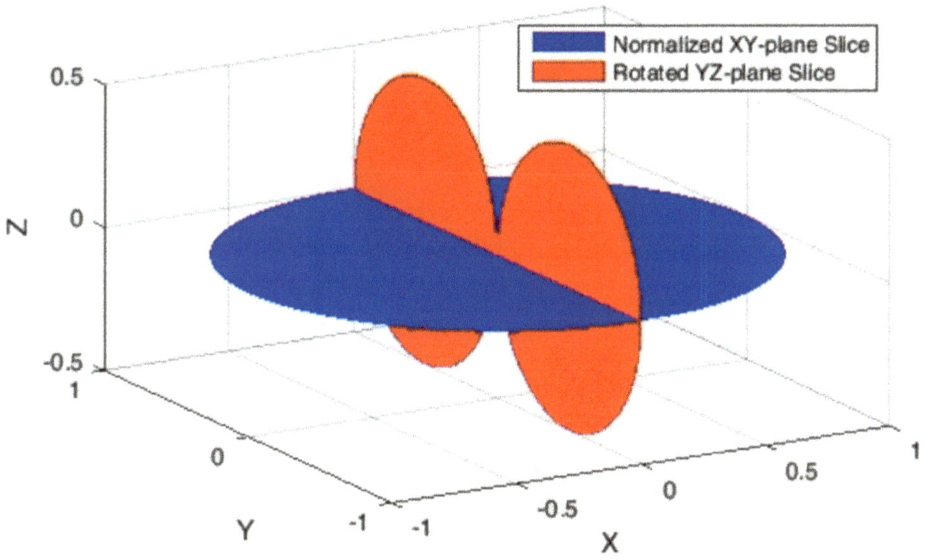

Figure 4.6 Elevation (E-plane) and azimuth (H-plane) patterns of a half-wave dipole.

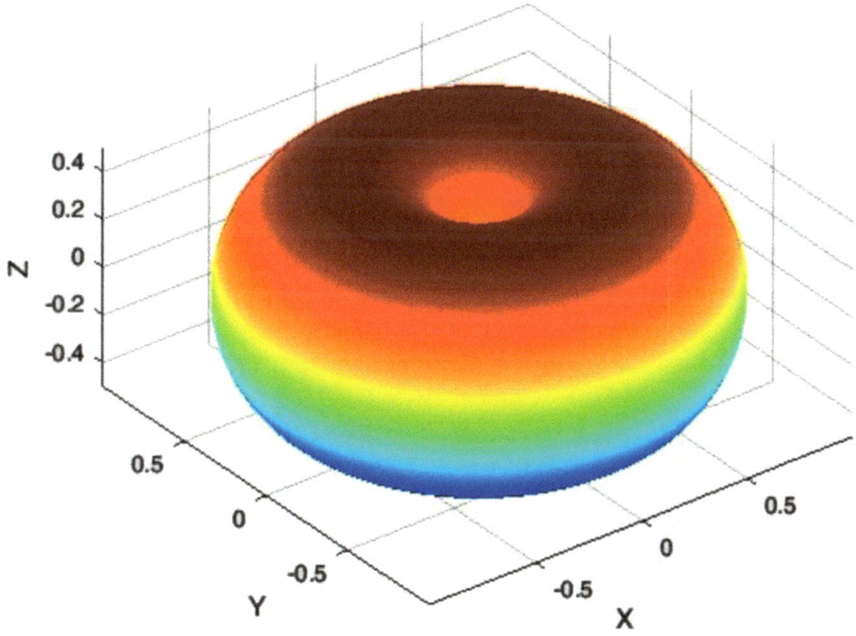

Figure 4.7 The 3D radiation pattern of a half-wave dipole.

```
% - A 3D surface plot of the radiation pattern
% - A slice at θ = 90 degrees (azimuth) for the XY-plane
% - A slice at ϕ = 90 degrees (elevation) for the YZ-plane
% - A combined 3D plot with both XY-plane and YZ-plane slices.

clc;
clear;
```

```
% Define theta and phi
theta = linspace(0, 2*pi, 360); % Azimuth angle (0 to 2*pi)
phi = linspace(0, pi, 180);     % Elevation angle (0 to pi)

[Theta, Phi] = meshgrid(theta, phi);

% Elevation (E-plane) pattern
E_pattern = (sin(Phi)); % E-plane radiation pattern
X_E = E_pattern .* sin(Phi) .* cos(Theta); % Convert to Cartesian
Y_E = E_pattern .* sin(Phi) .* sin(Theta);
Z_E = E_pattern .* cos(Phi);

% Azimuth (H-plane) pattern
H_pattern = abs(cos(Theta)); % H-plane radiation pattern
X_H = H_pattern .* sin(Phi) .* cos(Theta); % Convert to Cartesian
Y_H = H_pattern .* sin(Phi) .* sin(Theta);
Z_H = H_pattern .* cos(Phi);

% Plot Figure: Original 3D Radiation Pattern
figure;
surf(X_E, Y_E, Z_E, 'EdgeColor', 'none', 'FaceAlpha', 1.0);
colormap(jet);
title('3D Half-wave Dipole Radiation pattern');
xlabel('X');
ylabel('Y');
zlabel('Z');
axis equal;
grid on;
colorbar;

% Plot Figure: Slice with θ fixed at 90 degrees (XY-plane)
theta_fixed = pi/2;  % θ = 90 degrees
phi_fixed = linspace(0, pi, 180);  % Varying ϕ for this slice

% Calculate the slice at θ = 90 degrees (XY-plane)
E_pattern_fixed_theta = sin(phi_fixed);  % E-plane pattern at
fixed θ = 90 degrees
X_E_fixed_theta = E_pattern_fixed_theta .* sin(phi_fixed); % X
and Y components
Y_E_fixed_theta = E_pattern_fixed_theta .* cos(phi_fixed); % Y
and Z components
Z_E_fixed_theta = zeros(size(X_E_fixed_theta));  % No Z component
for the XY-plane slice

% Shift X coordinates by -0.5
X_E_fixed_theta_shifted = X_E_fixed_theta - 0.5;

% Normalize the XY plot range to [-1, +1]
X_min = min(X_E_fixed_theta_shifted);
X_max = max(X_E_fixed_theta_shifted);
```

```
Y_min = min(Y_E_fixed_theta);
Y_max = max(Y_E_fixed_theta);

% Normalize X and Y to be within [-1, +1]
X_E_fixed_theta_normalized = -1 + (X_E_fixed_theta_shifted - X_
min) * (1 - (-1)) / (X_max - X_min);
Y_E_fixed_theta_normalized = -1 + (Y_E_fixed_theta - Y_min) * (1
- (-1)) / (Y_max - Y_min);

% Plot the normalized H-plane slice at θ = 90 degrees in 3D
(XY-plane) with blue color
figure;
fill3(X_E_fixed_theta_normalized, Y_E_fixed_theta_normalized,
Z_E_fixed_theta, 'b', 'FaceAlpha', 1.0);  % Fill inside with
blue color
title('Normalized (Azimuth) H-plane Radiation Pattern');
xlabel('X');
ylabel('Y');
zlabel('Z');
grid on;
axis equal;

% Plot Figure: Slice with ϕ fixed at 90 degrees (YZ-plane)
phi_fixed = pi/2;  % ϕ = 90 degrees
theta_fixed = linspace(0, 2*pi, 360);  % Varying θ for this slice

% Calculate the slice at ϕ = 90 degrees (YZ-plane)
H_pattern_fixed_phi = abs(cos(theta_fixed));  % H-plane pattern
at fixed ϕ = 90 degrees
X_H_fixed_phi = zeros(size(H_pattern_fixed_phi)); % No X
component for the YZ-plane slice
Y_H_fixed_phi = H_pattern_fixed_phi .* sin(theta_fixed); %
Y component
Z_H_fixed_phi = H_pattern_fixed_phi .* cos(theta_fixed); %
Z component

% Rotate the YZ-plane slice by 90 degrees around the X-axis
theta_rotation = pi/2;  % 90 degrees
Y_H_rotated = Y_H_fixed_phi .* cos(theta_rotation) - Z_H_fixed_
phi .* sin(theta_rotation);
Z_H_rotated = Y_H_fixed_phi .* sin(theta_rotation) + Z_H_fixed_
phi .* cos(theta_rotation);

% Plot the filled rotated H-plane slice at ϕ = 90 degrees in 3D
(YZ-plane) with red color
figure;
fill3(X_H_fixed_phi, Y_H_rotated, Z_H_rotated, 'r', 'FaceAlpha',
1.0);  % Fill inside with red color
title('Elevation E-plane Radiation Pattern');
xlabel('X');
```

```
ylabel('Y');
zlabel('Z');
grid on;
axis equal;

% Combine XY-plane and rotated YZ-plane slices into a single
3D plot
figure;
hold on;

% Plot normalized XY-plane slice
fill3(X_E_fixed_theta_normalized, Y_E_fixed_theta_normalized,
Z_E_fixed_theta, 'b', 'FaceAlpha', 1.0, 'EdgeColor', 'none'); %
Blue

% Plot rotated YZ-plane slice
fill3(X_H_fixed_phi, Y_H_rotated, Z_H_rotated, 'r', 'FaceAlpha',
1.0, 'EdgeColor', 'none'); % Red

% Plot Figure: Combine XY-plane and YZ-plane radiation patterns
title('Combined Normalized XY-plane and YZ-plane
Radiation Patterns');
xlabel('X');
ylabel('Y');
zlabel('Z');
grid on;
axis equal;
view(-30, 15); % 3D view

% Extend the range of coordinates in X and Y from -1 to +1
xlim([-1, 1]);  % Extend X-axis range
ylim([-1, 1]);  % Extend Y-axis range

legend('Normalized XY-plane Slice', 'Rotated YZ-plane Slice');
hold off;
```

The power flux intensity, $\boldsymbol{P_r/d\Omega}$, works out as:

$$\frac{\boldsymbol{P_r}}{d\Omega} = \frac{1}{2}\boldsymbol{r}^2 \operatorname{Re} \boldsymbol{E} \times \boldsymbol{H}^* = \boldsymbol{I}_0^2 \frac{\boldsymbol{Z}_0}{8\pi^2}\left(\frac{\cos\left(\frac{\pi}{2}\cos\theta\right)}{\sin\theta}\right)^2 \tag{4.9}$$

$$P_r = \boldsymbol{I}_0^2 \frac{\boldsymbol{Z}_0}{8\pi^2} \int_0^{2\pi}\int_0^{\pi} \left(\frac{\cos\left(\frac{\pi}{2}\cos\theta\right)}{\sin\theta}\right)^2 \sin\theta \, d\theta \, d\varphi \tag{4.10}$$

Equation (4.10) can only be numerically solved. It follows that:

$$P_r = 36.568 I_0^2 \tag{4.11}$$

Now substituting (4.9) and (4.11) into the directivity function (3.36) yields the half-wave dipole:

$$D(\theta,\varphi) = 1.64\left(\frac{\cos\left(\frac{\pi}{2}\cos\theta\right)}{\sin\theta}\right)^2 \tag{4.12}$$

It is worth noting that the maximum directivity, D, in decibels yields 10 log (1.64) = 2.15 dB. For a short current filament, D works out as 10log (1.5) = 1.76 dB, and for an s.i.r., it works out as D = 10log(1) = 0 dB. We can also measure directivity or gain in dBd units. So an antenna with a gain of 10 dB is the same as 7.85 dBd, where the former gain is relative to an s.i.r. and the latter equal gain is relative to a half-wave dipole.

It is now possible to find the value of the radiation resistance, R_a, by substituting (4.11) into (3.43), for the half-wave dipole results, and $R_a = 73.136\Omega$.

It is worth mentioning that an empirical spiral impedance diagram was published in [2] for a dipole of a length-to-diameter ratio equal to 60, from which four resonant dipole lengths were shown at $l = 0.435\lambda$, 0.87λ, 1.31λ, and 1.75λ. These gave rise to real values of input impedance, $R = 67\Omega$, 336Ω, 95Ω, and 237Ω, respectively. It was noticed that the dipole lengths, $l = 0.435\lambda$ and $l = 1.31\lambda$, give rise to odd resonance, and $l = 0.87\lambda$ and $l = 1.75\lambda$ gave rise to even resonance. Odd and even resonance is associated with the number of half sinusoids of distributed current along the length of the dipole. Odd resonance results into the maximum current value at the center feed and even resonance results in an associated minimum current. Hence, the resistance values for odd resonance lengths are relatively lower than those for even resonance lengths.

4.4 Mutual Impedance of Coupled Parallel Dipoles

The mutual impedance, Z_{ba}, of two coupled dipole antennas, shown in Figure 4.8, is defined as the negative of the ratio of the emf V_{ba}, induced in dipole b to the current, i_a, flowing in dipole a. The mutual impedance can hence be expressed as:

$$Z_{ba} = -\frac{V_{ba}}{i_a} \tag{4.13}$$

If both dipoles are symmetrical, then by reciprocity, $Z_{ab} = Z_{ba}$, where:

$$Z_{ab} = -\frac{V_{ab}}{i_b} \tag{4.14}$$

The mutual impedance, in Figure 4.8, depends on the variation of spatial distance *d*. Figure 4.9 demonstrates the mutual resistance and reactance variation with the spatial distance between the two dipoles. Both curves vary on both sides of the zero impedance axis.

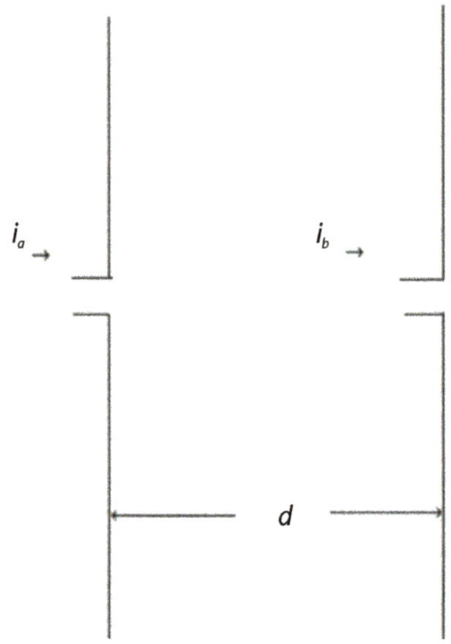

Figure 4.8 Two half-wave dipoles spaced by a distance, *d*, comparable to λ.

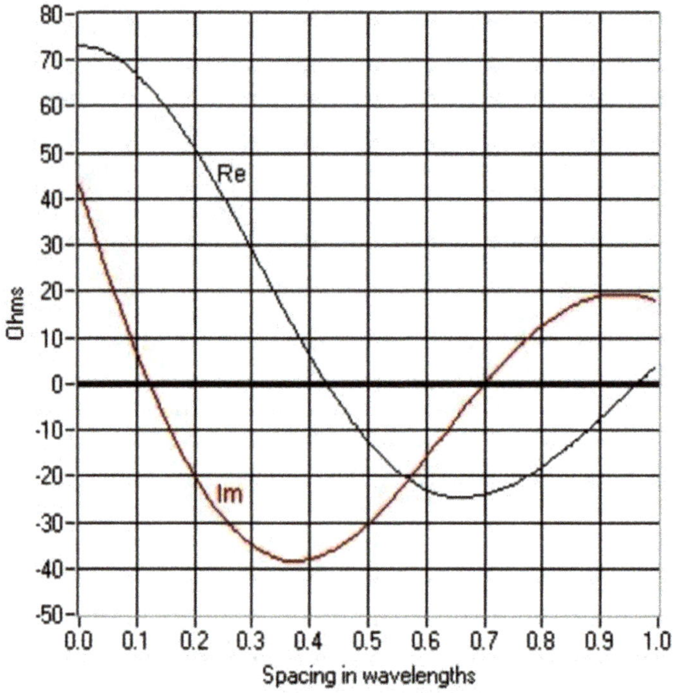

Figure 4.9 Mutual impedance of parallel nonstaggered half-wave dipoles. (Source: Wikimedia Commons.)

It is very common and practical to emphasize that not all elements of an array of multiple dipoles need to be energized with current sources. Usually, one element is energized with a source, the driver, while the others, the parasitic elements, are not. A simple characterization of two elements, shown in Figure 4.8, follows, with one dipole energized, while the other is a parasitic element:

$$V_a = i_a Z_a + i_b Z_{ab} \tag{4.15}$$

$$V_b = 0 = i_a Z_{ab} + i_b Z_b \tag{4.16}$$

where Z_a is the self-impedance of dipole a, Z_b is the self-impedance of dipole b, and Z_{ab} is the mutual impedance between the two dipoles, which varies with d, as shown in Figure 4.8.

For a symmetrical pair of dipoles, the mutual impedance is best defined as follows:

$$Z_{ab} = -\frac{E_b}{i_a} = -\frac{E_a}{i_b} \tag{4.17}$$

where E_b is the emf induced in dipole b due to current i_a in dipole a, and E_a is the emf induced in dipole z due to current i_b in dipole b. The negative sign is due to the opposite variations between the input voltage and the emf. The self-impedances, Z_a and Z_b, are simply V_a/i_a and V_b/i_b, respectively, where V_a and V_b are the input voltages of the respective dipoles.

We demonstrate an array of three dipoles that are separated by $\lambda/2$, as shown in Figure 4.10, from which the maximum gain of the array, relative to an isolated dipole, is to be calculated.

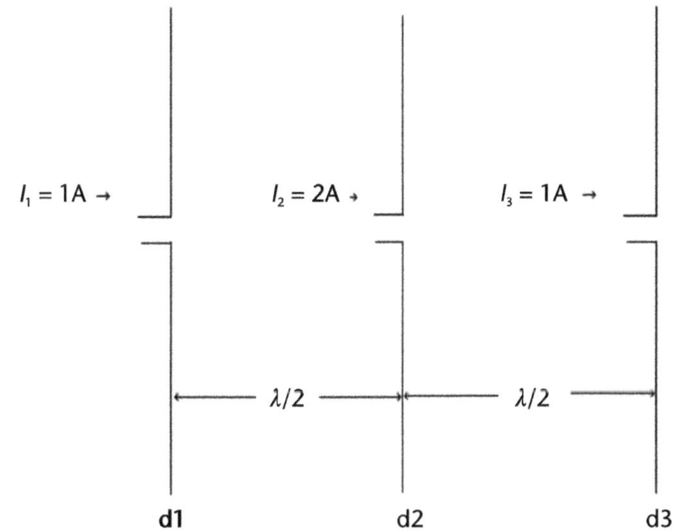

Figure 4.10 An array of three dipoles spaced by a distance, $\lambda/2$, with currents of unequal magnitude and all current in-phase.

Given that the self-impedance, for each dipole, $Z_{11} = Z_{22} = Z_{33} = 73\Omega$, while the mutual impedance, $Z_{12} = Z_{32} = -12.7\Omega$, and $Z_{13} = 3.8\Omega$, the following matric equation is proposed:

$$[V] = [Z][I] \tag{4.18}$$

Expanding the following equations give the following:

$$V_1 = Z_{11}I_1 + Z_{12}I_2 + Z_{13}I_3 \tag{4.19}$$

$$V_2 = Z_{21}I_1 + Z_{22}I_2 + Z_{23}I_3 \tag{4.20}$$

$$V_3 = V_1 \tag{4.21}$$

The input impedance, Z_{in1}, can be derived and calculated by substituting the given values in (4.19). Hence,

$$Z_{in1} = Z_{11} + Z_{12}\frac{I_2}{I_1} + Z_{13} = Z_{13} = 51.4\Omega \tag{4.22}$$

Similarly, Z_{in2} can be derived and calculated by substituting the provided values in (4.20). Hence,

$$Z_{in2} = Z_{22} + 2Z_{12}\frac{I_1}{I_2} + Z_{13} = 60.3\Omega \tag{4.23}$$

Now, the input power, P_a, to the array can be calculated as:

$$P_a = \frac{1}{2}I^2 Z_{in\_total} = \frac{1}{2}(51.4 + 4 \times 60.3 + 51.4) = \frac{344}{2}\text{W} \tag{4.24}$$

The input power P_d to a reference half-wave dipole of a standard input impedance = 73Ω to produce the same field should be energized with a total current of 4A. Hence,

$$P_d = \frac{1}{2}\left(I_{total}^2 \times 73\right) = \frac{1168}{2}\text{W} \tag{4.25}$$

Consequently, the maximum gain of the array works out as $P_d/P_a = 3.4 = 5.3$ dB.

At this point, multiple antenna elements and the radiation analysis of antenna arrays will be discussed, in detail, in the next chapter.

Solved Example

Example 4.1 A broadcast antenna, which is 5m long, transmits at 30 MHz.

(a) Determine what type of an antenna is used. What is your reason?

(b) Determine the value of the radiation resistance of the broadcast antenna.

(c) Calculate the length of another antenna, radiating at 30 MHz, with a radiation resistance equal to 1/10 of the radiation resistance of the previous antenna.

(d) Calculate the directivity of the new antenna (i.e., the antenna in item c) at $\theta = 30°$.

Solution 4.1

(a) Determine the type of antenna used and the reason.

The wavelength is calculated as:

$\lambda = c/f = 3 \times 10^8/30 \times 10^6 = 10\text{m}$

Since the antenna length is 5m, we compare it to the wavelength: $l = \lambda/2$.

This indicates that the antenna is a half-wave dipole antenna because its length is half the wavelength, which is a common design choice for efficient radiation.

(b) Determine the radiation resistance of the broadcast antenna. For a half-wave dipole, the typical radiation resistance is: $R_r \approx 73\Omega$.

(c) Calculate the length of another antenna radiating at 30 MHz, with a radiation resistance equal to 1/10 of the previous antenna.

The new radiation resistance should be:

$R_r' = 73/10 = 7.3\Omega$

For a thin cylindrical dipole, radiation resistance is approximately given by:

$R_r' = 7.3 = 80 \times \pi^2 \times (l/\lambda)^2 = 80 \times 9.8696 \times (1/10)^2$

$l = 0.96\text{m}$

(d) Calculate the directivity of the new antenna at $\theta = 30°$.

$$D(\theta,\varphi) = 1.5x\sin^2\theta = 1.5 \times 0.25 = 0.375$$

Example 4.2 Develop a MATLAB code that provides a detailed visualization of the impedance behavior of the half-wave dipole antenna over a frequency range between 90 MHz and 110 MHz. Utilize the following model in your program and plot the solution.

Proposed Model The input impedance $Z_{\text{in}}(f)$ of a half-wave dipole is modeled as:

$$Z_{\text{in}}(f) = R + jX(f)$$

where $R \approx 73\Omega$ is the radiation resistance at resonance and $X(f) = 70 \cdot (\sin(2\pi f/f_0 - 1))$ is the reactance, where $f_0 = 100$ MHz is the resonant frequency.

This model assumes that the antenna is resonant at 100 MHz and reactance changes with frequency.

Solution 4.2

```
% MATLAB program to plot the impedance behavior of the half-wave
dipole antenna over a frequency range between 90 MHz and 110 MHz.
% Constants
f0 = 100e6;  % Resonant frequency (100 MHz)
R = 73;      % Radiation resistance (in Ohms)
f_min = 90e6;  % Minimum frequency for bandwidth analysis
f_max = 110e6; % Maximum frequency for bandwidth analysis
frequencies = linspace(f_min, f_max, 1000); % Frequency range

% Reactance model
X = 70 * sin(2 * pi *( frequencies / f0 - 1));

% Impedance model
Z_in = R + 1i * X;

% Plot the real and imaginary parts of the impedance
figure;
subplot(2, 1, 1);
plot(frequencies / 1e6, real(Z_in), 'DisplayName', 'Real Part
(R)');
hold on;
plot(frequencies / 1e6, imag(Z_in), 'k--', 'DisplayName',
'Imaginary Part (X)');
title('Input Impedance of Half-Wave Dipole Antenna');
xlabel('Frequency (MHz)');
ylabel('Impedance (Ohms)');
legend('show');  % Show legend for the first subplot
grid on;

% Plot the magnitude of the impedance
subplot(2, 1, 2);
plot(frequencies / 1e6, abs(Z_in), 'k', 'DisplayName',
'|Z_{in}|');
title('Magnitude of Input Impedance');
xlabel('Frequency (MHz)');
ylabel('abs(Zin)(Ohms)');
legend('show');  % Show legend for the second subplot
grid on;
```

Figure 4.11 demonstrates the variation of half-wave dipole impedance with frequency from 90 MHz to 110 MHz.

Problems for Students

4.1 A half-wave dipole antenna operates at a frequency of 100 MHz. Calculate the following:
   (a) The physical length of the dipole;
   (b) The radiation resistance of the dipole;

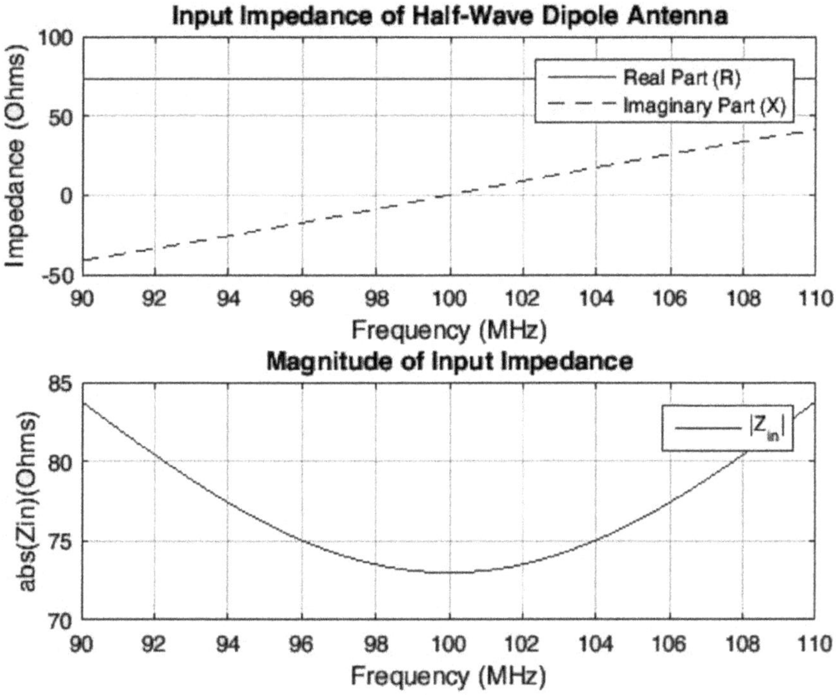

Figure 4.11 Variation of half-wave dipole impedance with frequency.

(c) The directivity of the dipole antenna.

4.2 Two identical dipoles, separated by $\lambda/2$, have impedance of $75 + j0\Omega$ and $25 + j25\Omega$, respectively, and the mutual impedance is $50 + j0\Omega$. The first dipole is driven by a voltage of 25V, while the second is a parasitic element. Determine the currents in the two radiators, in magnitude and phase.

4.3 Two identical monopoles, separated by $\lambda/2$ have a mutual impedance of $25\sqrt{2}\,\exp(j\pi/4)\Omega$. Knowing that Z_{in} of a monopole is one-half the value of a half-wave dipole and one of the monopoles is driven by a voltage of 12V, while the other one is a parasitic element (i.e., not driven). Determine the currents in the two monopoles in magnitude and in phase.

4.4 Derive and develop a MATLAB code to plot the electric and magnetic radiation far fields of a full-wave dipole. Discuss the main features of these patterns and explain the null directions and maximum radiation directions.

4.5 A half-wave dipole antenna has a feed impedance of $73 + j42.5\Omega$. Design a quarter-wave transformer, at $f = 1$ GHz, to match this antenna to a 50Ω transmission line. Calculate the characteristic impedance, Z_0, and the physical length of the transformer.

4.6 A half-wave dipole antenna is designed for use at a frequency of 2.4 GHz.

(a) Calculate the resonant frequency of the antenna in terms of its physical length. What is the corresponding wavelength for the antenna at this frequency?

(b) Assume that the antenna operates with a typical quality factor (Q) of 30. Calculate the bandwidth of the antenna using the relationship: bandwidth $= f_r/Q$ where f_r is the resonant frequency.

(c) If the antenna has a loss resistance of 10Ω and the radiation resistance is 73Ω, calculate the efficiency of the antenna. Discuss the effect of losses on antenna performance.

4.7 Consider a half-wave dipole antenna placed above a perfect ground plane.

(a) Calculate the input impedance of the dipole antenna placed above the ground plane. Compare this with the impedance of a free-standing dipole antenna and explain the differences.

(b) Given that the dipole is above a ground plane, determine how the radiation pattern is modified, particularly in the lower hemisphere (below the antenna). How does the ground plane affect the directivity and gain of the antenna?

(c) If the dipole is placed at a height of 0.25λ above the ground, calculate the approximate effective area of the antenna. How would this change if the height were increased to 0.5λ?

Problems for Instructors

4.1 Advanced Derivation of Input Impedance: Derive the input impedance of a half-wave dipole antenna using the Pocklington integral equation. Explain the impact of antenna thickness on input impedance and discuss the approximations used in the analysis.

Note that the Pocklington integral equation describes the current distribution $I(z)$ on a thin wire antenna. For a straight, thin wire antenna of length L, the equation is given by

$$\frac{d^2 I(z)}{dz^2} + k^2 I(z) = -j\omega\mu J(z)$$

where $k = 2\pi/\lambda$ is the wavenumber, ω is the angular frequency, μ is the permeability of the medium, and $J(z)$ is the impressed current source (voltage excitation).

4.2 Design Optimization for Monopole Antenna: Design a monopole antenna with a ground plane to achieve a specific resonant frequency. Include the effects of ground plane size, material conductivity, and antenna height. Compare the radiation pattern with that of a half-wave dipole.

4.3 Mutual Coupling Analysis for Array Design: Given two parallel half-wave dipole antennas placed at varying distances, compute the mutual impedance using induced EMF methods. Analyze how mutual coupling affects array performance in terms of radiation pattern, impedance matching, and beamforming.

4.4 Nonideal Conditions in Practical Dipole Design: Investigate the performance of a half-wave dipole antenna in a lossy medium, such as seawater or ionized plasma. Include the impact of material conductivity, relative permittivity, and magnetic permeability on radiation efficiency and pattern.

4.5 Comparison of Monopole and Dipole Antennas in Urban Environments: Conduct a performance comparison between a half-wave dipole and a monopole antenna in an urban setting. Discuss the influence of nearby structures on antenna gain, directivity, and bandwidth.

4.6 Experimental Measurement of Mutual Impedance: Devise an experimental setup to measure the mutual impedance of two coupled parallel dipoles. Outline the key parameters to measure, such as spacing, frequency, and orientation, and provide a detailed explanation of the measurement techniques.

4.7 Multidipole System Performance Evaluation: Analyze the performance of a three-element dipole array system. Compute the mutual impedance between all elements and evaluate how element spacing affects the overall radiation pattern, gain, and input impedance of the system.

References

[1] Brown, G. H., and O. M. Woodward, "Experimentally Determined Impedance Characteristics of Cylindrical Antennas," *Proceedings of the IRE*, Vol. 33, No. 4, 1945, pp. 257–262.

[2] Kraus, J. D., *Antennas*, Second Edition, New York: McGraw-Hill, 1988, p. 374.

Selected Bibliography

Carr, J., and G. Hippisley, *Practical Antenna Handbook*, Fifth Edition, New York: McGraw-Hill/TAB Electronics, 2011.

Friesen, D. J., *Introduction to Antennas and RF Propagation Analysis*. Norwood, MA: Artech House, 2021.

Geyi, W., *Foundations of Antenna Radiation Theory: Eigenmode Analysis*, New York: John Wiley & Sons, 2023.

Stutzman, W. L., and G. A. Thiele, *Antenna Theory and Design*, New York: John Wiley & Sons, 2012.

CHAPTER 5

Radiation from Multiple Elements

5.1 Introduction

In this chapter, we analyze the radiation characteristics of multiple antenna elements, focusing specifically on arrays of thin, linear antennas. We begin with a systematic approach to determining the far-field radiation pattern and gain of these arrays, which are foundational in understanding how antenna arrays behave in practice. We introduce the fundamental concept of array design by considering the simplest form, an array consisting of two driven half-wavelength elements. The term "element" refers to the basic unit antenna that forms the building block of any array structure. These elements are assumed to be thin and linear, which simplifies the analysis while still providing a robust foundation for more complex configurations.

We then extend our discussion to more sophisticated antenna array structures, such as uniform linear arrays (ULAs), end-fire arrays, uniform circular arrays (UCAs), planar arrays, phased arrays, and conformal arrays. Each of these configurations has distinct radiation characteristics that make them optimal for a variety of applications, including wireless communication, radar systems, satellite communication, and the evolving needs of next-generation networks. The ULA has maximum radiation at 90°, which is perpendicular to the array, making it ideal for broadside configurations. In contrast, the end-fire array directs maximum radiation at 0° or 180°, requiring phase progression ($\beta = \pm kd$) for optimal performance. We also examine the UCA, which supports omnidirectional or steerable beamforming, allowing for dynamic beam control. The planar array enables 2D beam control for both azimuth and elevation, which is particularly useful in applications that require coverage across a wide range of directions. The phased array enables electronic beam steering by adjusting phase shifts, providing rapid and precise control over the direction of the radiation beam. The conformal array is specifically designed to be mounted on curved surfaces, offering specialized solutions for complex applications such as aircraft and spacecraft communications. Each of these array types exhibits unique features that make them suitable for specific applications, ranging from wireless communication to radar systems, satellite communication, and emerging 5G/6G networks.

These arrays differ in their element arrangements, radiation patterns, and how they handle beam shaping and directivity, which are essential for improving performance in modern communication systems. A key concept that we explore in this chapter is pattern multiplication, a principle that simplifies the prediction of the total radiation pattern of an array. According to this principle, the overall radiation pattern of an array can be approximated as the product of the individual radiation patterns of each element. This concept is a powerful tool for predicting the performance of complex arrays without needing to perform exhaustive calculations for every possible configuration.

One of the main objectives in designing antenna arrays is to control the shape and direction of the beam, enhancing gain while minimizing unwanted radiation such as back radiation. We will also delve into methods for calculating gain using radiation pattern approximation, providing insights into how practical array designs achieve high directivity and efficiency. The gain of an antenna array is directly linked to its ability to focus energy in the desired direction, and techniques for improving this property are crucial for meeting the performance requirements of modern communication systems.

The field of antenna array design is rapidly evolving, particularly with the advent of 5G and 6G technologies, which demand more efficient, versatile, and high-performance antennas. As such, we will also explore emerging antenna technologies that are pushing the boundaries of what is possible in antenna design. For instance, the reconfigurable dipole array (RDA) integrates tunable elements, such as varactor diodes and micro-electro-mechanical systems (MEMS) switches, enabling dynamic beamforming and frequency adjustments in real time for adaptive wireless systems. The metamaterial-enhanced dipole array (MDA) employs advanced metamaterials to achieve enhanced gain, bandwidth, and beamforming characteristics, making it ideal for satellite communication and stealth applications. The fractal dipole array (FDA) utilizes self-similar fractal structures to create compact, wideband arrays suitable for IoT, wearable technology, and radar systems. The sparse dipole array (SDA), optimized with machine learning algorithms, can reduce the number of elements while maintaining high performance, making it a promising solution for 5G/6G networks. The hybrid optical-RF dipole array (HORDA) blends RF and photonic technologies to facilitate ultrahigh-speed data transmission in terahertz/millimeter-wave applications. Lastly, the 3D-printed dipole array (3D-PDA) uses additive manufacturing techniques to create lightweight, customizable antennas for use in aerospace and wearable devices.

This chapter provides both a theoretical foundation and a practical framework for understanding antenna arrays. It lays the groundwork for future advancements in antenna array technology, offering a comprehensive overview of current designs, as well as cutting-edge trends poised to shape the future of communication systems. As such, it serves as an essential resource for anyone interested in the rapidly advancing field of antenna technology.

5.2 Radiation from an Array of Two Driven Elements

Consider two identical radiating elements that are identical and normal to the page. The elements are spaced by a distance d and are assumed to be radiating a magnetic field at some angle φ. Figure 5.1 shows elements 1 and 2 radiating E_1 and E_2, respectively. The total radiation of the two elements is the vector sum of both field radiations. The magnitude of each field is proportional to the current that energizes the component. Hence, E_1 is proportional to I_1 and E_2 is proportional to

I_2. There would be a total phase difference, ψ, between the fields E_1 and E_2, where it has two components:

- The phase difference, α, between the currents I_1 and I_2, which energize the radiating elements.
- The phase difference due to the path difference, $kd\cos(\varphi)$, where $k = 2\pi/\lambda$.

Hence, the total phase difference:

$$\Psi = \alpha + kd\cos(\phi) \tag{5.1}$$

A vector diagram can be constructed for the total field $\mathbf{E}_t$ (Figure 5.2), where:

$$\mathbf{E}_t = \mathbf{E}_1 + \mathbf{E}_2 \tag{5.2}$$

If $\psi = 0, 2\pi, 4\pi$...and so forth and $|E_1| = |E_2|$, then $\mathbf{E}_1$ and $\mathbf{E}_2$ add in phase and $|E_t| = 2|E_1|$.

If $\psi = \pi, 3\pi, 5\pi$... and so forth and $|E_1| = |E_2|$, then $\mathbf{E}_1$ and $\mathbf{E}_2$ subtract in anti-phase and $|E_t| = 0$.

Let us consider the simple case where the two elements are isotropic point sources of the same amplitude. The two-point sources in Figure 5.1 are separated by a distance of d and located symmetrically with respect to the origin of the x-coordinate system. The origin is taken as a reference for phase. At a distant point, r in the direction φ, the field from source 1 is retarded by $\psi/2$, while the field from source 2 is advanced by $\psi/2$. The total field is then:

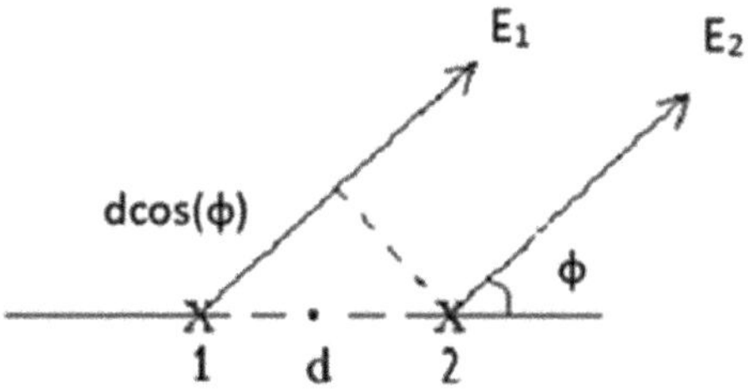

Figure 5.1 An array with two identical radiators separated by a distance *d*.

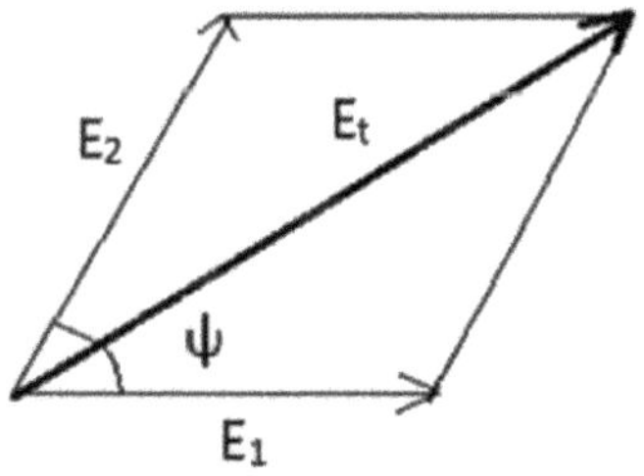

Figure 5.2 A vector diagram of the total field $\mathbf{E}_t$.

$$E_t = E_0 e^{-j\frac{\Psi}{2}} + E_0 e^{j\frac{\Psi}{2}} = 2E_0 \cos\left(\frac{\Psi}{2}\right) \tag{5.3}$$

The effect of changing the current and spatial phase on the radiation pattern of the double-element array can be shown. Figure 5.3 demonstrates normalized radiation patterns with various effect of the current phase and spatial phase from a double element antenna array.

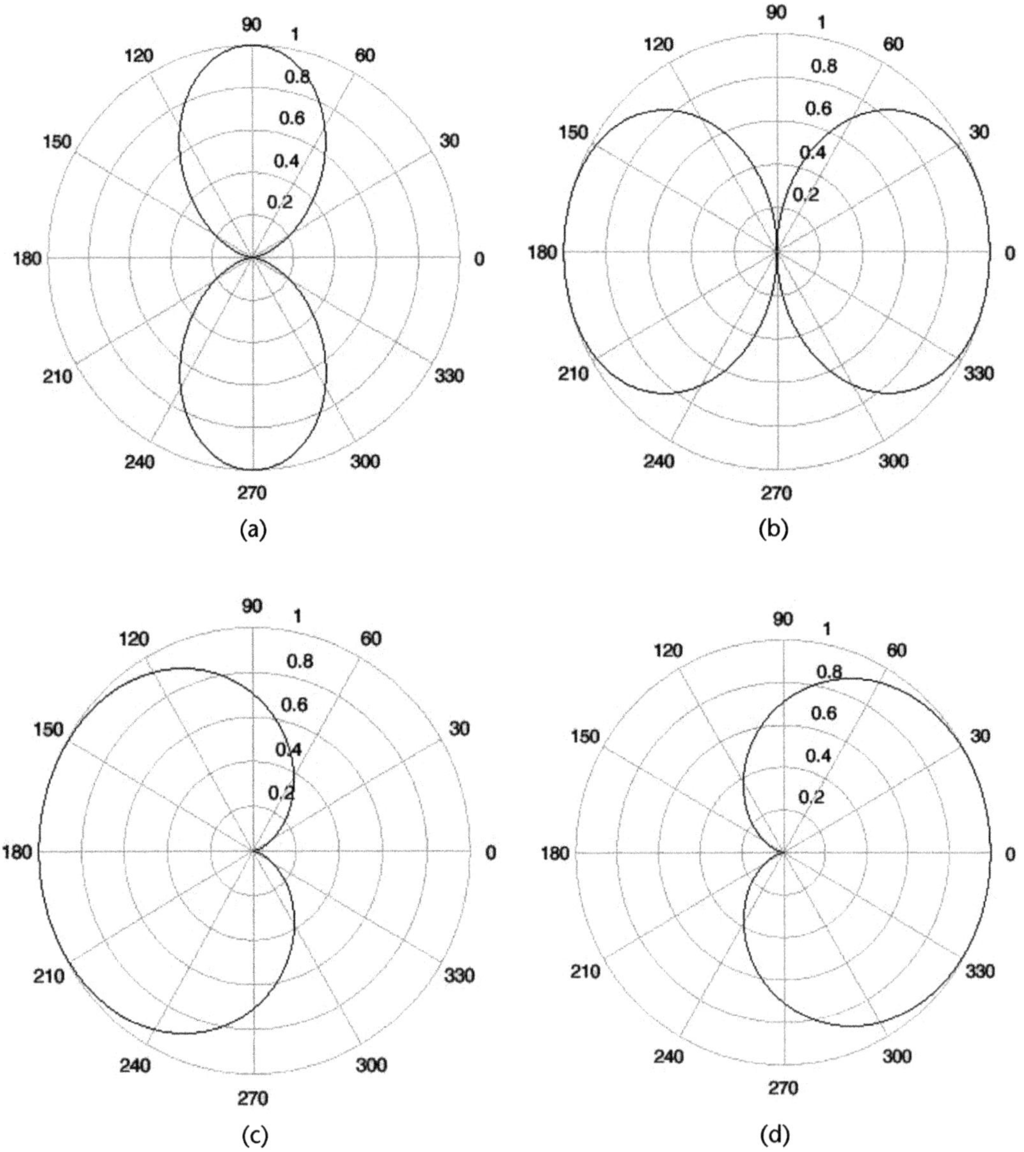

Figure 5.3 Normalized radiation patterns with various effects of the current phase and spatial phase from a double-element antenna array. (a) $\Psi = 0 + \pi \cos(\varphi)$, (b) $\Psi = \pi + \pi \cos(\varphi)$, (c) $\Psi = \pi/2 + \pi/2 \cos(\varphi)$, (d) $\Psi = -\pi/2 + \pi/2 \cos(\varphi)$, (e) $\Psi = \pi/4 + 5\pi/4 \cos(\varphi)$, (f) $\Psi = 0 + 2\pi \cos(\varphi)$, (g) $\Psi = -\pi/2 + 2\pi \cos(\varphi)$, and (h) $\Psi = \pi + 2\pi \cos(\varphi)$.

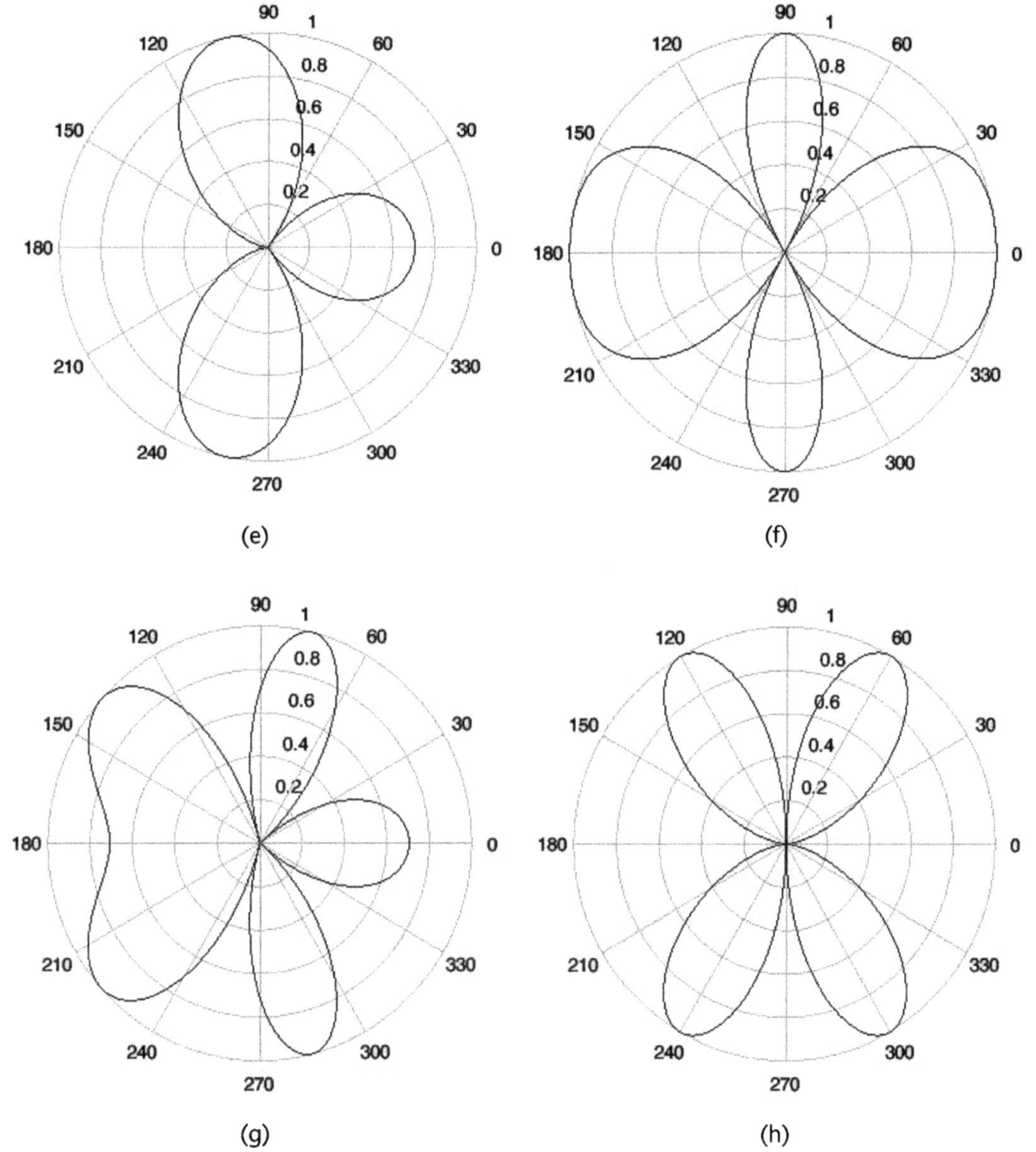

Figure 5.3 *(Continued)*

5.3 Theory of Linear Arrays

Consider a linear array of n isotropic elements with uniform separation distance d, as shown in Figure 5.4. It shows that there is a progressive phase shift of $kd\cos\varphi$ due to spacing. If at a distance r, the field due to i_0 is H_0/unit current, then the total field, H_t, would be:

$$H_t = H_0\left\{i_0 + i_1 e^{jkd\cos\varphi} + i_2 e^{j2kd\cos\varphi} + \cdots + i_{n-1} e^{j(n-1)kd\cos\varphi}\right\} \tag{5.4}$$

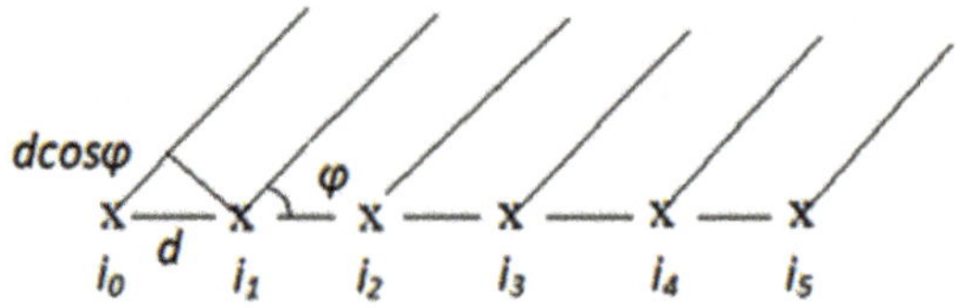

Figure 5.4 Antenna array with *n*-driven identical elements and uniform separation *d*.

For a ULA, the following conditions must be satisfied:

- Currents are all at the same magnitude, that is,

$$|i_0| = |i_1| = |i_2| = |i_{n-1}|$$

- There is a uniform progressive phase shift of currents along the array, where i_1 leads i_0 by phase angle α, i_2 leads i_0 by phase angle 2α, and i_{n-1} leads i_0 by phase angle $(n-1)\alpha$, that is,

$$i_1 = i_0 e^{j\alpha},\ i_2 = i_0 e^{j2\alpha} i_{n-1} = i_0 e^{j(n-1)\alpha}$$

It follows that:

$$H_t = H_0 \left\{ \begin{array}{c} i_0 + i_0 e^{j\alpha} e^{jkd\cos\varphi} + i_0 e^{j2\alpha} e^{j2kd\cos\varphi} + \cdots \\ + i_0 e^{j(n-1)\alpha} e^{j(n-1)kd\cos\varphi} \end{array} \right\} \tag{5.5}$$

Rearranging (5.5) yields:

$$H_t = H_0 i_0 \left\{1 + e^{j\Psi} + e^{j2\Psi} + \cdots + e^{j(n-1)\Psi}\right\} = H_0 i_0 F(\varphi) \tag{5.6}$$

where ψ is expressed in (5.1) and $F(\varphi)$ is called the array factor, which gives the radiation pattern.

$F(\varphi)$ is a geometrical progression of the form:

$$F(\varphi) = 1 + a + a^2 + a^3 + \cdots a^{n-1} \tag{5.7}$$

where $a = e^{j\Psi}$, $a^2 = e^{j2\Psi}$, $a^{n-1} = e^{j(n-1)\Psi}$ ... and so forth.

Multiplying both sides with a yields:

$$aF(\varphi) = a + a^2 + a^3 + \cdots a^{n-1} + a^n \tag{5.8}$$

Subtracting (5.8) from (5.7) gives after substituting back, a, in exponential form:

$$F(\varphi) = \frac{1 - e^{jn\Psi}}{1 - e^{j\Psi}} \tag{5.9}$$

The magnitude of the antenna array factor:

$$|F(\varphi)| = \left|e^{j(n-1)\frac{\Psi}{2}}\right| \left|\frac{e^{-jn(\psi/2)} - e^{jn(\psi/2)}}{e^{-j(\psi/2)} - e^{j(\psi/2)}}\right| \tag{5.10}$$

Simplifying yields:

$$|F(\varphi)| = \left|\frac{\sin\left(\frac{n\Psi}{2}\right)}{\sin\left(\frac{\Psi}{2}\right)}\right| \tag{5.11}$$

Figure 5.5 represents the antenna array factor, $F(\varphi)$, which can be generated from the product of $|\sin(n\psi/2)|$ and $1/|\sin(\psi/2)|$. The magnitude of the array actor as $\psi/2 \to 0$ works out as n, which equals the number of the array elements. Furthermore, the shape of the curve depends only on n, hence called the uniform radiation pattern for the uniform linear array. It provides all information about the radiation pattern. The maximum radiation of the main beam occurs at $\psi/2 = 0$ and $|F|_{\psi/2\to 0} = n$. Zero radiation occurs at $\psi/2 = \pm\pi/n, \pm 2\pi/n, \pm 3\pi/n, \pm 4\pi/n$. The sidelobe maxima occur at $\psi/2 = \pm 3\pi/2n, \pm 5\pi/2n, \pm 7\pi/2n$. The magnitude ratio of the first, second, or third sidelobe is relative to the maximum magnitude of the main lobe and can be worked out from the array factor expressed in (5.11). So the first sidelobe maximum works out: $A_1 = |\sin(5*(3\pi/10))/\sin(3\pi/10)| = 1.236$, and, hence, $A_1/5 = 0.247$. Similarly, the second sidelobe maximum works out: $A_2 = |\sin(5*(5\pi/10))/\sin(5\pi/10)| = 1$, and, hence, $A_2/5 = 0.2$.

Consider an array of six driven identical elements with zero current phase, $\alpha = 0$, and spatial distance between successive elements, $d = \lambda/2$. Substitution into (5.1) yields $\psi = \pi\cos(\varphi)$ or $\psi/2 = \pi\cos(\varphi)/2$. The range of $\psi/2$ can be defined from $\pi/2$ to $-\pi/2$ by substituting $\varphi = 0, \pi$, respectively. Accordingly, the array factor with six driven elements where ($\alpha = 0$, $d = \lambda/2$) is shown in Figure 5.6. There

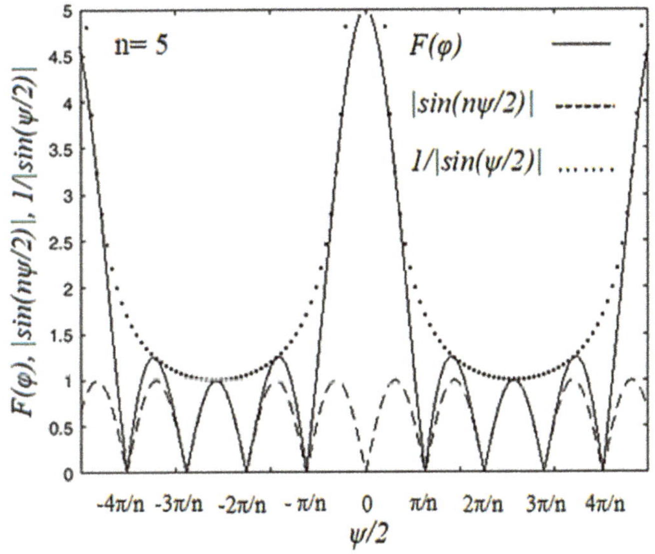

Figure 5.5 Antenna array factor, $F(\varphi)$, with n driven elements.

are two sidelobes on each side of the main lobe. The normalized sidelobe ratio is 1:1.414/6:1.035/6, which is 1:0.236:0.173. The main beam occurs when $\psi/2 = \pi \cos(\varphi)/2 = 0$, from which $\varphi = 90°$. Furthermore, zero radiation occurs at $\psi/2 = \pm\pi/6$, $\pm\pi/3$, $\pm\pi/2$, from which $\varphi = 70.5°$, $109.5°$, $48.2°$, $131.8°$, $0°$, and $180°$, respectively. Maximum radiation occurs at $\psi/2 = \pm\pi/4$ and $\pm5\pi/12$, from which $\varphi = 60°$, $120°$, $33.6°$, and $146.4°$, respectively. For completion of the radiation space between $\varphi = \pi$

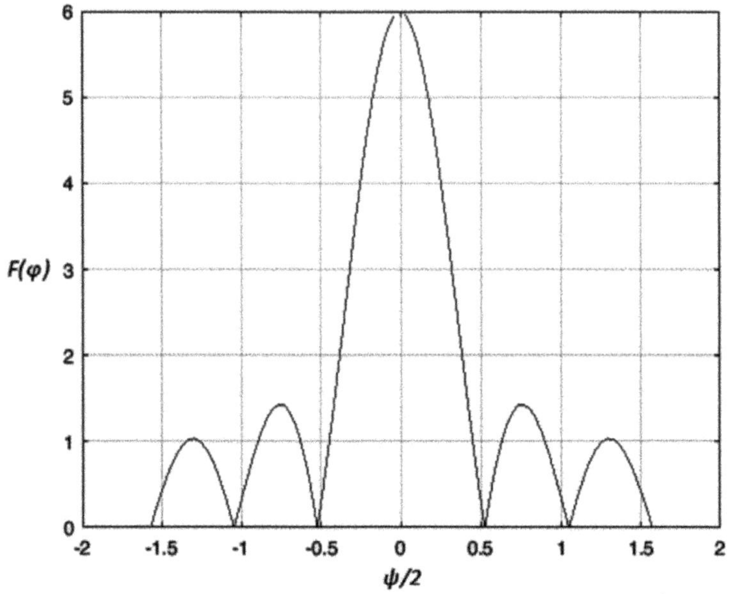

Figure 5.6 Antenna array factor, $F(\varphi)$, with six driven elements ($\alpha = 0$, $d = \lambda/2$).

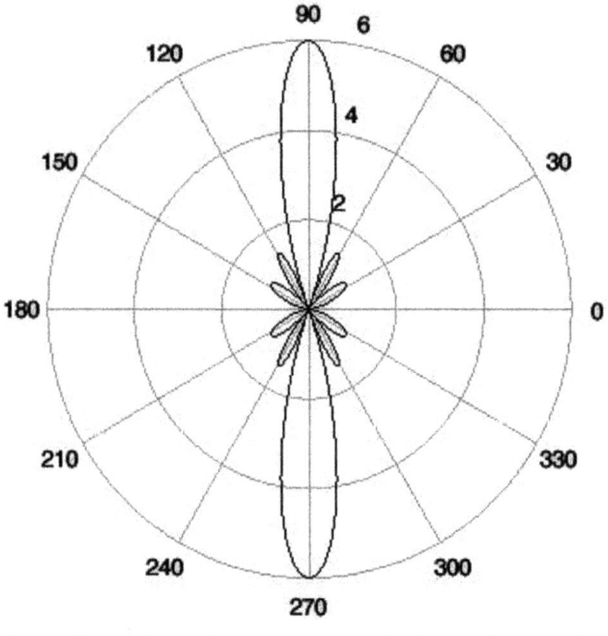

Figure 5.7 Radiation pattern of an antenna array with six driven elements ($\alpha = 0$, $d = \lambda/2$).

and 2π, the radiation pattern is simply a mirror image of that throughout the space between $\varphi = 0$ and π. The beamwidth between the first two nulls is 39.5°. Figure 5.7 demonstrates the radiation pattern of an antenna array with six driven elements.

The effect of the current phase progression on the radiation pattern can be observed, if the previous example was considered with $\alpha = \pi/2$, instead of $\alpha = 0$. Substitution into (5.1) yields $\psi = \pi/2 + \pi\cos(\varphi)$ or $\psi/2 = \pi/4 + \pi\cos(\varphi)/2$. The range of $\psi/2$ can be defined from $3\pi/4$ to $-\pi/4$ by substituting $\varphi = 0$, and π, respectively. Accordingly, the array factor with six driven elements where ($\alpha = \pi/2$, $d = \lambda/2$) is shown in Figure 5.8. There are two sidelobe maxima on one side and a single sidelobe maximum on the other side of the main lobe. The normalized sidelobe ratio is 1:0.236:0.173:0.173:0.236. The main beam occurs when $\psi/2 = \pi/4 + \pi\cos(\varphi)/2 = 0$, from which $\varphi = 120°$. Furthermore, zero radiation occurs at $\psi/2 = \pm\pi/6$, $\pi/3$, $\pi/2$, $2\pi/3$ from which $\varphi = 146°$, 99.6°, 80.4°, 60°, and 33.6°, respectively. Maximum radiation occurs at $\psi/2 = \pm\pi/4$, $5\pi/12$, $7\pi/12$, and $9\pi/12$ from which $\varphi =$ 90°, 180°, 70.5°, 48.2°, and 0°, respectively. Again, for completion of the radiation space between $\varphi = \pi$ and 2π, the radiation pattern is simply a mirror image of that throughout the space between $\varphi = 0$ and π. The beamwidth between the first two nulls is 46.4°. Figure 5.9 demonstrates the radiation pattern of the antenna array with six driven elements.

5.4 Pattern Multiplication

We can simply apply the principle of pattern multiplication, if the pointed sources, originally assumed in the calculation of the array factor, are now changed for other radiators such as identical half-wave dipoles or even nonisotropic array radiators Figure 5.10(a–c) demonstrates the pattern multiplication process of a half-wave dipole, an array factor, and a total radiation pattern, respectively.

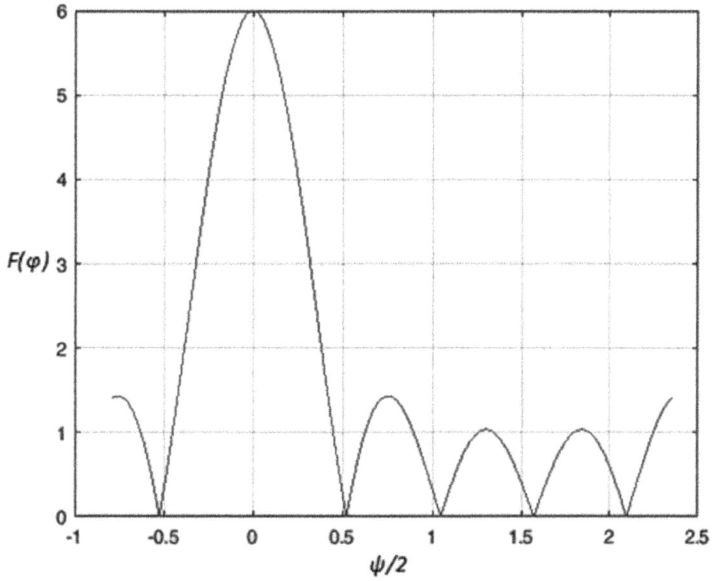

Figure 5.8 Antenna array factor, $F(\varphi)$, with six driven elements ($\alpha = \pi/2$, $d = \lambda/2$).

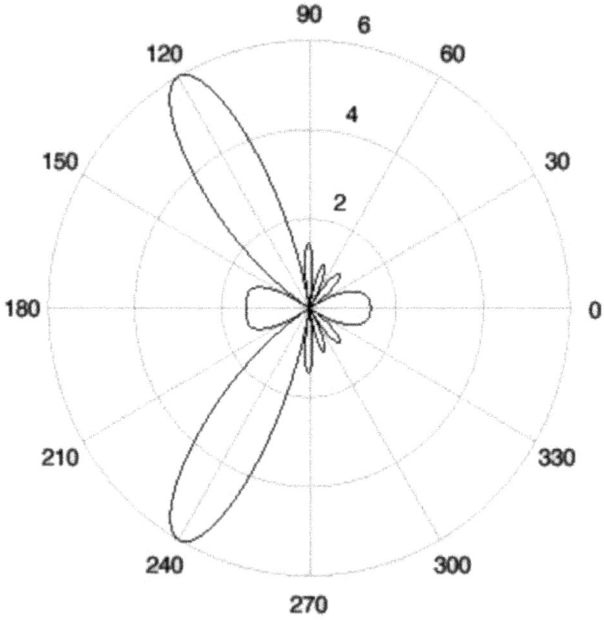

Figure 5.9 Radiation pattern of an antenna array with six driven elements ($\alpha = \pi/2$, $d = \lambda/2$).

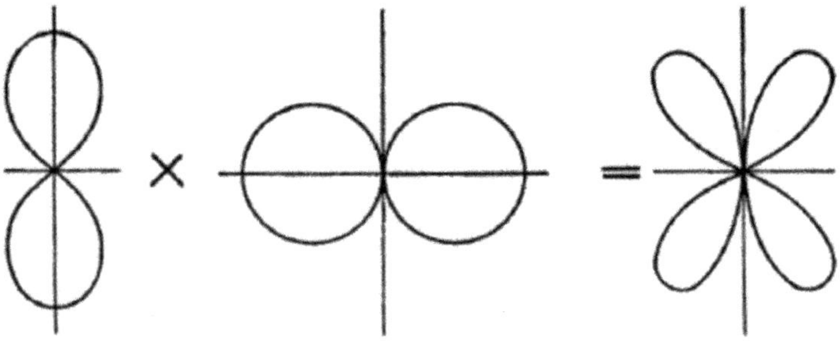

(a) (b) (c)

Figure 5.10 Pattern multiplication: (a) nonisotropic source (half-wave dipole), (b) an array factor, and (c) total radiation pattern.

We now demonstrate an antenna array of $n = 2$, $\alpha = \pi$, and $d = \lambda/2$. The radiation elements are now identical half-wave dipoles instead of spherical isotropic radiators. The array factor is the radiation field pattern with s.i.r. elements or point sources. By applying the principle of pattern multiplication, the total radiation field of the antenna array can be produced, as shown in Figure 5.11, where the total radiation pattern is shown with 4 lobes.

Another example is an antenna array with $n = 4$, $\alpha = 2\pi/3$, and $d = \lambda/2$. Figure 5.12 demonstrates the total radiation field of the antenna array that can be produced by applying the mentioned principle of pattern multiplication. The total radiation pattern shows an increased first sidelobe level and a reduced back lobe.

The following MATLAB code demonstrates the total radiation field plots by multiplying the half-wave dipole pattern with the array factor. Figures 5.11 and 5.12 were produced, accordingly.

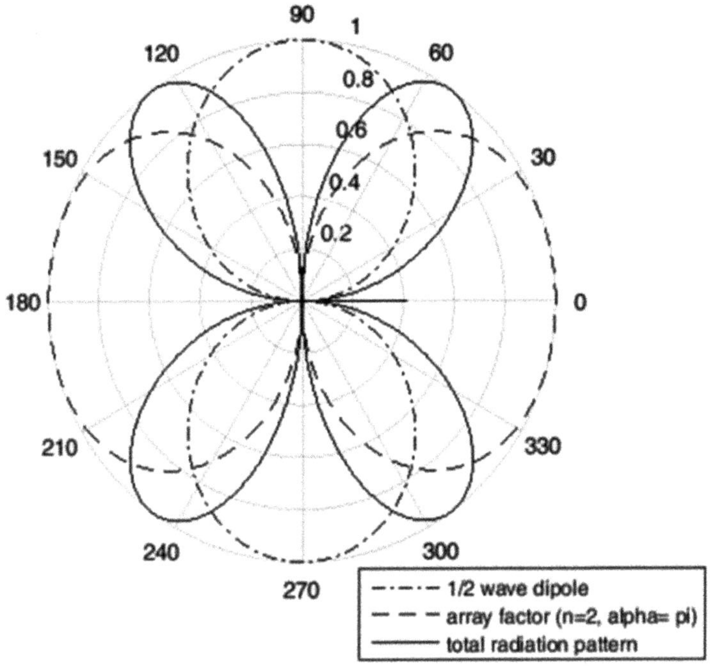

Figure 5.11 Normalized radiation patterns of an antenna array where ($n = 2$, $\alpha = \pi$, $d = \lambda/2$) of identical half-wave dipole radiators.

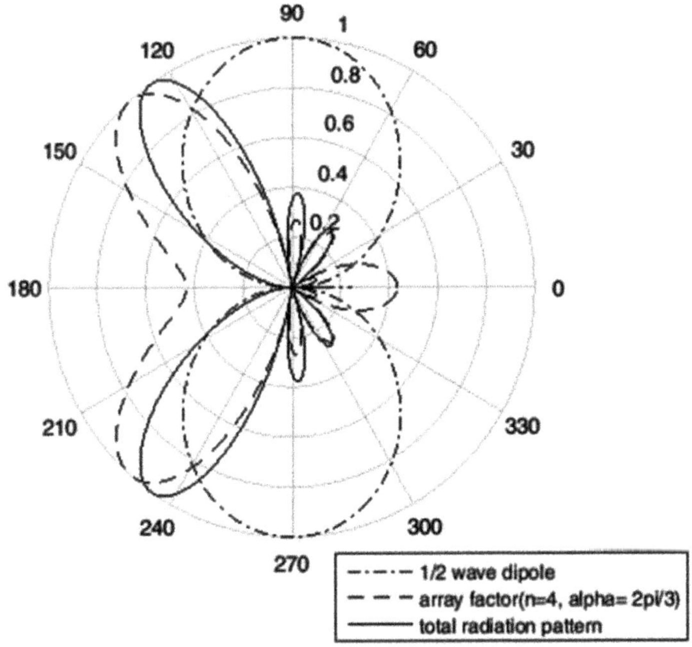

Figure 5.12 Normalized radiation patterns of an antenna array where ($n = 4$, $\alpha = 2\pi/3$, $d = \lambda/2$) of identical half-wave dipole radiators.

MATLAB Program 2

```
% MATLAB program to compute and polar plot the total radiation
field prodiced from an array of n half-wave dipoles.

% Define the range of theta from 0 to 2*pi for a complete plot
theta = linspace(0, 2*pi, 1000);

% Compute the electric field pattern E(theta)
E_theta = abs(cos((pi/2) * cos(theta)) ./ sin(theta));

% Handle the singularity at theta = 0 and theta = pi by setting
those values to zero
singularity_indices = (theta == 0 | theta == pi);
E_theta(singularity_indices) = 0;

% Normalize the electric field to its maximum value
E_theta = E_theta / max(E_theta);

% Create the polar plot
figure;
polar(theta, E_theta, 'k-.');
hold on

% Compute and Polar Plot the antenna array factor of (n, α) for
d= λ/2
% Parameters
n = 4;   % Number of elements
alpha =  2*pi/3 ;  % Phase difference between elements
d = 0.5;  % Element spacing in wavelengths
lambda = 1;  % Wavelength
k = 2 * pi / lambda; % Wave number

% Define the range of phi
phi = linspace(0, 2 * pi, 1000); % phi from 0 to 2*pi
% Calculate Ψ
Psi = alpha + k * d * cos(phi);

% Calculate |F(ϕ)|
F_phi = abs(sin(n * Psi / 2) ./ sin(Psi / 2));
% Normalize |F(ϕ)|
F_phi = F_phi / max(F_phi);

% Polar plot |F(ϕ)|
polar(phi, F_phi, 'k--');
hold on

% Calculate the total radiation field pattern
Amp= E_thet.*F_phi;
Amp= Amp/max(Amp);
polar(phi, Amp, 'k-')
```

5.5 Gain Calculation by Radiation Pattern Approximation

Real antennas do not have an ideal radiation distribution. Energy varies with angular displacement and losses occur due to sidelobes. However, if the radiation pattern is measured, and the beamwidth is determined, then gain can be calculated by simply approximating the radiation pattern, using a rectangular area, as shown in Figure 5.13. The model assumes that the radiated energy is completely concentrated in the main lobe and hence there would be no sidelobes. It further assumes that all radiated energy is within a half-width of the antenna (outside these –3-dB boundaries, there is no energy radiation) and, within those boundaries, the energy is uniformly distributed. Assuming that the antenna pattern is uniform, the maximum directivity is equal to the area of the isotropic sphere divided by the sector (cross-section) area.

From Figure 5.13, where $a = r \sin\theta$, $b = r \sin\varphi$, and $area = ab = r^2 \sin\theta \sin\varphi$, and $\theta = BW_\theta$ and $\varphi = BW_\varphi$ (BW is the 3-dB beamwidth), we can substitute into the following equation to find the maximum directivity.

$$D = \frac{\text{Area of sphere}}{\text{Area of antenna pattern}} = \frac{4\pi r^2}{r^2 \sin\theta \sin\varphi} = \frac{4\pi}{\sin\theta \sin\varphi} \tag{5.12}$$

For small angles $\sin\theta = \theta$, $\sin\varphi = \varphi$ in radians and

$$D_{\text{max}} = \frac{4\pi}{BW_\theta Bw_\varphi \text{ (degrees)}} \frac{360 * 360}{2\pi * 2\pi} = \frac{41{,}253}{BW_\theta Bw_\varphi \text{ (degrees)}} \tag{5.13}$$

If the average antenna efficiency factor, η, is considered, equal to 0.7, then the gain:

$$G = \eta D_{\text{max}} = \frac{28{,}877}{BW_\theta Bw_\varphi \text{ (degrees)}} \tag{5.14}$$

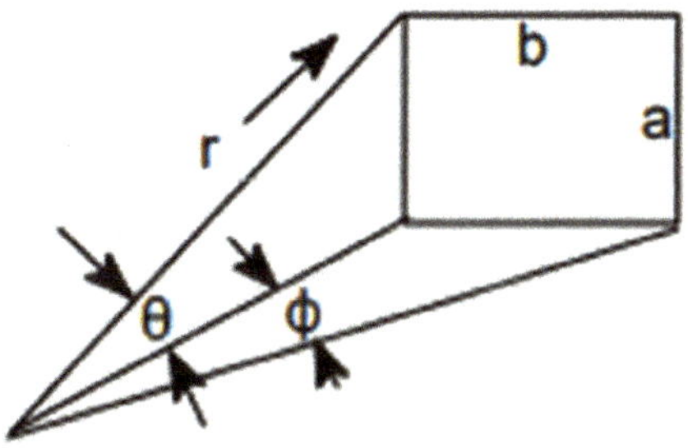

Figure 5.13 Approximating the antenna pattern as a rectangular area.

5.6 Suppression of Back Radiation

It is possible to suppress the back radiation of an array antenna, thus doubling its gain by 3 dB. This is useful in cases where the reception or transmission of a signal is meant to be unidirectional, thus saving the back radiation from being wasted. One technique is to place two parallel arrays, $\lambda/4$ apart while maintaining the phase between the currents i_1 and i_2 in quadrature (i.e., $\alpha = \pi/2$). This is similar to the case of the two-element array, shown in Figure 5.3(c, d), where the back radiation is suppressed and the radiation pattern appears solely unidirectional, as in Figure 5.14.

Another technique is to place a parallel reflector at a distance of $\lambda/4$ apart. From transmission line principles, an electromagnetic wave traveling a quarter-wavelength distance towards a short load would affect the voltage phase by 180° causing the direction of the wave voltage to flip. In addition, traveling a quarter-wavelength distance towards the reflector and back would incur an additional phase change of 180°. The total phase change would thus be 360° and the reflected wave would ultimately travel with the forward wave, in phase, thus suppressing the backward radiation and enhancing the forward radiated signal by 3 dB, as shown in Figure 5.15.

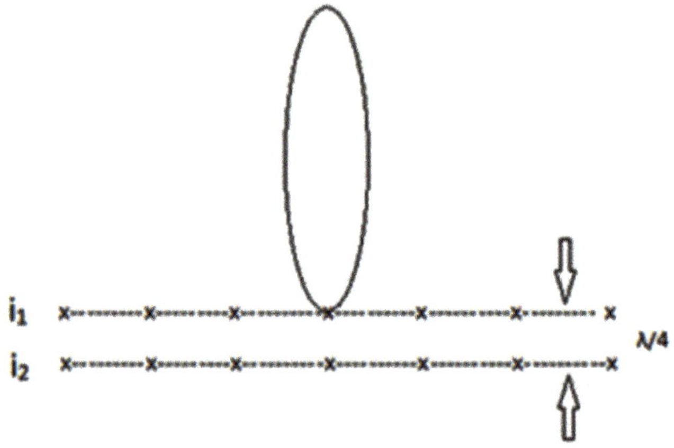

Figure 5.14 Two parallel arrays with currents in quadrature and spaced by $\lambda/4$.

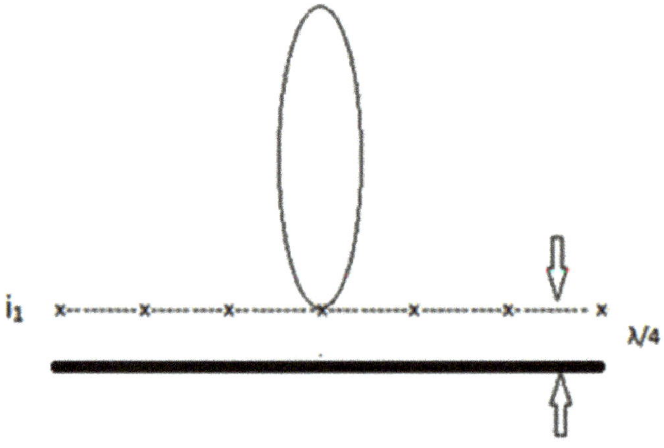

Figure 5.15 A reflector in parallel with the antenna array and spaced by $\lambda/4$.

5.7 Other Types of Antenna Arrays

While our study has primarily focused on ULAs in broadside and end-fire configurations, there are several other types of antenna arrays with distinct characteristics and applications. These arrays differ in their element arrangements, phase distributions, and beamforming capabilities. For example, UCAs enable 360° beam steering, planar arrays offer control in 2D, and phased arrays achieve electronic beam steering without mechanical movement. Additionally, conformal arrays are designed to fit curved surfaces for specialized applications.

Table 5.1 provides an overview of these antenna arrays, along with their array factor expressions and key properties. A detailed explanation of the variables used in the expressions follows the table.

The variable definitions are:

AF = array factor;

N = number of elements in the array;

M = number of elements in the x-direction (for planar arrays);

k = wave number $k = 2\pi/\lambda$ is the wavelength;

d = element spacing in a linear array;

d_x, d_y = element spacing along x and y axes (for planar arrays);

θ = elevation angle in spherical coordinates;

ϕ = azimuth angle in spherical coordinates;

R = radius of circular array;

Table 5.1 Array Factor Expressions for Different Types of Antenna Arrays

Array Type	*Array Factor Expression*	*Array Characteristics*
Uniform linear array (ULA) (broadside)	$AF = \sum_{n=0}^{N-1} e^{jkdn\cos\theta}$	Maximum radiation at 90° (perpendicular to the array); no progressive phase shift
End-fire (ULA)	$AF = \sum_{n=0}^{N-1} e^{j(nkd\cos\theta + n\beta)}$	Maximum radiation at 0° or 180°, requires phase progression $\beta = \pm kd$
Uniform circular array (UCA)	$AF = \sum_{n=0}^{N-1} e^{jkR\cos(\theta-\theta_n)+jn\beta}$	Omnidirectional or steerable beamforming; elements arranged in a circle of radius R
Planar array	$AF = \sum_{m=0}^{M-1}\sum_{n=0}^{N-1} e^{jk(md_x\sin\theta\cos\varphi + nd_y\sin\theta\sin\varphi)}$	Beam control in 2D (azimuth and elevation); elements arranged in a rectangular grid
Phased array	$AF = \sum_{n=0}^{N-1} e^{j(nkd\cos\theta + \psi_n)}$	Electronic beam steering by adjusting phase shifts ψ_n for each element
Conformal array	$AF = \sum_{n=0}^{N-1} e^{j(kr_n\cdot\hat{k} + \psi_n)}$	Mounted on curved surfaces for specialized applications; position vector r_n varies based on array shape

θ_n = angular position of the nth element in a circular array, $\theta_n = 2\pi n/N$;
β = progressive phase shift applied between elements;
ψ_n = custom phase shift for phased arrays;
r_n = position vector of element n in conformal arrays;
k = wave propagation vector;
d_n = element spacing in log-periodic arrays, follows a logarithmic rule.

5.8 Proposed Novel Antenna Arrays and Their Future Applications

5.8.1 Multibeam Reconfigurable Arrays

Multibeam reconfigurable arrays are essential for modern communication systems that require efficient management of multiple signal beams simultaneously. These arrays can dynamically adjust their radiation patterns, providing flexibility in targeting different users or devices without the need for physical repositioning. The significance of this technology lies in its ability to support dense, multi-user environments such as 5G networks and beyond. By enabling simultaneous data streams to various users, multibeam arrays enhance spectral efficiency and reduce interference. Moreover, they complement existing technologies such as phased arrays by introducing a new level of adaptability and spatial diversity. Their role in future applications, such as advanced satellite communication systems and autonomous vehicle networks, will further accelerate the development of high-speed, reliable communication services.

5.8.2 Space-Time Adaptive Arrays

Space-time adaptive arrays (STAA) provide advanced interference mitigation through spatial and temporal processing of incoming signals. This array technology can significantly improve the performance of wireless communication systems in environments prone to interference, such as dense urban settings or dynamic outdoor environments. STAA are ideal for applications that demand low-latency, high-capacity connections, such as 5G, IoT, and cognitive radio networks. By combining spatial beamforming with time-domain adjustments, STAA can offer superior signal clarity and robustness, even in the presence of noise and fading. This array technology has the potential to complement conventional antenna arrays by providing real-time optimization of signal reception, thus enabling a more efficient use of available spectrum.

5.8.3 Cognitive Radio Arrays

Cognitive radio arrays are designed to dynamically adjust their operating parameters based on real-time spectrum conditions. These arrays can sense the electromagnetic environment, identify underutilized frequency bands, and intelligently allocate them for communication, making them highly efficient for future spectrum management.

The importance of cognitive radio arrays lies in their ability to significantly increase spectral efficiency and reduce interference in crowded frequency bands. They could complement existing communication systems by offering flexible spectrum access for applications such as dynamic spectrum sharing in 5G, beyond 5G, and even in the emerging 6G networks. Cognitive radio arrays will be key to optimizing resources in the evolving wireless landscape, where demand for bandwidth continues to outstrip supply.

5.8.4 Ultrawideband Arrays

Ultrawideband (UWB) arrays provide significant advantages in terms of bandwidth and resolution, making them ideal for high-resolution imaging, radar systems, and wireless communication applications. By using very short pulses of radio waves, UWB arrays can detect fine details with high accuracy, even in cluttered or complex environments. Their applications are particularly promising in fields such as autonomous vehicle navigation, indoor positioning systems, and medical imaging. UWB arrays complement traditional narrowband antenna arrays by offering a broader frequency range and the ability to operate in a more crowded frequency spectrum. As demand for precise location-based services grows, UWB arrays will become a critical part of next-generation communication and sensing systems.

5.8.5 Metamaterial-Based Antenna Arrays

Metamaterial-based antenna arrays leverage engineered materials with properties not found in nature to achieve unique electromagnetic responses. These materials allow for highly compact, lightweight antennas with extraordinary directional properties, improved bandwidth, and reduced interference. Metamaterial arrays can outperform traditional antenna designs in terms of efficiency, directivity, and bandwidth. In applications such wireless communication, radar systems, and medical devices, metamaterial arrays hold the potential to reduce antenna size while improving performance. These arrays could significantly complement existing technologies by enabling more compact, efficient systems in devices where size and weight are crucial, such as in wearable technology and IoT devices.

5.8.6 Integrated Photonic-Antenna Arrays

Integrated photonic-antenna arrays combine the principles of optical and RF communication, creating an intersection between photonics and traditional electromagnetic systems. These arrays have the potential to support ultrahigh-speed data transfer and reduce latency by integrating optical waveguides with traditional antenna systems. As data rates continue to increase, the integration of photonic elements with antenna arrays can enable faster, more efficient data transmission for applications such as 6G networks, quantum communication, and secure satellite communication. By complementing existing RF-based communication systems with optical technologies, integrated photonic-antenna arrays will drive the development of next-generation, ultrahigh-performance communication networks.

Solved Examples

Example 5.1 Determine the main features of the following sets of two identical radiators: (a) separation $\lambda/2$, currents in quadrature; (b) separation $\lambda/4$, currents in antiphase; and (c) separation λ, currents in antiphase.

Solution 5.1 A simple MATLAB program can be easily developed to produce the plot, described in Figure 5.16.

Example 5.2 For a ULA of 6 elements, each separated by $\lambda/4$:

(a) Calculate the sidelobe level (SLL) relative to the main lobe.
(b) Discuss the impact of element spacing on the SLL and compare the results with a similar array where elements are separated by $\lambda/2$.

Solution 5.2 A MATLAB program is hereby developed to solve the problem and plot figures.

Figure 5.17 demonstrates the radiation patterns of an array factor of $n = 6$ with the two spacing variations, $d = \lambda/4$ and $\lambda/2$.

```
% Matlab program to compute and plot an array factor of n = 6
with two variations d1 = 0.25λ and d2 = 0.5λ.
n = 6;                    % Number of elements
d1 = 0.25;                % Element spacing in wavelengths (lambda/4)
```

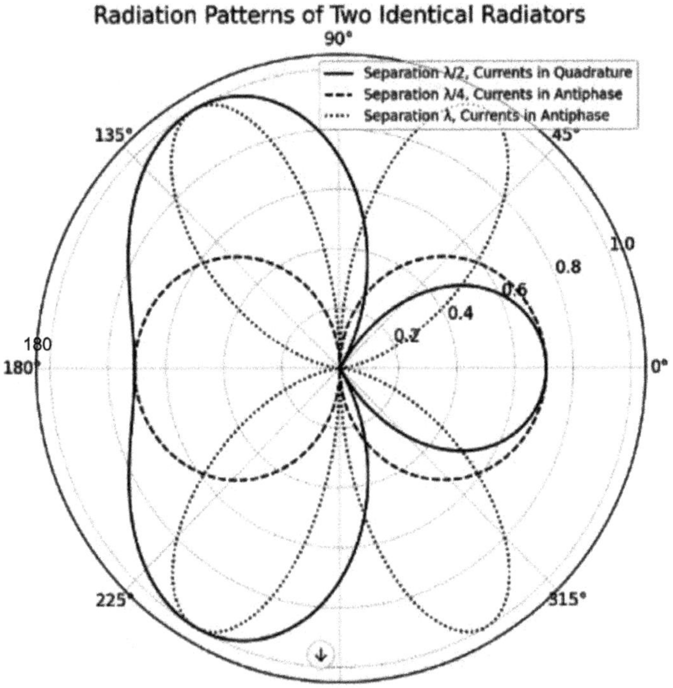

Figure 5.16 Radiation pattern of two identical radiators with different spacing and current phase.

```
d2 = 0.5;              % Element spacing in wavelengths (lambda/2)
lambda = 1;            % Wavelength
k = 2 * pi / lambda; % Wave number

% Define the range of theta
theta = linspace(0, 2 * pi, 1000); % theta from 0 to 2*pi

% Calculate array factors for both spacings
Psi1 = k * d1 * cos(theta);
Psi2 = k * d2 * cos(theta);

% Calculate |F(θ)| for both spacings
F_theta1 = abs(sin(n * Psi1 / 2) ./ sin(Psi1 / 2));
F_theta2 = abs(sin(n * Psi2 / 2) ./ sin(Psi2 / 2));

% Normalize |F(θ)|
F_theta1_norm = F_theta1 / max(F_theta1);
F_theta2_norm = F_theta2 / max(F_theta2);

% Find the index for theta = pi/2
[~, index_pi_over_2] = min(abs(theta - pi/2));

% Calculate the normalized array factors in decibels at theta =
pi/2
F_theta1_norm_db = 20 * log10(F_theta1_norm(index_pi_over_2));
F_theta2_norm_db = 20 * log10(F_theta2_norm(index_pi_over_2));

% Print the results
fprintf('F_theta1_norm at theta = pi/2: %.2f dB\n',
F_theta1_norm_db);
fprintf('F_theta2_norm at theta = pi/2: %.2f dB\n',
F_theta2_norm_db);
```

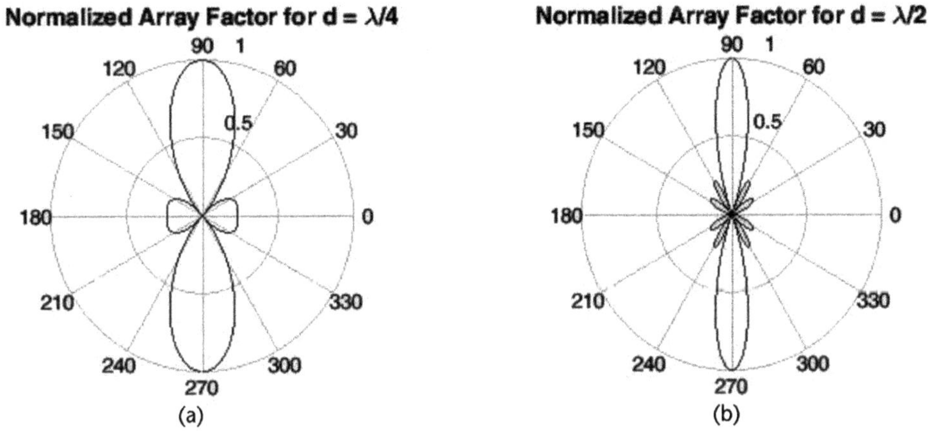

Figure 5.17 Radiation pattern of an array factor of $n = 6$ with two spacing variations, $d = \lambda/4$ and $\lambda/2$, showing SLL and HPBW. (a) SLL (first) = 0.2392 (–12.42 dB); (b) SLL (first) = 0.2392 (–12.42 dB), SLL (second) = 0.1727 (–15.25 dB), HPBW = 17°.

```
% Plotting
figure;
subplot(2,1,1);
polar(theta, F_theta1_norm, 'k-');
title('Normalized Array Factor for d = \lambda/4');

subplot(2,1,2);
polar(theta, F_theta2_norm, 'k-');
title('Normalized Array Factor for d = \lambda/2');

% Examine data for SLL and HPBW
[theta' F_theta1_norm' F_theta2_norm']
```

Problems for Students

5.1 A ULA has five radiators, separated by $\lambda/2$. Determine the salient features of the radiation pattern for: (a) currents in phase, and (b) π radians progressive phase shift.

5.2 Sketch the radiation pattern for a ULA of 8 elements, separated by $\lambda/2$, with currents in phase. What is the width of the main beam (to the first zeros)?

5.3 Design a ULA of 10 elements, each separated by $\lambda/2$, to achieve a main lobe beamwidth of 20°.
   (a) Calculate and plot the array factor and the resulting radiation pattern.
   (b) Explain the steps and the influence of the number of elements and element spacing on the beamwidth.

5.4 Consider a ULA of 12 elements, separated by $\lambda/2$, with current amplitudes tapered according to a cosine distribution (i.e., the current amplitude of each element follows a cosine function):
   (a) Calculate and plot the radiation pattern and compare it with a pattern where the currents are uniform.
   (b) Discuss the effects of tapering on the main lobe and sidelobes.

5.5 A ULA of 16 elements is designed with a spacing of $\lambda/2$.
   (a) Determine the phase shift required between adjacent elements to steer the main beam to an angle of 30° from the broadside.
   (b) Calculate and sketch the radiation pattern for this configuration.
   (c) Explain the principle of beam steering and its applications.

Problems for Instructors

5.1 Advanced Array Configuration Analysis: Derive the radiation pattern of a linear array of N equally spaced driven elements, considering both in-phase and out-of-phase excitations. Discuss how changing the interelement spacing affects the main lobe and sidelobe levels.

5.2 Array Factor Derivation: Given a linear array with two driven elements separated by a distance d, derive the array factor and the resultant radiation pattern for both constructive and destructive interference scenarios. Compare this with the pattern of a single element.

5.3 Gain Calculation for Complex Arrays: Calculate the gain of a linear array with *N* elements, given that the array pattern approximates a known radiation pattern. Use pattern multiplication techniques to find the overall gain, and discuss the effects of element spacing and excitation phase.
5.4 Back Radiation Suppression: Analyze the effectiveness of various techniques (such as the use of parasitic elements or reflector planes) for suppressing back radiation in a linear array. Provide a comparative analysis of the theoretical versus practical effectiveness of these techniques.
5.5 Pattern Multiplication Application: Given two antenna patterns (e.g., a single dipole and a parabolic reflector), use pattern multiplication to determine the combined pattern when they are used in an array configuration. Analyze the impact of this combination on overall array performance.
5.6 Impedance Matching in Arrays: Discuss methods for impedance matching in a linear array to minimize reflections and maximize power transfer. Provide an example involving a practical impedance matching network for a specific array configuration.
5.7 Practical Array Design Challenge: Design a linear array with a specified gain and beamwidth, and discuss the trade-offs between the number of elements, spacing, and array length. Include calculations for the required element excitation and spacing to meet the design specifications.

Selected Bibliography

Balanis, C. A., *Antenna Theory: Analysis and Design*, New York: John Wiley & Sons, 2015.

Dohmen, C., J. W. Odendaal, and J. Joubert, "Synthesis of Conformal Arrays with Optimized Polarization," *IEEE Transactions on Antennas and Propagation*, Vol. 55, No. 10, 2007, pp. 2922–2925.

Dutta, K. P., "Study of Broadside Linear Array Antenna with Different Spacing and Number of Elements," *International Journal of Advanced Engineering Research and Science*, Vol. 4, No. 5, 2017, p. 237181.

Kraus, J. D., *Antennas*, Second Edition, New York: McGraw-Hill, 1988.

Mailloux, R. J., *Phased Array Antenna Handbook*, Third Edition, Norwood, MA: Artech House, 2018.

Noordin, N. H., et al., "Uniform Circular Arrays for Phased Array Antenna," *2011 IEEE Loughborough Antennas & Propagation Conference*, November 2011, pp. 1–4.

Stutzman, W. L., and G. A. Thiele, *Antenna Theory and Design*, New York: John Wiley & Sons, 2012.

Tan, W., et al., "Analysis of Different Planar Antenna Arrays for mmWave Massive MIMO Systems," *2017 IEEE 85th Vehicular Technology Conference (VTC Spring)*, June 2017, pp. 1–5.

Visser, H. J., *Array and Phased Array Antenna Basics*, New York: John Wiley & Sons, 2006.

Yao, G. W., et al., "Research on a New Kind of High Directivity End-Fire Antenna Array," *Progress in Electromagnetics Research B*, Vol. 33, 2011, pp. 135–151.

CHAPTER 6

Frequency-Independent Antennas: Log-Periodic Dipole Array

6.1 Introduction

Before the 1950s, broadband antennas had limited bandwidth ratios, typically around 2:1, restricting their effectiveness in multifrequency applications. A significant breakthrough in antenna design during that period expanded bandwidth capability to 40:1 or more, leading to the development of frequency-independent antennas. Characterized by geometries defined through angular structures, these antennas became integral in applications spanning the 10 to 10,000-MHz range, including television broadcasting, point-to-point communications, and reflector and lens feed systems.

Log-periodic antennas, a key advancement in frequency-independent designs, are renowned for their ability to operate efficiently over an extensive bandwidth ratio, often reaching 40:1 or higher. They maintain relatively stable input impedance, gain, and directivity across their entire operating range, making them highly effective for a variety of applications. Their logarithmic structure, where element sizes and spacing vary progressively, enables consistent resonant behavior at multiple frequencies, ensuring predictable performance. These antennas are widely deployed in communication networks, broadcasting, radar systems, and spectrum monitoring, where wideband coverage is essential. However, achieving optimal performance requires meticulous design, including careful selection of element dimensions and spacing. One inherent trade-off is that broader bandwidth necessitates a larger physical structure, which can present challenges in space-constrained environments.

Among the most widely used designs within this category is the log-periodic dipole array (LPDA), along with variations such as the log-periodic spiral and log-periodic bowtie antennas. Unlike traditional narrowband antennas, log-periodic antennas feature a sequence of radiating elements arranged in a logarithmic pattern, allowing them to function efficiently across a broad frequency spectrum. The LPDA, in particular, is valued for its ability to provide stable impedance and predictable radiation characteristics over a wide range, making it an ideal choice for applications requiring frequency agility. The periodic variation in element size ensures continuous resonance, further enhancing operational efficiency. However, the trade-off remains—the wider the frequency coverage, the larger the physical footprint of the antenna, which can be a limiting factor in compact installations.

Log-periodic antennas play a crucial role in modern RF applications, offering a practical solution for systems requiring seamless operation across multiple frequencies without the need for multiple antennas. Their versatility simplifies system design and enhances operational flexibility, making them indispensable in environments where broad frequency coverage and reliability are critical. By optimizing design parameters, engineers can maximize their performance without significant degradation in impedance matching or radiation efficiency. While their size can be a limitation, their ability to provide stable broadband operation makes them an essential component in advanced communication and radar technologies.

6.2 Description of LPDA

Figure 6.1 shows a schematic diagram of an LPDA. The array constitutes of a sequence of side-by-side parallel linear half-wave dipoles, which are logarithmically scaled in their lengths (l_n), diameters (d_n), spacings (R_n), and gap spacings at dipole centers (s_n). The diagram shows that the LPDA dimensions decrease logarithmically as defined by the geometric ratio τ, where τ is called the scale factor.

The equation, which involves τ, is expressed as:

$$\tau = \frac{l_n}{l_{n+1}} = \frac{R_n}{R_{n+1}} = \frac{d_n}{d_{n+1}} = \frac{s_n}{s_{n+1}} \tag{6.1}$$

Another parameter, which is usually associated with a dipole array, is the spacing factor σ, where:

$$\sigma = \frac{R_{n+1} - R_n}{2l_{n+1}} \tag{6.2}$$

While an element in a linear array is usually driven by a source and the other elements act as parasites, all elements in an LPDA are connected to the source. One way to energize an LPDA is by mechanically criss-crossing or transposing the feed between adjacent elements, as shown in Figure 6.2, where an 180° is added to the terminal of each element. The mechanical phase reversal between these elements produces a phase progression, which beams the energy in the direction of the short elements. The most active elements for this feed arrangement are those that are near resonance with a combined radiation pattern towards the vertex of the array.

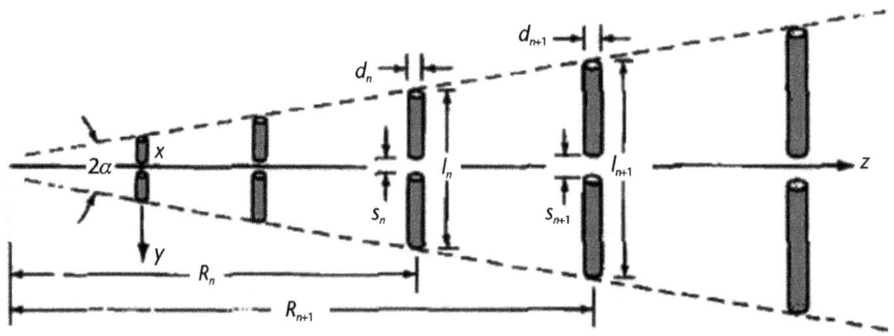

Figure 6.1 LPDA.

A two-conductor transmission line is considered a convenient feed arrangement for a criss-cross connection as shown in Figure 6.2. Using a coaxial cable instead, a practical approach to achieve the 180° phase reversal between adjacent elements can be realized by the cable connection, shown in Figure 6.3.

6.3 Design Equations of LPDA

A complete and practical design of LPDAs is originally attributed to Carrel [1], who proposed a set of curves to aid in the design process. The initial curves had directivities 1 to 2 dB too high, which have been realized and later reduced by an average of 1 dB as published by Balanis [2]. Carrel computed an empirical Figure, which shows the variation of relative spacing factor, σ, with the scale factor, τ, for a set of antenna directivities. As the scale factor and optimum spacing factor increase, the antenna directivity increases too. All designed dimensions of the LPDA can then be determined except for the center-to-center spacing s between the two rods of the feeder line. Carrel published another empirical Figure to determine s, which is essential for realizing Z_0; the characteristic impedance of the feeder line. It shows the variation of the normalized characteristic impedance, relative to R_{in} (R_{in} is usually 50Ω for a coaxial cable), of the feeder line to that of the dipole element.

Table 6.1 illustrates selected figure data of τ at optimum σ for constant directivities.

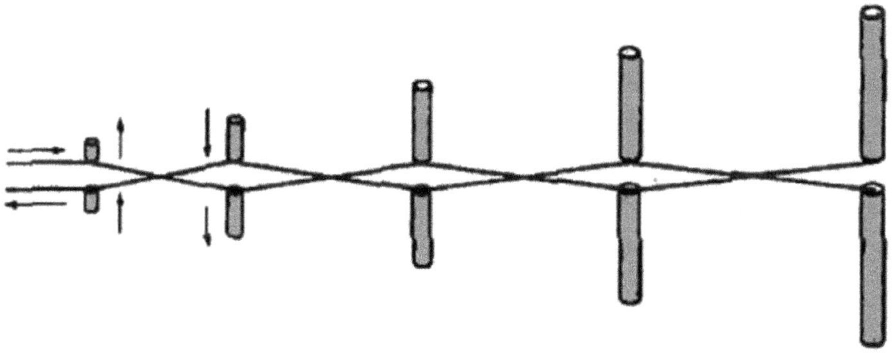

Figure 6.2 Criss-cross connection showing the direction of the current.

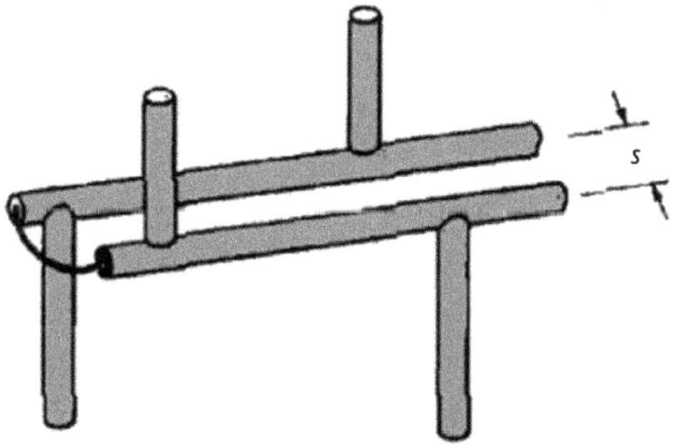

Figure 6.3 Coaxial connection showing the feeder lines spaced by s.

Table 6.1 Selected Figure Data Subset of Scale Factor τ at Optimum Spacing Factor σ for Constant Directivities

Directivity (dB)	*Scale Factor (τ)*	*Optimum Spacing Factor (σ)*
7.0	0.137	0.778
7.5	0.148	0.822
8.0	0.157	0.865
8.5	0.166	0.895
9.0	0.171	0.916
10.0	0.178	0.943
11.0	0.183	0.967

Source: [2].

The general configuration of a log-periodic array is described in terms of the design parameters τ, α, and σ, where:

$$\alpha = \tan^{-1}\left[\frac{1-\tau}{4\sigma}\right] \tag{6.3}$$

Carrel has introduced the following empirical equation to calculate the bandwidth of the active region, B_{ar}:

$$B_{ar} = 1.1 + 7.7(1-\tau)^2 \cot\alpha \tag{6.4}$$

In practice, a slightly larger bandwidth, the designed bandwidth B_s, is usually designed than that which is the desired bandwidth, B, such that:

$$B_s = BB_{ar} = B\left[1.1 + 7.7(1-\tau)^2 \cot\alpha\right] \tag{6.5}$$

where $B = f_{max}/f_{min}$.

The total length of the structure L, from the shortest l_{min} to the longest l_{max} element, is given by:

$$L = \frac{\lambda_{max}}{4}\left(1 - \frac{1}{B_s}\right)\cot\alpha \tag{6.6}$$

where $\lambda_{max} = 2l_{max} = v/f_{min}$.

The number of elements is determined by:

$$N = 1 + \frac{\ln(B_s)}{\ln(1/\tau)} \tag{6.7}$$

The center-to-center spacing s of the feeder line conductors can be determined by specifying the required input impedance, assumed to be real, and the diameter of the dipole elements and the feeder line conductors. For this, we first define an average characteristic impedance of the elements given by:

$$Z_a = 120\left[\ln\left(\frac{l_n}{d_n}\right) - 2.25\right] \tag{6.8}$$

For an ideal design, the length-to-diameter ratio should be the same for all elements of the array. However, the elements are practically divided into two or three groups (small, middle, and large elements) with all elements having the same diameter, in each group, but not the same length.

Table 6.2 describes selected values reflecting relative Z_0 versus relative Z_a for various relative spacing, σ', where $\sigma' = \sigma/\sqrt{\tau}$ = relative mean spacing, Z_a = average characteristic impedance of the elements, R_{in} = input impedance (real), and Z_0 = characteristic impedance of the feeder line.

The center-to-center spacing s between the two rods of the feeder line, each of identical diameter d, is determined by:

$$s = d\cosh\left(\frac{Z_0}{120}\right) \tag{6.9}$$

6.4 Design Procedure

In this section, we demonstrate the steps for designing an LPDA. An antenna is required to cover all the VHF TV channels, starting with $f_l = 54$ MHz and ending with $f_h = 216$ MHz. The desired directivity is 8 dB and the input impedance is 50Ω. The outside diameter (d_{max}) for the largest element and the feeder line is 1.9 cm, while the diameter for the smallest element (d_{min}) is 0.48 cm. These diameters yield identical l/d ratios for the largest and smallest elements.

The design solution is:

1. For $D_0 = 8$ dB, we can get from Table 6.4, the optimum value $\sigma = 0.157$ and the corresponding value $\tau = 0.865$.
2. Using (6.3), $\alpha = 12.13°$.
3. Using (6.4), $B_{ar} = 1.753$.
4. Using (6.5), $B_s = (216/54) \times 1.753 = 7.01$.
5. Using (6.6), $L = 5.541$m.

Table 6.2 Selected Figure Data Subset of Z_0/R_{in} at Z_a/R_{in} for Various Relative Spacing, σ'

	$\sigma' = 0.06$	$\sigma' = 0.1$	$\sigma' = 0.16$	$\sigma' = 0.24$
Z_a/R_{in}	Z_0/R_{in}	Z_0/R_{in}	Z_0/R_{in}	Z_0/R_{in}
1	4.335	2.772	1.972	1.603
2	2.444	1.739	1.442	1.271
4	1.596	1.333	1.182	1.113
6	1.374	1.210	1.102	1.052
7	1.309	1.161	1.079	1.036

Source: [2].

6. Using (6.7), $N = 14.43$ (14 or 15 elements).
7. Let $N = 14$, $\lambda_{max} = v/f_{min} = 3 \times 10^8/54 \times 10^6 = 5.5555$m, and $l_{max} = l_{14} = \lambda_{max}/2 = 2.7777$m, and then $l_{13} = \tau l_{14}$, $l_n = \tau l_{n+1}$.
8. Calculate $R_{14} = (l_{14}/2)/\tan(\alpha) = 6.4620$m. $R_{13} = \tau R_{14}$, $R_n = \tau R_{n+1}$.

   $L' = R_{14} - R_1 = 5.4812$m, which is very close to $L = 5.541$m, calculated from (6.6). Hence, $N = 14$ satisfies the design and there is no need to increase N by one more element.

   Table 6.4 shows all calculated lengths of radial distances of LPDA dipoles in meter units and in reference to Figure 6.1.
9. $\sigma' = \sigma/\sqrt{\tau} = 0.157/\sqrt{0.865} = 0.169$.
10. Now $l_{max}/d_{max} = 277.77$ cm/1.9 cm = 146.195.
11. Using 6.8, $Z_a = 120[\ln(146.195) - 2.25] = 328.19\Omega$.
12. Thus, $Z_a/R_{in} = 6.564$.
13. From Table 6.5 and the calculated values for Z_a/R_{in} and σ', $Z_0/R_{in} \approx 1.2$, as a maximum estimate. Hence, $Z_0 = 60\Omega$.
14. Finally, $s = 1.9 \times \cosh(60/120) = 2.14$ cm. This allows for a separation of $2.14 - 1.9 = 0.24$ cm, between the conducting surfaces of the feedline rods.

The design of an LPDA antenna involves careful consideration of various factors to ensure optimal performance. One important aspect is achieving impedance matching across the desired frequency range. Impedance matching is critical for efficient power transfer between the antenna and the transmission line or system. While it is challenging to achieve perfect impedance matching across the entire bandwidth, designing for an expanded bandwidth allows for a reasonable impedance match over a wider frequency range. This helps to minimize signal reflections and maximize power transfer. Additionally, the LPDA antenna's element lengths are carefully designed to create a periodic structure that exhibits the desired broadband characteristics. By considering element spacing, dipole tapering, and element shaping, the antenna can effectively transfer power across the desired frequency range, maximizing the signal strength and minimizing reflections. Thus, a comprehensive design approach that accounts for impedance matching and other design parameters ensures the LPDA antenna's effective performance over the intended bandwidth.

Table 6.3 Lengths of LPDA Dipoles

l_{14}	l_{13}	l_{12}	l_{11}	l_{10}	l_9	l_8
2.7777	2.4027	2.0784	1.7978	1.5551	1.3451	1.1635
l_7	l_6	l_5	l_4	l_3	l_2	l_1
1.0064	0.8706	0.7530	0.6514	0.5634	0.4874	0.4216

Table 6.4 Lengths of Radial Distances in Reference to Figure 6.1

R_{14}	R_{13}	R_{12}	R_{11}	R_{10}	R_9	R_8
6.4620	5.5896	4.8350	4.1823	3.6177	3.1293	2.7068
R_7	R_6	R_5	R_4	R_3	R_2	R_1
2.3414	2.01253	1.7519	1.5154	1.3108	1.1338	0.9808

Solved Examples

Example 6.1 For a log-periodic dipole array antenna with 12 elements and a given element spacing σ of 0.2, and a scaling factor τ of 0.85:

(a) Estimate the main lobe direction.

(b) Discuss the expected radiation pattern and beamwidth.

Solution 6.1

```
% Matlab program to compute radiation pattern of LPDA
% Parameters
tau = 0.85;
sigma = 0.2;
N = 12;
f = 1e9;        % Frequency in Hz
c = 3e8;        % Speed of light in m/s
lambda = c / f;
k = 2 * pi / lambda;

% Element lengths and positions
L = lambda * (tau .^ (0:N-1));
d = L(1:end-1) * sigma;
r = [0, cumsum(d)];

% Assuming equal current distribution and no phase shift
for simplicity
I = ones(1, N);
phi = zeros(1, N);

% Calculate Array Factor
theta = linspace(0, 2*pi, 360);
AF = zeros(size(theta));

for n = 1:N
   AF = AF + I(n) * exp(1j * (k * r(n) * cos(theta) + phi(n)));
end

AF = abs(AF);

% Normalize Array Factor
AF = AF / max(AF);

% Calculate Radiation Pattern of a Single Dipole
E_element = cos((pi/2) * cos(theta)) ./ sin(theta);
E_element(theta == 0 | theta == pi) = 0; % Avoid division by zero

% Total Radiation Field
E_total = AF .* E_element;
```

```
% Normalize Total Radiation Field
E_total = abs(E_total / max(E_total));

% Plot Radiation Pattern
polar(theta, E_total);
title('Radiation Pattern of LPDA');
HPBW= 38.60
N= 12, τ = 0.85, σ = 0.2
```

Figure 6.4 demonstrates the radiation pattern of an LPDA antenna with 12 elements, $\sigma = 0.2$, and $\tau = 0.85$.

Example 6.2 Modify the code in Example 6.1 to test the effect of changing τ or σ on the radiation pattern.

Solution 6.2 It is noticed that reducing τ broadens the pattern while increasing σ narrows it.

Figure 6.5 demonstrates the effect of changing σ and τ on the radiation pattern of an LPDA antenna with 12 elements.

Example 6.3 Develop a MATLAB program to analyze the performance of an LPDA antenna by computing and plotting:

(a) The reflection coefficient ($|\Gamma|$) in dB scale as a function of frequency.
(b) The gain (in dB) as a function of frequency.

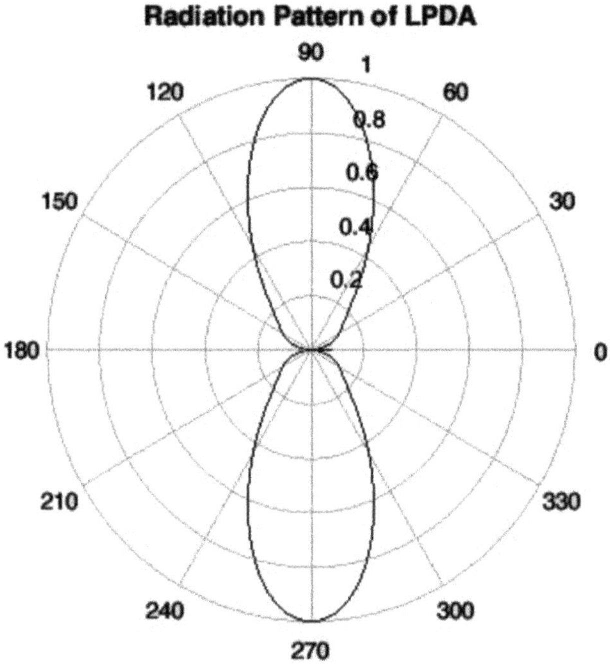

Figure 6.4 Radiation pattern of an LPDA antenna with 12 elements, $\sigma = 0.2$, and $\tau = 0.85$.

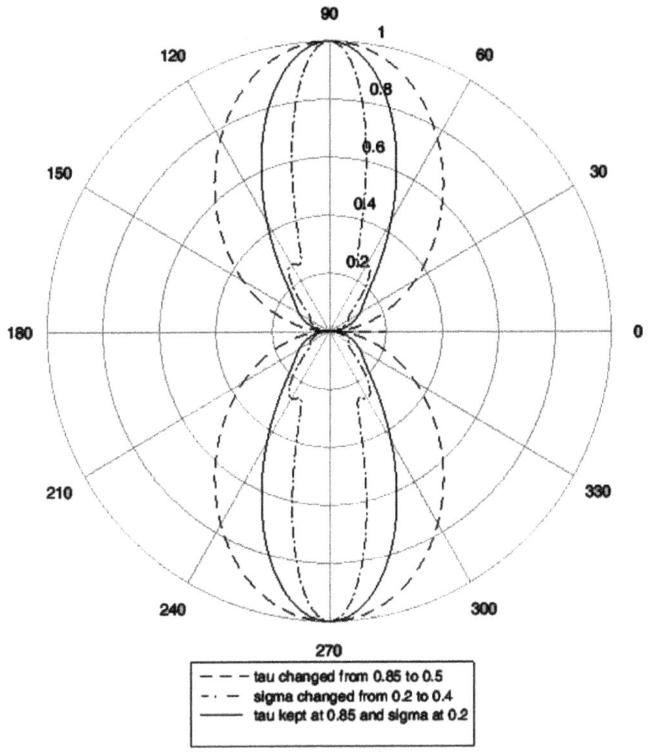

Figure 6.5 Effect of changing σ and τ on the radiation pattern of an LPDA antenna with 12 elements.

The analysis will be conducted over a frequency range of 0.5 GHz to 1.5 GHz. The antenna's input impedance and gain will be approximated using empirical models relevant to LPDA antennas.

Solution 6.3 We will use the following formulas in the development of the MATLAB code.

1. Reflection coefficient ($|\Gamma|$): The reflection coefficient quantifies the impedance mismatch between the LPDA antenna and the transmission line. It is given by:

$$\Gamma = Z_{\text{antenna}} - Z_{\text{feed}}/Z_{\text{antenna}} + Z_{\text{feed}}$$

   where Z_{antenna} is the impedance of the LPDA antenna and Z_{feed} is the characteristic impedance of the feedline (typically 50Ω).
2. LPDA antenna impedance model: Let the impedance of an LPDA antenna vary with frequency and is approximated using a frequency-dependent scaling model:

$$Z_{\text{antenna}} = Z_{\text{feed}} \times (f/f_{\text{ref}})^{\alpha} \times L_{\text{coil}}$$

   where f is the operating frequency, $f_{\text{ref}} = 1$ GHz (a reference frequency for scaling), $\alpha = 0.5$ (empirical scaling factor for impedance variation), and L_{coil}

= $1 + 0.1\sin(2\pi f/f_{ref})$ is a small correction factor modeling the loading coil effects in the LPDA.

3. Gain estimation: The gain of an LPDA antenna depends on the number of active elements in the array, which increases as frequency increases. A simplified empirical relation is used:

$$N_{elements} = 5 + 10\log_{10}(f/f_{min})$$

where $N_{elements}$ is the estimated number of active dipoles at frequency f and $f_{min} = 0.5$ GHz (lowest operating frequency).

The gain (in dB) is given by:

$$Gain_{dB} = 10\log_{10}(N_{elements})$$

This equation follows the principle that, as more dipoles become active, the gain increases logarithmically.

The following MATLAB code plots the variation of (a) reflection coefficient and (b) gain with frequency of an LPDA with 10 elements as shown in Figure 6.6.

```
% MATLAB Script for Analyzing Reflection Coefficient and Gain of
a Log-Periodic Dipole Array (LPDA)
% Frequency range: 0.5 GHz to 1.5 GHz
% Plots: Reflection Coefficient (dB) vs Frequency, Gain (dB)
vs Frequency

% Clear workspace and command window
clc; clear; close all;

% Define frequency range using a logarithmic scale
f_min = 0.5e9;  % Minimum frequency (0.5 GHz)
f_max = 1.5e9;  % Maximum frequency (1.5 GHz)
frequencies = logspace(log10(f_min), log10(f_max), 200); % 200
frequency points

% Define characteristic impedance of the feedline
Z_feed = 50; % Feedline impedance (typically 50 Ohms)

% Initialize arrays to store results
Gamma = zeros(1, length(frequencies));  % Reflection coefficient
(linear scale)
Gain_dB = zeros(1, length(frequencies)); % Antenna gain in dB

% Loop over each frequency and calculate the reflection
coefficient and gain
for i = 1:length(frequencies)
    f = frequencies(i); % Current frequency value
    f_ref = 1e9;  % Reference frequency (1 GHz, adjustable
for LPDA)
```

(a)

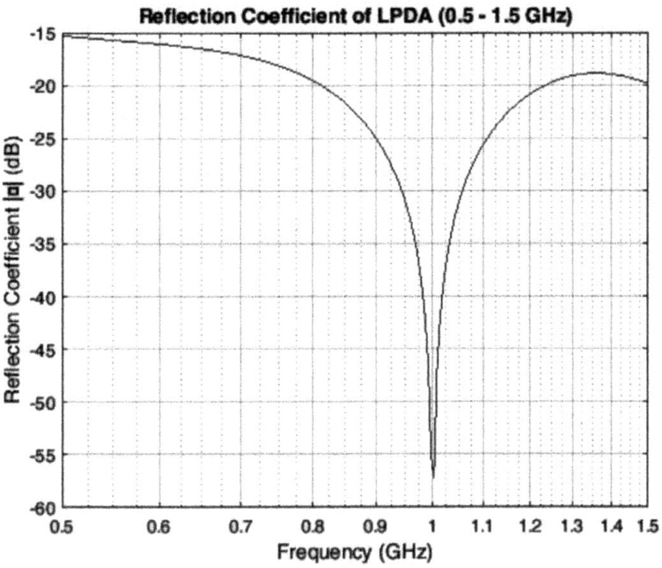

(b)

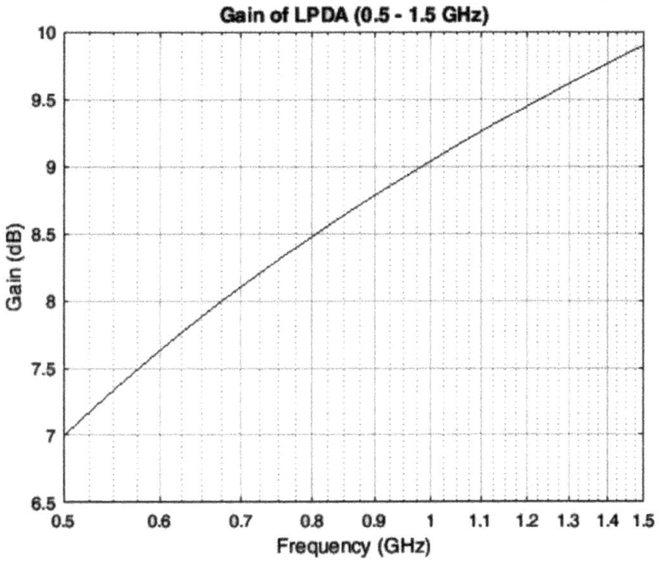

Figure 6.6 Variation of: (a) reflection coefficient and (b) gain with frequency of an LPDA with 10 elements.

```
    alpha = 0.5;  % Impedance scaling factor for LPDA

    % Simulating the effect of loading coils using a
sinusoidal variation
    Lpda_coil_effect = 1 + 0.1 * sin(2 * pi * f / f_ref);

    % Calculate antenna impedance (simplified empirical model)
    Z_antenna = Z_feed * (f / f_ref) ^ alpha * Lpda_coil_effect;
```

```
    % Calculate the reflection coefficient (linear scale)
    Gamma(i) = abs((Z_antenna - Z_feed) / (Z_antenna + Z_feed));

    % Estimate Gain (dB) based on LPDA characteristics
    % The number of active elements increases with frequency
    N_elements = 5 + 10 * (log10(f / f_min)); % Example function
for element activation
    Gain_dB(i) = 10 * log10(N_elements); % Convert to dB scale
end

% Convert Reflection Coefficient to dB scale
Gamma_dB = 20 * log10(Gamma);

% Plot Reflection Coefficient (in dB) vs Frequency
figure;
semilogx(frequencies / 1e9, Gamma_dB, 'k'); % Logarithmic x-axis
xlabel('Frequency (GHz)'); % X-axis label
ylabel('Reflection Coefficient |Γ| (dB)'); % Y-axis label
title('Reflection Coefficient of LPDA (0.5 - 1.5 GHz)'); %
Plot title
grid on;
% Plot Gain vs Frequency
figure;
semilogx(frequencies / 1e9, Gain_dB, 'k'); % Logarithmic x-axis
xlabel('Frequency (GHz)'); % X-axis label
ylabel('Gain (dB)'); % Y-axis label
title('Gain of LPDA (0.5 - 1.5 GHz)'); % Plot title
grid on;
```

Problems for Students

6.1 Given a log-periodic dipole array antenna with the following parameters:
- The shortest dipole element length $l_1 = 0.05$m.
- The scaling factor $\tau = 0.9$.
- The spacing factor $\sigma = 0.18$.
- Number of elements $N = 10$.

Calculate:
(a) The length of the longest dipole element.
(b) The spacing between the first and second dipole elements.
(c) The overall length of the antenna.

6.2 An LPDA antenna operates over a frequency range from 300 MHz to 3 GHz. Given that the shortest dipole element corresponds to the highest frequency and the longest dipole element corresponds to the lowest frequency, calculate:
(a) The length of the shortest dipole element.
(b) The length of the longest dipole element.

6.3 Given a log-periodic dipole array with the following parameters:
- Shortest dipole element length $l_1 = 0.06$m.
- Scaling factor $\tau = 0.85$.
- Spacing factor $\sigma = 0.15$.

- Number of elements $N = 15$.

Calculate the input impedance at the feed point of the antenna at a frequency of 1 GHz. Assume that the characteristic impedance of each dipole element is 73W.

6.4 An LPDA antenna is designed with 10 elements, a scaling factor τ of 0.88, and a spacing factor σ of 0.18. The antenna is used to operate over a frequency range from 200 MHz to 2 GHz. Calculate the expected gain of the antenna at the following frequencies: (a) 300 MHz, (b) 1 GHz, and (c) 1.5 GHz.

6.5 An LPDA antenna is designed to operate between 500 MHz and 5 GHz. The array consists of 12 elements with a scaling factor $\tau = 0.85$ and a spacing factor $\sigma = 0.16$.
   (a) Calculate the total length of the array.
   (b) Determine the physical location (distance from the feed point) of the dipole corresponding to the geometric center frequency (1.58 GHz).
   (c) Assuming that the conductor losses account for a reduction of 15% in efficiency, calculate the antenna's effective gain at 2 GHz if the theoretical gain is 8.5 dB.

6.6 An LPDA antenna with 15 elements is designed with a scaling factor $\tau = 0.9$ and a spacing factor $\sigma = 0.2$.
   (a) Using the design parameters, estimate the half-power beamwidth (HPBW) of the antenna at 1 GHz. Assume an ideal uniform current distribution.
   (b) To suppress the sidelobes, a nonuniform amplitude excitation is applied across the dipoles. Discuss the trade-offs in terms of radiation efficiency and directivity due to this modification.

Problems for Instructors

6.1 LPDA Fundamentals: Explain the concept of frequency independence in the context of LPDA. Calculate the range of operating frequencies for an LPDA designed with a longest element of 1m and a shortest element of 0.5m. Assume a speed of light, $c = 3 \times 10^8$ m/s, and define the frequency range using the fundamental LPDA design principles discussed in the chapter.

6.2 Element Lengths and Spacing: Given an LPDA with 10 elements and a frequency range from 100 MHz to 1 GHz, derive the lengths of each element using the logarithmic periodicity principles. Also, calculate the spacing between elements based on a specified scaling factor $\tau = 0.9$ and angular configuration $\alpha = 0.12$ radians. Discuss how these values impact the antenna's overall performance and bandwidth.

6.3 Design Equation Applications: Using the design equations provided in Section 6.3, design an LPDA that operates within the frequency range of 200 MHz to 2 GHz. Calculate and specify:
   (a) The number of elements required;
   (b) The length of the longest and shortest elements;
   (c) The scaling factor τ;
   (d) The required gain at the center frequency (determine feasible gain values for LPDA within this range). Discuss how variations in τ and α might alter the design and performance characteristics.

6.4 Design Procedure and Parameter Sensitivity: Following the design procedure in Section 6.4, develop an LPDA for the frequency range of 300 MHz to 3 GHz with a targeted gain of 7 dBi. Calculate each element's length and spacing. Then analyze how the performance changes (e.g., gain, directivity, bandwidth) if:
  (a) The scaling factor τ is increased by 10%.
  (b) The spacing angle α is decreased by 0.02 radians. Explain how adjustments in τ and α can serve as tuning parameters in LPDA design.

6.5 Pattern Characteristics: An LPDA designed for operation over a 10:1 bandwidth ratio shows specific radiation pattern characteristics across its operating range. For a specified LPDA configuration, plot the expected radiation patterns at the minimum, mid-point, and maximum frequencies of its operational range. Discuss the stability of the radiation patterns as frequency changes and the implications for applications requiring uniform pattern coverage.

6.6 Real-World Application Scenario: Consider an LPDA designed for UHF and VHF television broadcasting in rural areas where line of sight to the broadcast source is challenging. Based on typical LPDA design parameters for such applications, calculate an appropriate element length and spacing configuration for a frequency range of 200 to 800 MHz. Additionally, evaluate the LPDA's performance in terms of directivity and gain, comparing it to other possible antenna choices (e.g., Yagi-Uda).

6.7 MATLAB Simulation of LPDA Design: Using MATLAB, simulate an LPDA designed to cover the frequency range from 50 MHz to 500 MHz with a scaling factor of $\tau = 0.85$ and an element spacing angle $\alpha = 0.1$ radians. Write a MATLAB script to:
  (a) Plot the gain versus frequency for the LPDA.
  (b) Visualize the radiation pattern at three different frequencies within the operational range. Discuss any insights you gain from the simulation regarding the LPDA's performance over its frequency range.

References

[1] Carrel, R. L., "Analysis and Design of the Log-Periodic Dipole Antenna," PhD dissertation, Electrical Engineering Department, University of Illinois, University Microfilms, Inc., Ann Arbor, MI, 1961.

[2] Balanis, C. A., *Antenna Theory: Analysis and Design*, Second Edition, New York: John Wiley & Sons, 1997, p. 561 (Figure 11.13) and p. 553 (Figure 11.14).

Selected Bibliography

Abdelhak, B., et al., "High Gain and Wide Beam Printed LPDA with Improved Bandwidth Based on Multiple Scaling Factors," *2024 IEEE International Telecommunications Conference (ITC-Egypt)*, July 2024, pp. 85–90.

Casula, G. A., et al., "Design of a Printed Log-Periodic Dipole Array for Ultra-Wideband Applications," *Progress in Electromagnetics Research C*, Vol. 38, 2013, pp. 15–26.

Chu, Q. X., X. R. Li, and M. Ye, "High-Gain Printed Log-Periodic Dipole Array Antenna with Parasitic Cell for 5G Communication," *IEEE Transactions on Antennas and Propagation*, Vol. 65, No. 12, 2017, pp. 6338–6344.

Isbell, D., "Log Periodic Dipole Arrays," *IRE Transactions on Antennas and Propagation*, Vol. 8, No. 3, 1960, pp. 260–267.

Kyei, A., D. U. Sim, and Y. B. Jung, "Compact Log-Periodic Dipole Array Antenna with Bandwidth-Enhancement Techniques for the Low Frequency Band," *IET Microwaves, Antennas & Propagation*, Vol. 11, No. 5, 2017, pp. 711–717.

CHAPTER 7

Rectangular and Circular Aperture Antennas

7.1 Introduction

Aperture antennas are a class of antennas that utilize an opening in a conducting surface to transmit or receive electromagnetic waves. These apertures, typically found in metallic plates or dielectric materials, serve as pathways for electromagnetic energy propagation. Among the most common aperture antennas is the parabolic reflector, often referred to as a parabolic dish. This design leverages a curved reflective surface to direct and focus electromagnetic waves towards a focal point, where an antenna element, such as a feed horn or dipole, is positioned to facilitate transmission or reception.

One of the key advantages of aperture antennas is their ability to provide high directivity and gain. The parabolic reflector, for example, efficiently concentrates radiated energy into a narrow beam, making it ideal for applications requiring long-range communication or precise signal reception. Due to their high performance, aperture antennas are widely employed in satellite communications, wireless networks, radar systems, and radio telescopes. Their ability to operate across a broad spectrum, from microwave to optical wavelengths, further enhances their versatility.

Several variations of aperture antennas exist, each with distinct structural and operational characteristics. Slot antennas, for instance, utilize narrow openings in conducting surfaces to radiate electromagnetic waves, while horn antennas incorporate flared structures to guide and direct radiation efficiently. Among these, horn antennas stand out due to their exceptional directivity, broad bandwidth, low radiation losses, and minimal cross-polarization effects. Additionally, they exhibit high power handling capabilities, making them suitable for demanding applications such as radar, radio astronomy, and high-frequency communication systems. Their metallic construction also grants them superior durability and resistance to environmental challenges, setting them apart from alternatives such as microstrip antennas, which rely on thin substrates.

Despite the numerous advantages of horn antennas, the selection of an appropriate aperture antenna ultimately depends on the specific requirements of the application. Factors such as size, weight, cost, and operational frequency influence the choice between horn antennas, slot antennas, and microstrip designs. Each type presents unique trade-offs, necessitating careful consideration to ensure optimal performance in a given system.

7.2 The H-Plane Sectoral Horn

The geometry of this horn type is such that it is flared in the H-plane (Figure 7.1), but is kept fixed in the E-plane, to the narrow dimension of the waveguide feed.

For the analysis and subsequent construction of the H-plane sectoral horn, the two dimensions, A and R_H, are shown in Figure 7.2.

$$l_H^2 = R_0^2 + \left(\frac{A}{2}\right)^2 \tag{7.1}$$

$$\alpha_H = \tan^{-1}\left(\frac{A}{2R_0}\right) \tag{7.2}$$

$$R_H = (A - a)\sqrt{\left(\frac{I_H}{A} - \frac{1}{4}\right)} \tag{7.3}$$

Hence, the required dimensions for the construction of the H-plane sectoral horn are R_H and A.

The tangential field arriving at the input of the horn is composed of the transverse field components of the waveguide dominant mode TE_{10}:

$$E_y(x) = E_0 \cos\left(\frac{\pi}{a}x\right)e^{-j\beta_g z} \tag{7.4}$$

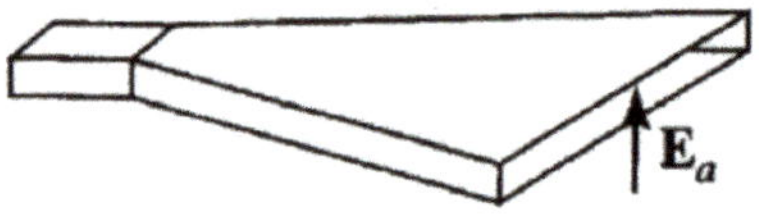

Figure 7.1 The H-plane sectoral horn.

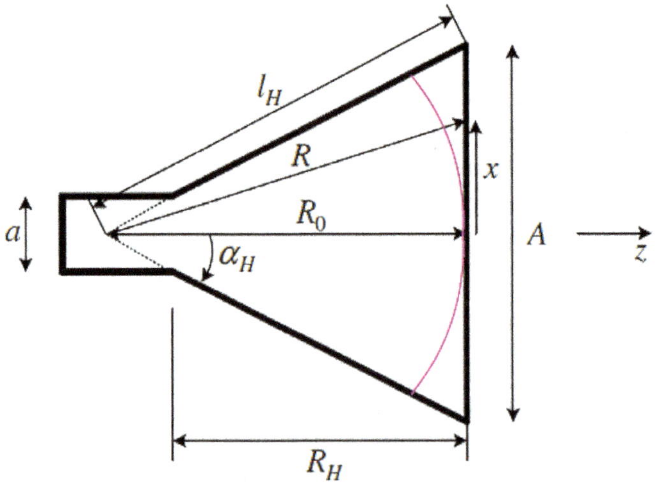

Figure 7.2 H-plane (*x*-*z*) cut of an H-plane sectoral horn.

$$H_x(x) = -\frac{E_y(x)}{Z_g} \tag{7.5}$$

where

$$Z_g = \frac{\eta}{\sqrt{1-(\lambda_0/2a)^2}} \tag{7.6}$$

is the wave impedance of the TE_{10} waveguide mode.

$$\beta_g = \beta_0\sqrt{1-\left(\frac{\lambda_0}{2a}\right)^2} \tag{7.7}$$

is the propagation constant of the TE_{10} mode.

Here, $\beta_0 = \omega\sqrt{\mu\varepsilon} = 2\pi/\lambda_0$, and λ is the free-space wavelength. The field that is illuminating the aperture of the horn is essentially a spatially expanded version of the waveguide field. Note that the wave impedance of the flared waveguide (the horn) gradually approaches the intrinsic impedance of open space η, as A (the H-plane width) increases.

The waves arriving at the horn aperture are not in phase due to the different path lengths from the horn apex. The aperture phase variation is given by $e^{j\beta(R-R0)}$.

Since the aperture is not flared in the y-direction, the phase is uniform along y. We first approximate the path of the wave in the horn:

$$R = \sqrt{R_0^2 + x^2} = R_0\sqrt{1+\left(\frac{x}{R_0}\right)^2} \approx R_0\left[1+\frac{1}{2}\left(\frac{x}{R_0}\right)^2\right] \tag{7.8}$$

The last approximation holds if $x << R_0$. Then we can assume that:

$$R - R_0 \approx \frac{1}{2}\frac{x^2}{R_0} \tag{7.9}$$

Using (7.9), the field at the aperture is approximated as:

$$E_{ay}(x) \approx E_0 \cos\left(\frac{\pi}{A}x\right)e^{-j\frac{\beta}{2R_0}x^2} \tag{7.10}$$

The far field can now be calculated from the following integral expression:

$$F(P) = -\varepsilon\frac{e^{-j\beta r}}{4\pi r}\iint E_{ay}(x)e^{j\beta a_r\cdot r}\,dx\,dy = -\varepsilon\frac{e^{-j\beta r}}{4\pi r}I_y^E(\theta,\varphi) \tag{7.11}$$

where

$$\begin{aligned} a_r \cdot r &= \left(\sin\vartheta\cos\varphi\, a_x + \sin\vartheta\sin\varphi\, a_y\right)\cdot\left(x a_x + y a_y\right) \\ &= x\sin\vartheta\cos\varphi + y\sin\vartheta\sin\varphi \end{aligned} \tag{7.12}$$

The double integral can be further expanded as:

$$I_y^E(\theta,\varphi) = \int_{-A/2}^{A/2} E_0\cos\left(\frac{\pi}{A}x\right)e^{-j\frac{\beta}{2R_0}x^2}\, e^{j\beta x\sin\theta\cos\varphi}\,dx \int_{-b/2}^{b/2} e^{j\beta y\sin\theta\sin\varphi}\,dy \tag{7.13}$$

The second integral can be analytically solved to yield:

$$\int_{-b/2}^{b/2} e^{j\beta y\sin\theta\sin\varphi}\,dy = b\frac{\sin(\beta b\sin\theta\sin\varphi/2)}{(\beta b\sin\theta\sin\varphi/2)} \tag{7.14}$$

The first integral, $I(\theta, \varphi)$, is cumbersome and can be numerically solved:

$$I(\theta,\varphi) = \int_{-A/2}^{A/2} E_0\cos\left(\frac{\pi}{A}x\right)e^{-j\frac{\beta}{2R_0}x^2}\, e^{j\beta x\sin\theta\cos\varphi}\,dx \tag{7.15}$$

The amplitude pattern of the H-plane sectoral horn can be obtained as:

$$E(\theta,\varphi) = \frac{1+\cos\theta}{2}\cdot\frac{\sin\left(\frac{\beta b}{2}\sin\theta\sin\varphi\right)}{\left(\frac{\beta b}{2}\sin\theta\sin\varphi\right)}\cdot I(\theta,\varphi) \tag{7.16}$$

The principal-plane patterns consequently follow:

E-plane pattern ($\varphi = 90°$)

$$F_E(\theta) = \frac{1+\cos\theta}{2}\cdot\frac{\sin\left(\frac{\beta b}{2}\sin\theta\right)}{\left(\frac{\beta b}{2}\sin\theta\right)} \tag{7.17}$$

H-plane pattern ($\varphi = 0°$)

$$F_H(\theta) = \frac{1+\cos\theta}{2}\cdot I(\theta,\varphi) \tag{7.18}$$

The following program is written in MATLAB, which plots the two principal-plane radiation patterns, H-plane and E-plane, in polar diagrams with a logarithmic scale, as shown in Figure 7.3. The range of the scale is limited to a 40-dB range, and, hence, values in the computed arrays, less than the minimum within the considered range, were forced to the minimum value. The user has the choice to use the accurate or the approximate integrand formula as shown in the program code.

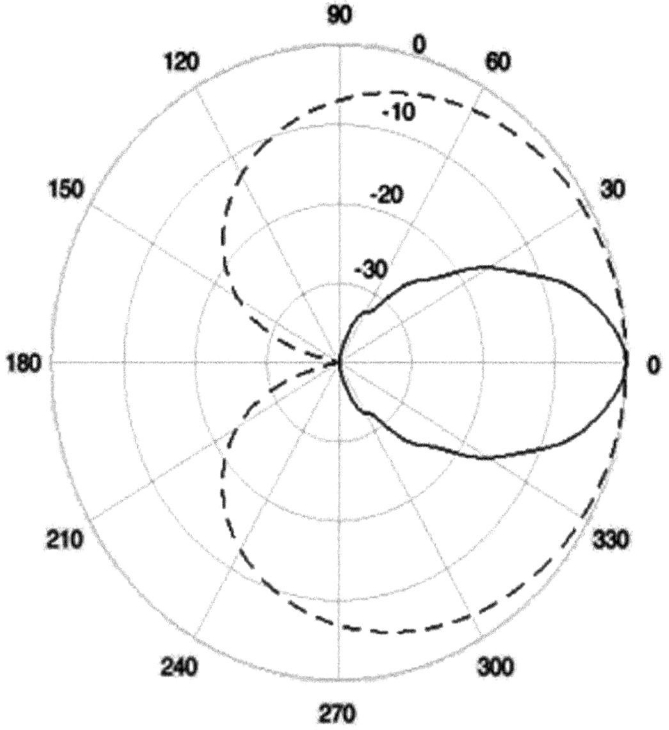

Figure 7.3 Principal-plane radiation patterns of H-plane sectoral horn. H-plane (——), E-plane (-----); half-power beamwidth (HPBW) (H-plane) = 20°, HPBW (E-plane) = 116°.

MATLAB Program 3

```
% Matlab program to compute the principle plane
radiation patterns
% of an H-plane sectoral horn.
% Horn dimensions: R0=6 lam, A= 5.5 lam.
% Waveguide dimensions: a=0.5 lam, b= 0.25 lam.
% Flare angle = 49.25^0
% where wavelength, lam,  is defined in meter (@ f= 10 GHz)
lam=3*10^-2;

% Define the function.
% Accurate formula for the H-plane integrand
integrand = @(x, theta) cos(pi*x/(5.5*lam)) .* exp(-1j*2*pi/
lam*sqrt((6*lam)^2 + x.^2)) .* exp(1j*2*pi/lam*x.*sin(theta));

% Set the integration limits and theta steps.
x_lower = -2.75*lam;
x_upper = 2.75*lam;
theta_step = pi/180;

% Perform the integration.
theta = -pi/2:theta_step:pi/2;
result = zeros(size(theta)); % Initialize result vector
```

```
for i = 1:length(theta)
  result(i) = (1+cos(theta(i)))/2.*integral(@(x)integrand(x,
theta(i)), x_lower, x_upper);
end

% Plot the H-plane (phi=0 degree) polar diagram.
H= abs(result)/max(abs(result));
H(1:8)= 0;
H(174:181)=0;
H= 20*log10(H);
H= H+40;
polar (theta, H, 'k');
hold on

% Plot the E-plane (phi=90 degree) polar diagram.
theta = -pi:pi/180:pi;
F= abs((1+cos(theta))/2.*sin(pi/4*sin(theta))./
(pi/4*sin(theta)));
F= 20*log10(F);
F(1:12)= -40;
F(350:361)= -40;
F= F+40;
polar (theta, F, 'k--')
```

The directivity of the H-plane sectoral horn is calculated by the general directivity expression for apertures.

$$D_0 = \frac{4\pi}{\lambda^2} \cdot \frac{\left|\iint E_a \, ds\right|^2}{\iint |E_a|^2 \, ds} \tag{7.19}$$

The integral in the denominator is proportional to the total radiated power.

$$\iint |E_a|^2 \, ds = \int_{-b/2}^{b/2} \int_{-A/2}^{A/2} |E_0|^2 \cos^2\left(\frac{\pi}{A} x\right) dx \, dy = |E_0|^2 \frac{Ab}{2} \tag{7.20}$$

The integral in the numerator of (7.19) is solved where the field is substituted with its phase approximated as in (7.10). The final result is:

$$D_H = \frac{4\pi}{\lambda^2} \varepsilon_t \varepsilon_{ph}^H (Ab) \tag{7.21}$$

where $\varepsilon_t = 8/\pi^2$ is the factor that explicitly shows the aperture efficiency associated with the aperture cosine taper.

The factor $\varepsilon^H_{ph}$is the aperture efficiency associated with the aperture phase distribution. $\varepsilon^H_{ph} = 0.79$ for optimal directivity.

A family of universal directivity curves $(\lambda/b)D_H$ versus A/λ, with R_0 being a parameter is given in [1].

From these curves, it is obvious that, for a given axial length R_0 and at a given wavelength, there is an optimal aperture width A corresponding to the maximum directivity.

It can be shown that the optimal directivity is obtained if the relation between A and R_0 is:

$$\frac{A}{\lambda} = \sqrt{3\frac{R_0}{\lambda}} \tag{7.22}$$

Table 7.1 tabulates the dimensions of an H-plane horn for optimal directivity.

7.3 The E-Plane Sectoral Horn

The geometry of this horn type is such that it is flared in the E-plane, but is kept fixed in the H-plane, to the broad dimension of the waveguide feed.

The analysis follows the same steps as in the previous section. The field at the aperture is approximated by (7.23), which compares to (7.10).

$$E_{ay} \approx E_0 \cos\left(\frac{\pi}{a}x\right) e^{-j\frac{\beta}{2R_0}y^2} \tag{7.23}$$

The far field can now be calculated from the integral expressed in (7.11), where the double integral can be further expanded as:

$$I_y^E(\theta,\varphi) = \int_{-a/2}^{a/2} E_0 \cos\left(\frac{\pi}{a}x\right) e^{j\beta x \sin\theta\cos\varphi}\, dx \int_{-B/2}^{B/2} e^{-j\frac{\beta}{2R_0}y^2} e^{j\beta y \sin\theta \sin\varphi}\, dy \tag{7.24}$$

Table 7.1 Values of H-Plane Horn Aperture Dimensions for Optimal Directivity

R_0 *(in λ)*	$\frac{A}{\lambda} = \sqrt{3\frac{R_0}{\lambda}}$ *for optimal directivity*	$\frac{\lambda}{b}D_H = \frac{32}{\pi}\varepsilon_{ph}^H\left(\frac{A}{\lambda}\right)$ *where $\varepsilon^H_{ph} = 0.79$, for optimal directivity*
6	4.242	34.134
8	4.898	39.413
10	5.477	44.072
12	6.000	48.281
15	6.708	53.978
20	7.745	62.323
30	9.486	76.332
50	12.247	98.550
75	15.000	120.703
100	17.320	139.371

Figures 7.4 shows an E-plane sectoral horn, while Figure 7.5 demonstrates a cut of the horn in the y-z plane, for analysis purpose.

The first integral can be analytically solved to yield:

$$\int_{-a/2}^{a/2} E_0 \cos\left(\frac{\pi}{a}x\right) e^{j\beta x \sin\theta\cos\varphi}\,dx = \frac{\cos\left(\frac{\beta a}{2}\sin\theta\cos\varphi\right)}{\left(\frac{\pi}{a}+\beta\sin\theta\cos\varphi\right)} + \frac{\cos\left(\frac{\beta a}{2}\sin\theta\cos\varphi\right)}{\left(\frac{\pi}{a}-\beta\sin\theta\cos\varphi\right)} \tag{7.25}$$

The second integral, $I(\theta, \varphi)$, is cumbersome and can be numerically solved:

$$I(\theta,\varphi) = \int_{-B/2}^{B/2} e^{-j\frac{\beta}{2R_0}y^2} e^{j\beta y \sin\theta\sin\varphi}\,dy \tag{7.26}$$

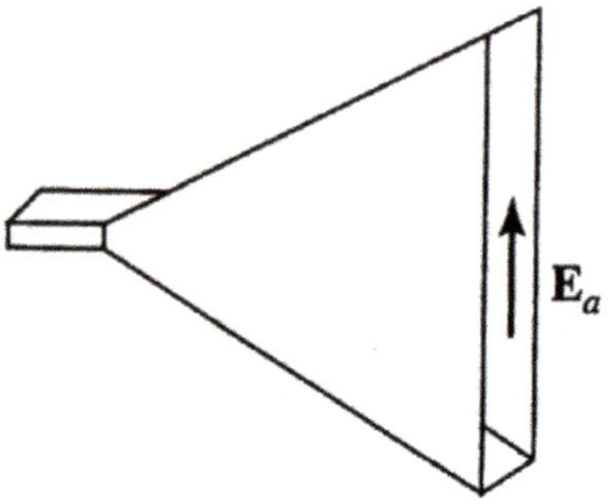

Figure 7.4 E-plane sectoral horn.

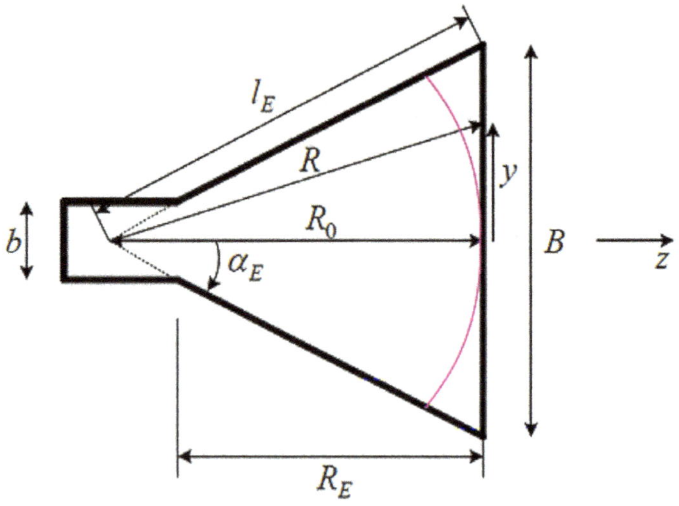

Figure 7.5 E-plane (y-z) cut from an E-plane sectoral horn.

The amplitude pattern of the E-plane sectoral horn can be obtained as:

$$E(\theta,\varphi)=\frac{1+\cos\theta}{2}\cdot\left(\frac{\cos\left(\frac{\beta a}{2}\sin\theta\cos\varphi\right)}{\left(\frac{\pi}{a}+\beta\sin\theta\cos\varphi\right)}+\frac{\cos\left(\frac{\beta a}{2}\sin\theta\cos\varphi\right)}{\left(\frac{\pi}{a}-\beta\sin\theta\cos\varphi\right)}\right)\cdot I(\theta,\varphi) \tag{7.27}$$

The principal-plane patterns consequently follow:

H-plane pattern ($\varphi = 0°$)

$$F_E(\theta)=\frac{1+\cos\theta}{2}\cdot\left(\frac{\cos\left(\frac{\beta a}{2}\sin\theta\right)}{\left(\frac{\pi}{a}+\beta\sin\theta\right)}+\frac{\cos\left(\frac{\beta a}{2}\sin\theta\right)}{\left(\frac{\pi}{a}-\beta\sin\theta\right)}\right) \tag{7.28}$$

E-plane pattern ($\varphi = 90°$):

$$F_H(\theta)=\frac{1+\cos\theta}{2}\cdot I(\theta,\varphi) \tag{7.29}$$

MATLAB Program 4 computes the principal radiation patterns and plots the E-plane and H-plane of the E-plane sectoral horn, as shown in Figure 7.6.

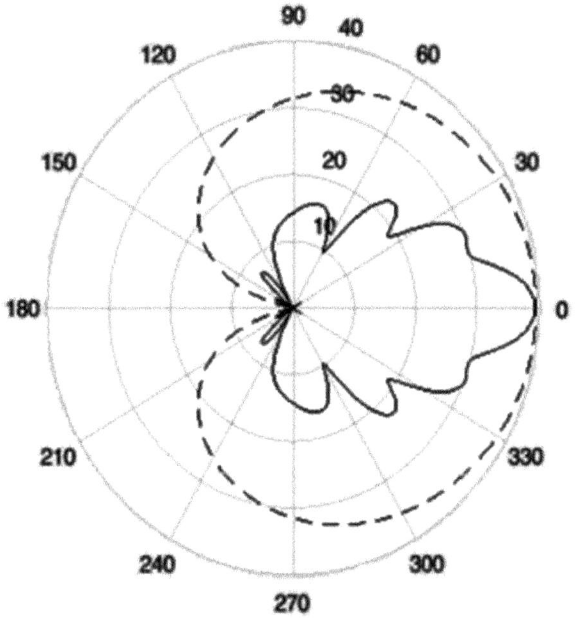

Figure 7.6 Principal-plane radiation patterns of E-plane sectoral horn. E-plane (——), H-plane (----); HPBW (E-plane) = 16°, HPBW (H-plane) = 102°.

MATLAB Program 4

```
% Matlab program to compute the principal plane
radiation patterns
% of an E-plane sectoral horn.
% Horn dimensions: R0=6 lam, B=3.5 lam.
% Waveguide dimensions: a=0.5 lam, b= 0.25 lam.
% Flare angle = 32.52^0
% where wavelength, lam, is defined in meter (@ f= 10 GHz)
lam= 3*10^-2;

% Define the function.
% Accurate formula for the E-plane integrand
integrand = @(y, theta) exp((-j*2*pi/lam).*(sqrt(y.^2 +(6*lam)^2)
-y.*sin(theta)));

% Set the integration limits and theta steps.
y_lower = -1.75*lam;
y_upper = 1.75*lam;
theta_step = pi/180;

% Perform the integration.
theta = 0:theta_step:2*pi;
result = zeros(size(theta)); % Initialize result vector
for i = 1:length(theta)
  result(i) = (1+cos(theta(i)))/2.*integral(@(y)integrand(y,
theta(i)), y_lower, y_upper);
end

% Plot the E-plane (phi=90 degree) polar diagram.
E= abs(result);
E= 20*log10(E);
E= E+60;
E(161:201)=0;
polar (theta, E, 'k');
hold on

% Plot the H-plane (phi=0 degree) polar diagram.
theta = -pi:pi/180:pi;
F1= (cos(pi/2*sin(theta))./(2*pi/lam+2*pi/lam.*sin(theta)));
F2= (cos(pi/2*sin(theta))./(2*pi/lam -2*pi/lam.*sin(theta)));
F= (F1+F2).* (1+cos(theta))/2;
H= 20*log10(F);
H= H+80;
H(1:12)=0;
H(350:361)=0;
polar (theta, H, 'k--')
```

The directivity of the E-plane sectoral horn is found in a manner analogous to the H-plane sectoral horn:

$$D_E = \frac{4\pi}{\lambda^2}\varepsilon_t \varepsilon_{ph}^E (aB) \tag{7.30}$$

where $\varepsilon_t = 8/\pi^2$ is the factor that explicitly shows the aperture efficiency associated with the aperture cosine taper and the factor ε^E_{ph} is the aperture efficiency associated with the aperture phase distribution. $\varepsilon^E_{ph} = 0.8$ for optimal directivity.

A family of universal directivity curves $(\lambda/a)D_E$ versus B/λ with R_0 being a parameter is given below.

The optimal relation between the flared height B and the horn apex length R_0 that produces the maximum possible directivity is:

$$\frac{B}{\lambda} = \sqrt{2\frac{R_0}{\lambda}} \tag{7.31}$$

Table 7.2 tabulates the dimensions of an E-plane horn for optimal directivity.

7.4 The Pyramidal Horn

The pyramidal (rectangular) horn is probably the most popular antenna in the microwave frequency ranges from approximately (1 GHz up to 18 GHz). The feeding waveguide is flared in both directions, the E-plane and the H-plane. The analysis results are combinations of the E-plane sectoral horn and the H-plane sectoral horn. The field distribution at the aperture is approximated as:

$$E_{ay} \approx E_0 \cos\left(\frac{\pi}{A}x\right) e^{-j\frac{\beta}{2}\left(\left(\frac{x}{R_E}\right)^2 + \left(\frac{y}{R_H}\right)^2\right)} \tag{7.32}$$

The E-plane principal pattern of the pyramidal horn is the same as the E-plane principal pattern of the E-plane sectoral horn. The same holds for the H-plane patterns of the pyramidal horn and the H-plane sectoral horn. The directivity of the

Table 7.2 Values of E-Plane Horn Aperture Dimensions for Optimal Directivity

R_0 *(in* λ*)*	$\frac{B}{\lambda} = \sqrt{2\frac{R_0}{\lambda}}$ *for optimal directivity*	$\frac{\lambda}{a}D_E = \frac{32}{\pi}\varepsilon^E_{ph}\left(\frac{B}{\lambda}\right)$ *where* $\varepsilon^E_{ph} = 0.8$*, for optimal directivity*
6	3.464	28.227
8	4.000	32.594
10	4.472	36.441
12	4.898	39.912
15	5.477	44.630
20	6.324	51.532
30	7.745	63.111
50	10.000	81.487
75	12.247	99.797
100	14.142	115.239

pyramidal horn can be found by introducing the phase efficiency factors of both planes and the taper efficiency factor of the H-plane:

$$D_P = \frac{4\pi}{\lambda^2}\varepsilon_t\varepsilon_{ph}^E\varepsilon_{ph}^H(AB) = \frac{4\pi}{\lambda^2}\varepsilon_{ph}^P(AB) \tag{7.33}$$

The gain of a horn is usually very close to its directivity because the radiation efficiency is very good (low losses). The directivity as calculated from (7.33) is very close to measurements. The above expression is a physical optics approximation, and it does not take into account only multiple diffractions and the diffraction at the edges of the horn arising from reflections from the horn interior. These phenomena, which are unaccounted for, lead to only minor fluctuations of the measured results about the prediction of (7.33). That is why horns are often used as gain standards in antenna measurements.

The optimal directivity of an E-plane horn is achieved when (7.31) is implemented, which also results in $\varepsilon^E_{ph} = 0.8$. The optimal directivity of an H-plane horn is achieved when (7.22) is implemented, which also results in $\varepsilon^H_{ph} = 0.79$. Thus, the optimal horn has a phase aperture efficiency of $\varepsilon^E_{ph}\varepsilon^H_{ph} = 0.632$.

The total aperture efficiency includes the taper factor, too. Hence, it works out as:

$$\varepsilon_{ph}^P = \frac{8}{\pi^2} * 0.632 = 0.512 \tag{7.34}$$

Therefore, the best achievable directivity for a rectangular waveguide horn is about one-half that of a uniform rectangular aperture.

7.4.1 Optimum Horn Design

Usually, the optimum (from the point of view of maximum gain) design of a horn is desired because it results in the shortest axial length. For a horn to be realizable, the following must be true:

$$R_E = R_H = R_P \tag{7.35}$$

In reference to Figures 7.2 and 7.6, the following equations from the two geometries can be derived:

$$\frac{R_0^H}{R_H} = \frac{A/2}{A/2 - a/2} = \frac{A}{A-a} \tag{7.36}$$

$$\frac{R_0^E}{R_E} = \frac{B/2}{B/2 - b/2} = \frac{B}{B-b} \tag{7.37}$$

The optimum-gain condition in the E-plane (7.31) is substituted into (7.37) to produce:

$$B^2 - bB - 2\lambda R_E = 0 \tag{7.38}$$

There is only one physically meaningful solution to:

$$B = \frac{1}{2}\left(b + \sqrt{b^2 - 8\lambda R_E}\right) \tag{7.39}$$

Similarly, the maximum-gain condition for the H-plane of (7.22) together with (7.36) yields:

$$R_H = \frac{A-a}{A}\left(\frac{A^2}{3\lambda}\right) = A\frac{(A-a)}{3\lambda} \tag{7.40}$$

Since $R_E = R_H$ must be fulfilled, (7.40) is substituted into (7.39), which gives:

$$B = \frac{1}{2}\left(b + \sqrt{b^2 + \frac{8A(A-a)}{3}}\right) \tag{7.41}$$

Substituting in the following horn's gain equation similar to (7.33) gives:

$$G = \frac{4\pi}{\lambda^2}\varepsilon_{ph}AB = \frac{4\pi}{\lambda^2}\varepsilon_{ph}A\frac{1}{2}\left(b + \sqrt{b^2 + \frac{8A(A-a)}{3}}\right) \tag{7.42}$$

This transforms into the following:

$$A^4 - aA^3 + \frac{3bG\lambda^2}{8\pi\varepsilon_{ph}}A - \frac{3G^2\lambda^4}{32\pi^2\varepsilon_{ph}^2} = 0 \tag{7.43}$$

An analytical solution of the above equation is cumbersome and a numerical solution can be found instead.

Another optimal solution may be found if a known R_P is desired. In this case, there would be no need for a nonlinear equation solution. The design procedure requires finding A and B from (7.22) and (7.31) and then substituting in the first part of the gain (7.42), where $\varepsilon_{ph} = 0.512$.

Let $R_P = 8\lambda$, and then A and B for an optimum horn design can be calculated from (7.22) and (7.23) and the results are $A = 4.9\lambda$ and $B = 4\lambda$, respectively. The gain from (7.42) works out as $G = 125.61$, equivalent to 21 dB. Similarly, and by applying (7.21) and (7.30), the gains for the H-sectoral horn and E-sectoral horn work out as 10.4 dB and 11.5 dB, respectively.

A MATLAB program was developed to simulate the principal E-plane and H-plane field patterns for a pyramidal horn. The integrands of both aperture fields were solved numerically and the field patterns were plotted, as shown in Figure 7.7.

MATLAB Program 5

```
% Matlab program to compute the principle plane
radiation patterns
% of a Pyramidal Horn.
```

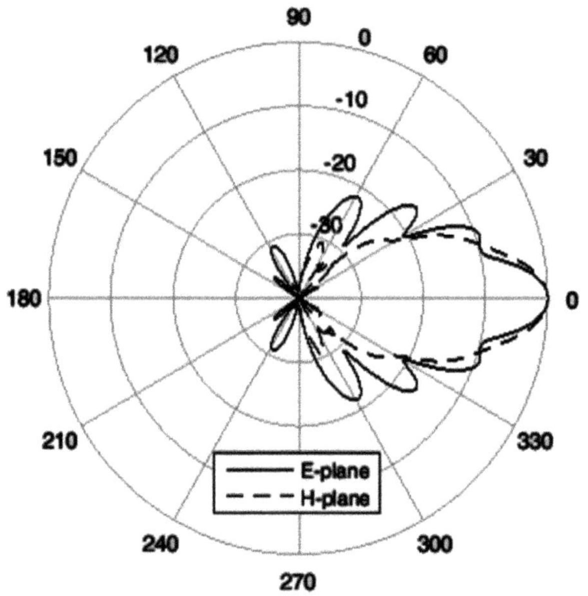

Figure 7.7 E-plane and H-plane field patterns of a pyramidal horn. ($R_P = 8\lambda$, $A = 4.9\lambda$, $B = 4\lambda$); HPBW (E-plane) = 14°, HPBW (H-plane) = 16°.

```
% Horn dimensions: R0=8 lam, A= 4.9 lam, B=4 lam.
% Waveguide dimensions: a=0.5 lam, b= 0.25 lam.
% where wavelength, lam, is defined in meter (@ f= 10 GHz)
lam= 3*10^-2;

% Define the function.
% Accurate formula for the E-plane integrand
integrand = @(y, theta) exp((-j*2*pi/lam).*(sqrt(y.^2 +(8*lam)^2)
-y.*sin(theta)));

% Set the integration limits and theta steps.
y_lower = -2*lam;
y_upper = 2*lam;
theta_step = pi/180;

% Perform the integration.
theta = 0:theta_step:2*pi;
result = zeros(size(theta)); % Initialize result vector
for i = 1:length(theta)
  result(i) = (1+cos(theta(i)))/2.*integral(@(y)integrand(y,
theta(i)), y_lower, y_upper);
end

% Plot the E-plane (phi=90 degree) polar diagram.
E= abs(result)/max(abs(result));
E= 20*log10(E);
E= E+40;
```

```
E(153:209)=0;
polar (theta, E, 'k');
hold on

% Accurate formula for the H-plane integrand
integrand = @(x, theta) cos(pi*x/(4.9*lam)) .* exp(-1j*2*pi/
lam*sqrt((8*lam)^2 + x.^2)) .* exp(1j*2*pi/lam*x.*sin(theta));

% Set the integration limits and theta steps.
x_lower = -2.45*lam;
x_upper = 2.45*lam;
theta_step = pi/180;

% Perform the integration.
theta = 0:theta_step:2*pi;
result = zeros(size(theta)); % Initialize result vector
for i = 1:length(theta)
  result(i) = (1+cos(theta(i)))/2.*integral(@(y)integrand(y,
theta(i)), x_lower, x_upper);
end

% Plot the H-plane (phi=90 degree) polar diagram.
H= abs(result)/max(abs(result));
H= 20*log10(H);
H= H+40;
H(86:99)=0;
H(149:213)=0;
polar (theta, H, 'k--');
```

Horn antennas operate well over a bandwidth of 50%. However, gain performance is optimal only at a given frequency. The gain increases with frequency, which is typical for aperture antennas. However, the increase shows saturation at higher frequencies. This is due to the decrease of the aperture efficiency, which is a result of an increased phase difference in the field distribution at the aperture.

7.5 The Circular Aperture

Conical horns, although widely applicable, have historically received less focus than symmetrical circular apertures. Circular apertures are more popular due to their simplicity, making them dominant in research and design. Conical horns' complex geometry poses challenges in analyzing, simulating, and using them practically. Understanding how electromagnetic waves behave in these structures requires advanced numerical methods, especially for modeling aperture phase distribution. Circular apertures serve as foundational elements in comprehending the intricate nature of conical horns. The distinctive characteristics of conical horns underscore the imperative need for expanded research endeavors to unlock their full potential and optimize performance across diverse applications.

7.5.1 The Uniform Aperture

A circular opening in an infinite ground plane, which is illuminated by a uniform plane wave, normally incident from behind, approximates the uniform circular aperture.

The idealized field distribution is described as:

$$E_a = E_0 \boldsymbol{a}_x, \quad \rho' \le a \tag{7.44}$$

The radiation integral is

$$I_x^E = E_0 \iint e^{j\beta \boldsymbol{a}_r \cdot \boldsymbol{r}'} ds' \tag{7.45}$$

where $\boldsymbol{r}' = \boldsymbol{a}_x \rho' \cos\varphi' + \boldsymbol{a}_y \rho' \sin\varphi'$ and $\boldsymbol{a}_r \cdot \boldsymbol{r}' = \rho' \sin\,\theta(\cos\varphi\ \text{cox}\varphi' + \sin\varphi \sin\varphi') = \rho' \sin\theta \cos(\varphi - \varphi')$.

Substituting back into (7.45) yields:

$$\begin{aligned} I_x^E &= E_0 \int_0^a \left[\int_0^{2\pi} e^{j\beta\rho' \sin\theta \cos(\varphi - \varphi')}\, d\varphi' \right] \rho'\, d\rho' \\ &= 2\pi E_0 \int_0^a \rho' J_0(\beta\rho' \sin\theta)\, d\rho' \end{aligned} \tag{7.46}$$

where $J_0(\ )$ is the Bessel function of the first kind of order zero.

The integral can be solved by applying the identity $\int x J_0(x) dx = x J_1(x)$. Hence:

$$I_x^E = 2\pi E_0 \frac{a}{\beta \sin\theta} J_1(\beta a \sin\theta) \tag{7.47}$$

where $J_1(\ )$ is the Bessel function of the first kind of order 1.

Since the aperture is mounted on a conducting plane, the general far-field solution can be written as:

$$E_\theta^{\text{far}}(\theta, \varphi) = j\beta \frac{e^{-j\beta r}}{4\pi r} \left(I_x^E \cos\varphi + I_y^E \sin\varphi \right) \tag{7.48}$$

$$E_\varphi^{\text{far}}(\theta, \varphi) = j\beta \frac{e^{-j\beta r}}{4\pi r} \cos\theta \left(I_y^E \cos\varphi - I_x^E \sin\varphi \right) \tag{7.49}$$

Principal-plane patterns

E-plane ($\varphi = 0°$):

$$E_\theta(\theta) = j\beta \frac{e^{-j\beta r}}{4\pi r} \pi a^2 E_0 \frac{2 J_1(\beta a \sin\theta)}{\beta a \sin\theta} \tag{7.50}$$

Hence, the normalized magnitude, $E_\theta(\theta)$:

$$E_\theta(\theta) = \frac{2J_1(\beta a \sin\theta)}{\beta a \sin\theta} = f(\theta) \tag{7.51}$$

H-plane ($\varphi = 90°$):

$$E_\varphi(\theta) = -j\beta \frac{e^{-j\beta r}}{4\pi r} \pi a^2 E_0 \frac{2\cos\theta J_1(\beta a \sin\theta)}{\beta a \sin\theta} \tag{7.52}$$

and the normalized magnitude, $E_\varphi(\theta)$:

$$E_\varphi(\theta) = \frac{2\cos\theta J_1(\beta a \sin\theta)}{\beta a \sin\theta} = \cos\theta \cdot f(\theta) \tag{7.53}$$

The following MATLAB program calculates and plots the far-field pattern of a circular radiating aperture, of $2a = 6\lambda$, mounted on a conducting plane, with the electric field in the x-direction, as shown in Figure 7.8.

MATLAB Program 6

```
% Radiation from a UNIFORM CIRCULAR APERTURE of 2a= 6 λ.
% Calculate the Bessel function of the first kind of order one,
J1( ).
z = -6*pi:0.1:6*pi;
J = besselj(1,z);
% z= 6pi*sin(theta)
J= J./z;
% zs= sin(theta)
zs= z./(6*pi);
% cs = cos(theta)
cs= sqrt(1-zs.^2);
H= 20*log10(2*cs.*abs(J));
E= 20*log10(2.*abs(J));
plot(zs, E, 'k -', zs, H, 'k --')
axis([-1 1 -50 0])
```

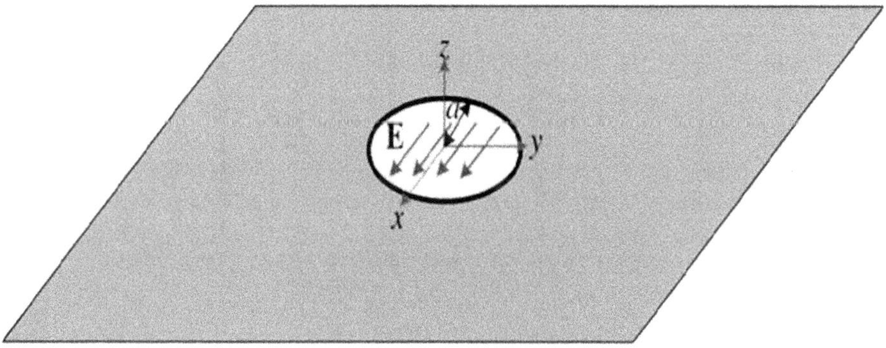

Figure 7.8 Circular aperture mounted on a conducting plane.

For a large aperture, where $a >> \lambda$, the cos θ factor, in the 3D pencil beam approaches unity. The half-power angle for the $f(\theta)$ factor is obtained at $\beta\, a \times \sin\theta \approx 1.6$. The HPBW is calculated as:

$$HPBW = 2\theta_{1/2} = 2\sin^{-1}\left(\frac{1.6}{\beta a}\right) \approx 2\left(\frac{1.6}{\beta a}\right)$$
$$= 1.02\frac{\lambda}{2a}\text{ rad} = 58.4\frac{\lambda}{2a}\text{ deg} \tag{7.54}$$

For example, if the diameter of the aperture is $2a = 6\lambda$ (Figure 7.9), then HPBW = 9.73°. The sidelobe level of any uniform circular aperture is −17.5 dB. Any uniform aperture has unity taper aperture efficiency, and its directivity can be found directly in terms of its physical area,

$$D_u = \frac{4\pi}{\lambda^2}A_p = \frac{4\pi}{\lambda^2}\pi a^2 \tag{7.55}$$

7.5.2 The Tapered Aperture

Many practical circular aperture antennas can be approximated as radially symmetric apertures with field amplitude distribution, which is tapered from the center toward the aperture edge.

The general radiation integral hence takes the following form:

$$I_x^E = 2\pi\int_0^a E_0(\rho')\rho' J_0(\beta\rho'\sin\theta)\,d\rho' \tag{7.56}$$

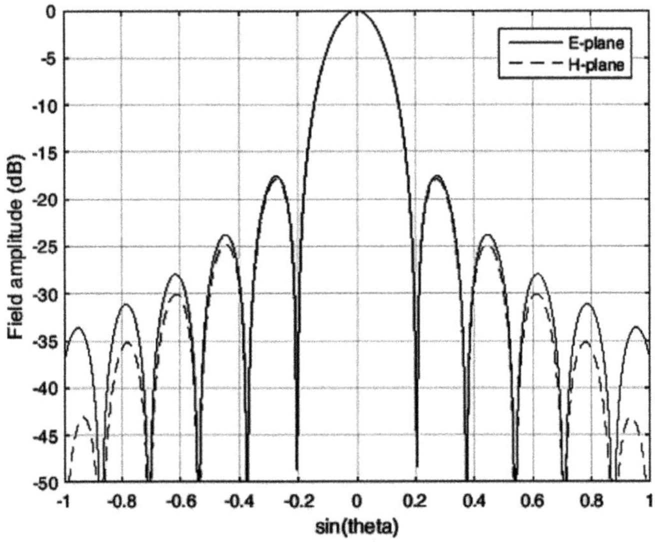

Figure 7.9 Far-field patterns of a uniform circular aperture of $2a = 6\lambda$ mounted on a conducting plane. (E-plane: ——, H-plane: ----.)

The following parabolic taper approximation of order n is commonly used:

$$E_a(\rho') = E_0\left\{1-\left(\frac{\rho'}{a}\right)^2\right\}^n \tag{7.57}$$

Figure 7.10 shows the tapered field amplitude distribution, as described by (7.57) for the two cases when $n = 1$ and $n = 2$.

Substituting (7.57) back into (7.56) yields:

$$I_x^E = 2\pi E_0\int_0^a\left\{1-\left(\frac{\rho'}{a}\right)^2\right\}^n(\rho')\rho' J_0(\beta\rho'\sin\theta)\,d\rho' \tag{7.58}$$

The following relation is used to solve (7.58). Hence:

$$\int_0^a\left(1-x^2\right)^n xJ_0(bx)\,dx = \frac{2^n n!}{b^{n+1}}J_{n+1}(b) \tag{7.59}$$

In our case, $x = \rho'/a$ and $b = \beta a\sin\theta$. Then

$$I_x^E = E_0\left(\frac{\pi a^2}{n+1}\right)f(\theta,n) \tag{7.60}$$

where the normalized field pattern is:

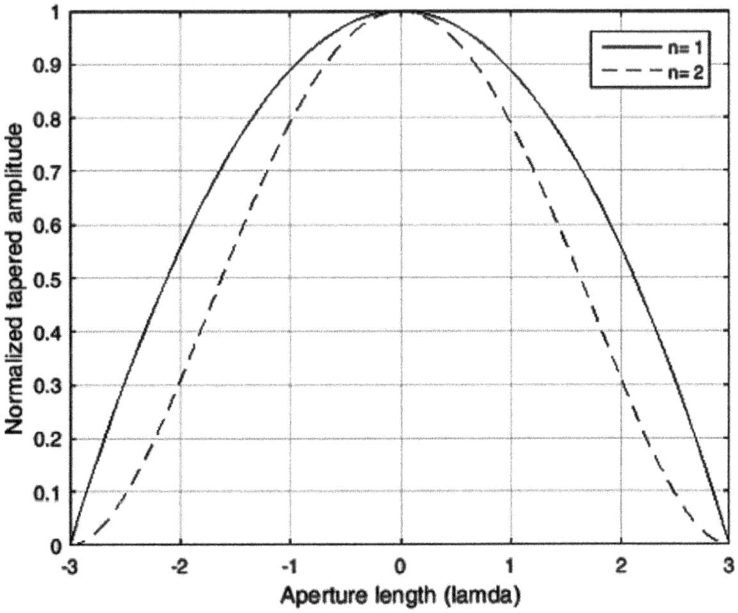

Figure 7.10 Parabolic tapered ($n = 1$) and parabolic squared tapered ($n = 2$) field amplitude distribution for circular aperture of $2a = 6\lambda$.

$$f(\theta, n) = \frac{2^{n+1}(n+1)!\, J_{n+1}(\beta a \sin\theta)}{(\beta a \sin\theta)^{n+1}}$$

For $n = 1$, the following expresses the normalized field of the parabolic tapered aperture, where J_2 is the Bessel function of the first kind of order 2.

$$f(\theta, 1) = \frac{8 J_2(\beta a \sin\theta)}{(\beta a \sin\theta)^2} \quad (7.61)$$

For $n = 2$, the following expresses the normalized field of the parabolic squared tapered aperture, where J_3 is the Bessel function of the first kind of order 3.

$$f(\theta, 2) = \frac{48 J_3(\beta a \sin\theta)}{(\beta a \sin\theta)^3} \quad (7.62)$$

MATLAB Program 7 calculates and plots the far field patterns, shown in Figure 7.11, of a parabolic tapered circular aperture with an aperture distribution, described by (7.61).

MATLAB Program 7

```
% Radiation from a PARABOLIC TAPERED CIRCULAR
% APERTURE of diameter 6 λ.
% Calculate the Bessel function of the first kind of order two.
```

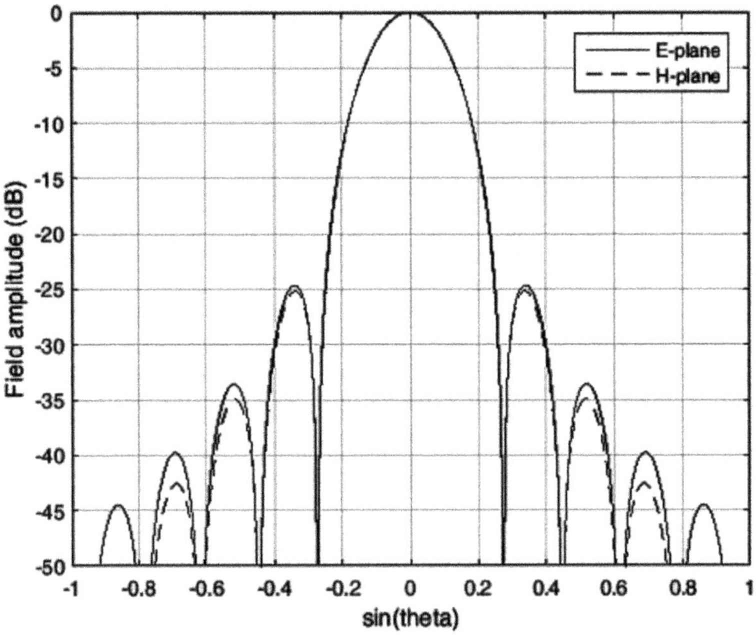

Figure 7.11 Far-field patterns of a parabolic tapered circular aperture of $2a = 6\lambda$. (E-plane: ——, H-plane: ----.)

```
z = -6*pi:0.1:6*pi;
J = besselj(2,z);
% z= 6pi*sin(theta)
J= J./z.^2;
% zs= sin(theta)
zs= z./(6*pi);
% cs = cos(theta)
cs= sqrt(1-zs.^2);
H= 20*log10(8.*cs.*abs(J));
E= 20*log10(8.*abs(J));
plot(zs, E, 'k -', zs, H, 'k --')
axis([-1 1 -50 0])
```

The half-power angle for the $f(\theta, 1)$ factor is obtained at $\beta a \cdot \sin\theta \approx 2$. The HPBW is calculated as:

$$\begin{aligned} HPBW &= 2\theta_{1/2} = 2\sin^{-1}\left(\frac{2}{\beta a}\right) \approx 2\left(\frac{2}{\beta a}\right) \\ &= 1.27\frac{\lambda}{2a} \text{ rad} = 72.7\frac{\lambda}{2a} \text{ deg} \end{aligned} \tag{7.63}$$

For example, if the diameter of the aperture is $2a = 6\lambda$, then HPBW = 12.11°. The sidelobe level of any uniform circular aperture is −24.6 dB.

For comparison, MATLAB Program 8 calculates and plots the far-field patterns, shown in Figure 7.12, of a parabolic squared tapered circular aperture with an aperture distribution, described by (7.62).

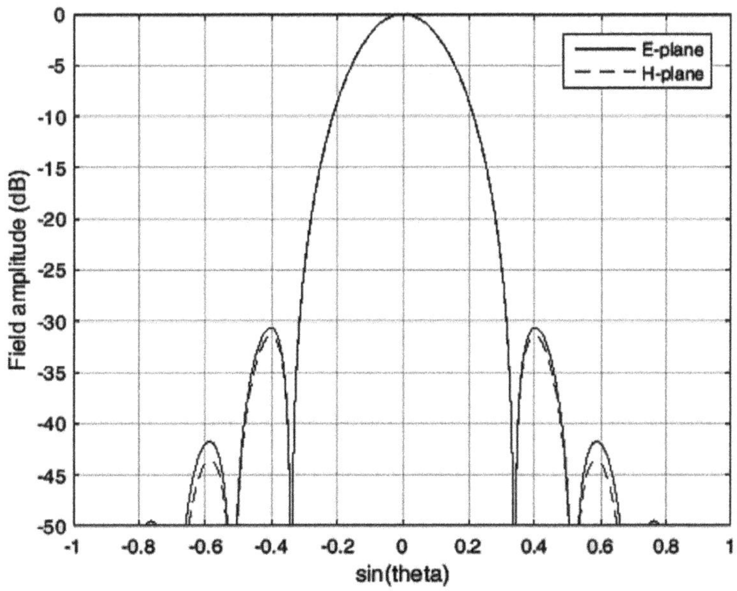

Figure 7.12 Far-field patterns of a parabolic squared tapered circular aperture of $2a = 6\lambda$. (E-plane: —, H-plane: ----.)

MATLAB Program 8

```
% Radiation from a PARABOLIC SQUARED TAPERED
% CIRCULAR APERTURE of diameter 6 λ.
% Calculate the Bessel function of the first kind of order three.
z = -6*pi:0.1:6*pi;
J = besselj(3,z);
% z= 6pi*sin(theta)
J= J./z.^3;
% zs= sin(theta)
zs= z./(6*pi);
% cs = cos(theta)
cs= sqrt(1-zs.^2);
H= 20*log10(48.*cs.*abs(J));
E= 20*log10(48.*abs(J));
plot(zs, E, 'k -', zs, H, 'k --')
axis([-1 1 -50 0])
```

The half-power angle for the $f(\theta, 2)$ factor is obtained at $\beta a \cdot \sin\theta \approx 2.32$. The HPBW is calculated as:

$$\begin{aligned} HPBW &= 2\theta_{1/2} = 2\sin^{-1}\left(\frac{2.32}{\beta a}\right) \approx 2\left(\frac{2.32}{\beta a}\right) \\ &= 1.47\frac{\lambda}{2a} \text{ rad} = 84.2\frac{\lambda}{2a} \text{ deg} \end{aligned} \tag{7.64}$$

For example, if the diameter of the aperture is $2a = 6\lambda$, then HPBW = 14.03°. The sidelobe level of any uniform circular aperture is −30.6 dB.

The parabolic taper (n = 1) provides lower sidelobes in comparison with the uniform distribution (n = 0), but it has a broader main beam. The first sidelobe is noticed to be 7 dB lower than that produced by the uniform aperture. Furthermore, the HPBW is noticed to be 24.5% broader than that from the uniform aperture. The parabolic taper (n = 2) provides even much lower sidelobes in comparison with the uniform distribution (n = 0). The first sidelobe is noticed to be 13 dB lower than that produced by the uniform aperture, while the HPBW is 44.2% broader than that produced by the uniform aperture. Hence, there is always a trade-off between low sidelobe levels and high directivity (small HPBW). The aperture taper efficiency, which estimates the effective radiating aperture relative to the physical aperture, can be calculated from the following formula:

$$\varepsilon_t = \frac{2n+1}{(n+1)^2} \tag{7.65}$$

Figure 7.13 shows changes of aperture taper efficiency, HPBW, and first sidelobe level with the n order of the parabolic taper function.

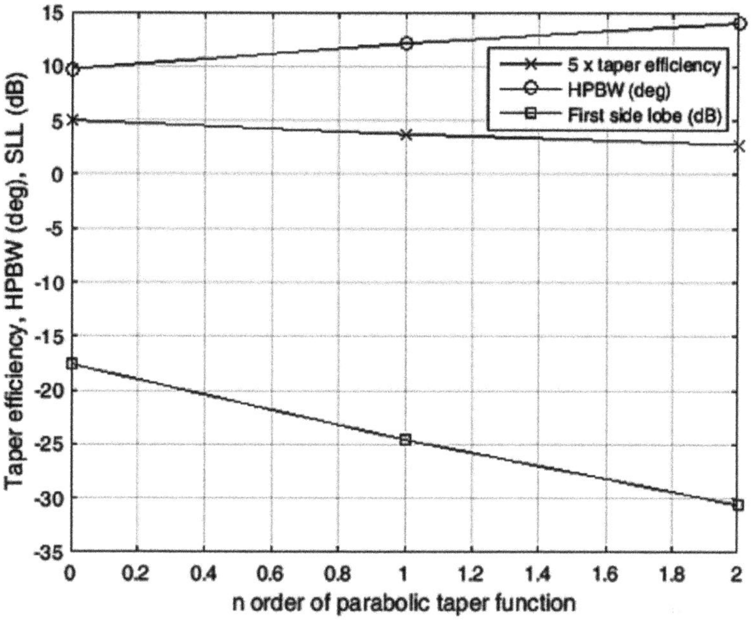

Figure 7.13 Aperture taper efficiency, HPBW (deg) and SLL (dB) versus nth order of parabolic taper function of a circular aperture ($2a = 6\lambda$).

Solved Examples

Example 7.1 Design a pyramidal horn, assuming that an aperture efficiency equals 0.7 with axial lengths $R_H = 20\lambda$ and $R_E = 30\lambda$ at the design frequency = 10 GHz. The dimensions of the waveguide feed were $a = 2.286$ cm and $b = 1.016$ cm.

(a) What are the aperture dimensions A and B, in centimeters?

(b) What are the beamwidths, BW_E and BW_H?

(c) What is the maximum gain of the pyramidal horn?

(d) If R_H became 35λ and R_E became 40λ, calculate the new aperture dimensions, beamwidths and maximum gain at the same frequency?

Solution 7.1

(a) $A = \sqrt{3\lambda R_H} = 23.23$ cm, $B = \sqrt{2\lambda R_E} = 13.41$ cm

(b) HPBWs are estimated using the following formulas:

$$HPBW_H = 78 \cdot \lambda/A = 10.07°, HPBW_E = 54 \cdot \lambda/B = 12.08°$$

(c) $G_{\max} = (4\pi/\lambda^2)\varepsilon^P_{ph}(AB) = 304.46$ (i.e., 24.83 dB)

(d) $A = 30.74$ cm, $B = 26.83$ cm

$$HPBW_H = 7.61°, HPBW_E = 6.03°$$

$G_{\max} = 806.1$ (i.e., 29.06 dB), assuming the same aperture efficiency

Example 7.2 A circular horn antenna has an aperture efficiency of 0.75 and operates at 8 GHz. The diameter of the aperture is 0.16m. Determine:

(a) The maximum directivity;
(b) The HPBW for a uniformly illuminated aperture;
(c) The HPBW for a parabolic tapered aperture;
(d) The HPBW for a parabolic square tapered aperture.

Solution 7.2

(a) $D_{\max} = (4\pi/\lambda_2)\varepsilon_a \pi a^2 = 134.75$ (i.e., 21.29 dB)
(b) HPBW = $58.4\lambda/2a = 13.68°$ (uniformly illuminated aperture)
(c) HPBW = $72.7\lambda/2a = 17.04°$ (parabolic tapered aperture)
(d) HPBW = $84.2\lambda/2a = 19.73°$ (parabolic square tapered aperture)

Example 7.3 A circular horn antenna is designed to operate at 12 GHz with an aperture diameter of 0.15m. Calculate:

(a) The maximum gain of the antenna;
(b) The 3-dB, 10-dB, and null-to-null beamwidths in degrees;
(c) The far-field distance.

Solution 7.3

(a) $D_{\max} = 284.24$ (i.e., 24.53 dB).
(b) A slight modification of MATLAB Program 7 would enable plotting the radiation field, demonstrated in Figure 7.14, of a parabolic tapered circular horn (aperture has $2a = 6.67\lambda$).

Upon examining the MATLAB data:

HPBW = 11.2°

10-dB beamwidth = 18.8°

Null-to-null beamwidth = 28.4°

(c) $R_{\text{far}} = 2D^2/\lambda = 2 \times 0.15^2/0.0225 = 2\text{m}$.

Problems for Students

7.1 Design a rectangular horn antenna to operate at 10 GHz. The desired gain is 20 dB. Determine the following:
(a) The dimensions of the horn (aperture width and height);
(b) The flare angles in the E-plane and H-plane;
(c) The beamwidths in the E-plane and H-plane.

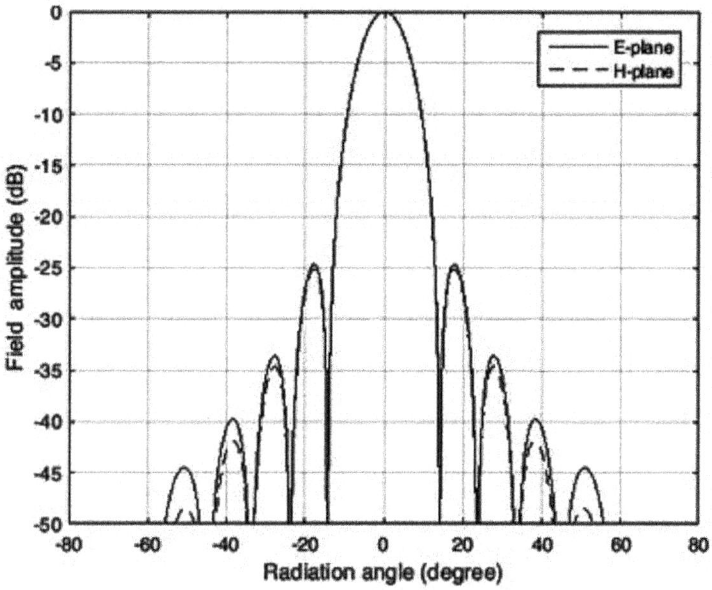

Figure 7.14 Radiation field of a parabolic tapered circular horn.

7.2 A circular horn antenna operates at a frequency of 15 GHz and has an aperture diameter of 0.1m. Calculate:
   (a) The directivity of the antenna;
   (b) The HPBW;
   (c) The near-field and far-field distances.

7.3 A rectangular horn antenna has an aperture efficiency of 0.6 and operates at 12 GHz. The aperture dimensions are 0.05m by 0.04m. Calculate:
   (a) The maximum directivity of the antenna;
   (b) The actual gain of the antenna.

7.4 Design a rectangular horn antenna with a desired directivity of 25 dB at a frequency of 5 GHz. Determine:
   (a) The required aperture dimensions (width and height);
   (b) The flare angles in the E-plane and H-plane;
   (c) The length of the horn;
   (d) The HPBW in both planes.

7.5 A rectangular horn antenna operates at 14 GHz and has aperture dimensions of 0.06m by 0.05m. Calculate:
   (a) The directivity of the antenna;
   (b) The E-plane and H-plane beamwidths;
   (c) The physical length of the horn, assuming a flare angle of 15° in both planes.

7.6 Design a circular horn antenna with an aperture efficiency of 0.85 to operate at 9 GHz. Determine:
   (a) The diameter of the aperture needed to achieve a gain of 18 dB;
   (b) The HPBW of the antenna;
   (c) The near-field region distance.

7.7 A rectangular horn antenna has a length of 50λ at 6 GHz. The dimensions of the waveguide feed are $a = 3.0$ cm and $b = 1.5$ cm. Calculate:
   (a) The aperture dimensions (width and height);
   (b) The E-plane and H-plane beamwidths;
   (c) The gain of the antenna;
   (d) The effect on the gain if the length is increased to 60λ.

Problems for Instructors

7.1 Fundamental Properties of Sectoral Horns: Explain the difference between the H-plane and E-plane sectoral horns in terms of their radiation characteristics and applications. Calculate the beamwidth and directivity of an H-plane sectoral horn with an aperture width of 0.3m, operating at a frequency of 10 GHz. Assume that the speed of light $c = 3 \times 10^8$ m/s. Discuss how these characteristics would change for an E-plane sectoral horn with the same aperture width.

7.2 Design of an H-Plane Sectoral Horn: Using the principles discussed in Section 7.2, design an H-plane sectoral horn with a targeted gain of 15 dBi at 8 GHz. Calculate the required aperture dimensions and explain how the design choices affect the directivity and beamwidth of the antenna. Further, discuss how adjusting the aperture size impacts the sidelobe levels and the overall efficiency of the horn.

7.3 E-Plane Sectoral Horn Gain and Beamwidth: For an E-plane sectoral horn antenna operating at 12 GHz, the aperture height is 0.2m, and the desired directivity is 18 dBi. Calculate the required length of the horn to achieve this directivity. Additionally, compute the beamwidth of the horn in the E-plane. Explain why the E-plane sectoral horn is preferred over other horn types in specific applications.

7.4 Analysis of Pyramidal Horn Design Parameters: Design a pyramidal horn antenna that operates at 5 GHz with an aperture efficiency of 70%. The desired gain is 20 dBi. Based on these requirements:
   (a) Calculate the aperture dimensions in both the E-plane and H-plane.
   (b) Determine the flare angles required to achieve the specified gain.
   (c) Explain the influence of flare angle adjustments on gain and beamwidth, and discuss any trade-offs that may arise in optimizing for compactness versus gain.

7.5 Circular Aperture Radiation Patterns: For a circular aperture antenna with a diameter of 0.5m operating at a frequency of 15 GHz, calculate the first null and HPBWs of the radiation pattern. Using this information, sketch the approximate radiation pattern of the circular aperture. Describe how increasing the aperture size affects the radiation pattern and explain why circular apertures are often chosen for high-directivity applications.

7.6 Comparison of Pyramidal and Sectoral Horn Antennas: Compare the design and performance of a pyramidal horn and a sectoral horn antenna for a satellite communication system operating at 14 GHz. Given the constraints of high directivity (20 dBi or above) and narrow beamwidth, recommend the more suitable antenna type for this application and justify your choice based on the calculated parameters (e.g., gain, beamwidth, aperture size). Additionally, discuss how the choice of antenna affects installation and pointing accuracy requirements.

7.7 MATLAB Simulation of Circular Aperture Antenna: Using MATLAB, simulate the radiation pattern of a circular aperture antenna with a diameter of 1m operating at 10 GHz. The simulation should:
  (a) Plot the far-field radiation pattern in the principal planes.
  (b) Calculate and display the first three null points in the pattern.
  (c) Estimate the aperture efficiency based on the simulated pattern. Discuss the observed pattern characteristics and how they align with theoretical expectations for a circular aperture antenna.

Reference

[1] Stutzman, W. L., and G. A. Thiele, *Antenna Theory and Design*, New York: John Wiley & Sons, 2012.

Selected Bibliography

Abhignya, G., et al., "Design, Fabrication and Testing of Pyramidal Horn Antenna," *International Journal of Engineering and Applied Sciences*, Vol. 2, No. 4, 2015, pp. 45–47.

Balanis, C. A., *Antenna Theory: Analysis and Design*, Second Edition, New York: John Wiley & Sons, 1997.

Fahama, H. S. N., "Design of a Conical Horn Antenna for Ultra-Wideband Applications," *Far East Journal of Electronics and Communications*, Vol. 28, 2024, pp. 21–28.

Jha, R. K., and J. A. Aas, "H-Plane Sectoral Dielectric Horn Antenna," *IEE-IERE Proceedings*, India, Vol. 15, No. 2, 1977, pp. 69–77.

Kobayashi, H., "Horn Antenna," in K. Yeap and K. Hirasawa, (eds.), *Analyzing the Physics of Radio Telescopes and Radio Astronomy*, Hershey, PA: IGI Global Scientific Publishing, 2020, pp. 144–177.

Narasimhan, M., and V. Rao, "Radiation from Wide-Flare Corrugated E-Plane Sectoral Horns," *IEEE Transactions on Antennas and Propagation*, Vol. 22, No. 4, 2003, pp. 603–608.

CHAPTER 8

Reflector Antennas

Reflector antennas are among the most widely used high-gain antennas, playing a crucial role in satellite communications, deep-space exploration, radio astronomy, radar systems, and terrestrial microwave links. Their ability to focus electromagnetic waves with high directivity makes them indispensable in applications requiring long-range and high-resolution performance. The simplest form of a reflector antenna consists of two primary components: a reflecting surface and a much smaller feed antenna positioned at the focal point of the reflector.

The most common reflector shape is the parabolic reflector, which provides excellent collimation of electromagnetic waves, ensuring that incident spherical wavefronts from a feed source transform into plane waves in the far field. This characteristic makes parabolic reflectors particularly advantageous for directional applications. Other common reflector configurations include cylindrical, corner, and spherical reflectors, each serving specific operational requirements.

More complex designs introduce a secondary reflector, known as a subreflector, positioned at the focal point to enhance performance. These are classified as dual-reflector antennas, with the most well-known types being the Cassegrain and Gregorian reflectors. The Cassegrain configuration features a concave subreflector that redirects energy from the primary feed toward the parabolic main reflector, achieving improved gain and compactness. The Gregorian system uses a convex subreflector to enhance performance while minimizing sidelobes and aberrations.

The analysis and design of reflector antennas rely on understanding aperture field distributions and their transformation into far-field radiation patterns. The aperture distribution method is a widely used analytical technique that considers the electric field in the aperture plane and its projection onto the effective aperture, enabling precise calculations of radiation patterns, gain, and efficiency.

One of the most critical parameters in reflector antenna design is gain, which directly affects communication range and signal strength. The gain of a reflector antenna is influenced by factors such as aperture efficiency, surface accuracy, feed illumination, and blockage effects. Optimizing these parameters ensures maximum performance, particularly in demanding applications such as deep-space communication and high-frequency radar systems.

This chapter provides a detailed examination of single parabolic reflectors, the aperture distribution method, and the gain of reflector antennas. Section 8.1 focuses on the single parabolic reflector, outlining its fundamental principles and design considerations. Section 8.2 delves into the aperture distribution method, covering the establishment of aperture fields, effective aperture projection, and electric far-field calculations. Section 8.3 explores gain calculations, emphasizing key factors that influence the overall efficiency and performance of reflector antennas. Section 8.4 compares analysis of reflector antennas with dipoles and horns. Section 8.5 explores real-world challenges in reflector antennas. Finally, Section 8.6 summarizes the main applications of reflector antennas.

8.1 Single Parabolic Reflector

The design of the parabolic reflector, shown in Figure 8.1, is based on optical geometry techniques (ignoring diffractions from the rim of the reflector), by choosing a plane perpendicular to the axis of the reflector through the focal point, as shown in Figure 8.2.

$$OP + PQ = 2f = \text{cons.} \tag{8.1}$$

$$PQ = r' \cos\theta \tag{8.2}$$

Substituting (8.2) into (8.1),

$$2f = r' + r'\cos\theta = r'(1+\cos\theta) \tag{8.3}$$

$$r' = \frac{2f}{(1+\cos\theta)} \tag{8.4}$$

From trigonometry, $(1 + \cos\theta) = 2\cos^2(\theta/2)$.
Therefore,

$$r' = f \cdot \sec^2\left(\frac{\theta}{2}\right) \tag{8.5}$$

Now the symmetry of the parabolic imposes no variation to the angle (φ) and so the rectangular coordinates can replace the spherical coordinates.

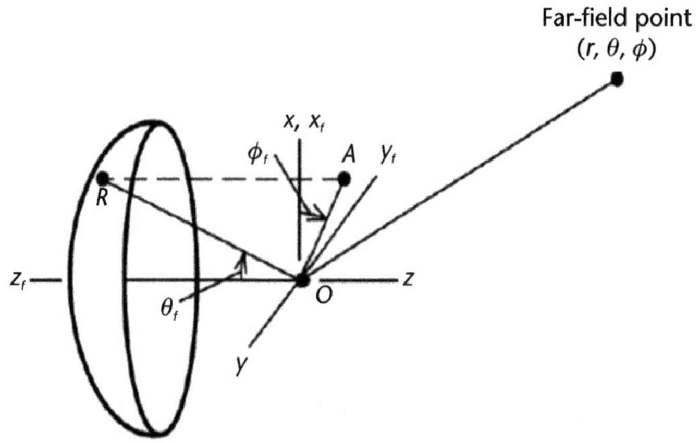

Figure 8.1 Parabolic reflector.

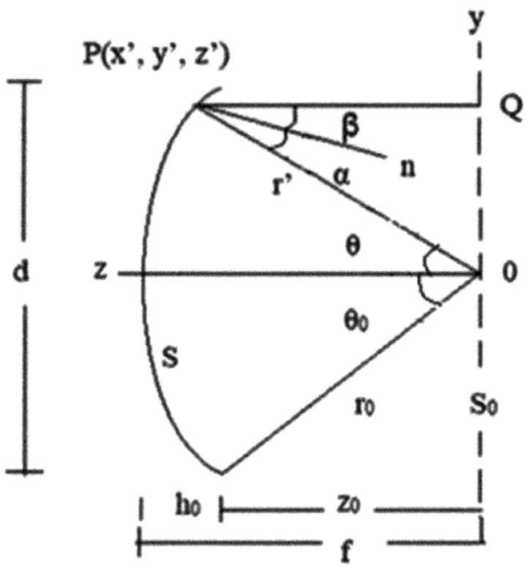

Figure 8.2 Cross-section of a parabolic reflector in the *yz*-plane.

$$r' + r'\cos\theta = \sqrt{x'^2 + y'^2 + z'^2} + z' = 2f \tag{8.6}$$

Therefore,

$$\sqrt{x'^2 + y'^2 + z'^2} = 2f - z' \tag{8.7}$$

$$x'^2 + y'^2 + z'^2 = 4f^2 + z'^2 - 4fz' \tag{8.8}$$

$$\rho'^2 = x'^2 + y'^2 = 4f\,(f - z') \tag{8.9}$$

Here, ρ' is the distance from a point Q to the focal point O, where Q is the projection of the point P on the reflector surface S onto the axis-orthogonal plane S_0, the aperture plane, at the focal point.

Another approach relates θ_0 to f and d, where:

$$\theta_0 = \tan^{-1}\left(\frac{d/2}{z_0}\right) \tag{8.10}$$

Reorganizing (8.9) for z_0 gives:

$$z_o = f - \frac{x_0^2 + y_0^2}{4f} = f - \frac{(d/2)^2}{4f} = f - \frac{d^2}{16f} \tag{8.11}$$

Substituting in (8.10) yields:

$$\theta_0 = \tan^{-1}\left(\frac{d/2}{f - \left(d^2/16f\right)}\right) = \tan^{-1}\left(\frac{f/2d}{(f/d)^2 - 1/16}\right) \tag{8.12}$$

Interestingly, (8.12) can be expressed in a different form. From trigonometry,

$$\tan\theta_0 = \frac{2\tan(\theta_0/2)}{1-\tan^2(\theta_0/2)} = \frac{2x}{1-x^2} \tag{8.13}$$

where $x = \tan\theta_0/2$.

Equating (8.13) to (8.12) gives:

$$\frac{1-x^2}{x} = \frac{1-(d/4f)^2}{(d/4f)} \tag{8.14}$$

Therefore,

$$x = \tan\frac{\theta_0}{2} = \frac{d}{4f} \tag{8.15}$$

$$f = \left(\frac{d}{4}\right)\cot\frac{\theta_0}{2} \tag{8.16}$$

We now develop a MATLAB program that plots normalized *p¢/d* to *z¢/d* for various values of *f/d* in (8.9).

Equation (8.9) has the following normalized form:

$$\left(\frac{\rho'}{d}\right)^2 = 4\left(\frac{f}{d}\right)\left(\frac{f}{d}-\frac{z'}{d}\right) \tag{8.17}$$

MATLAB Program 9

```
% Matlab program of parabolic antenna
% m = f/d where f is reflector's focus and d is diameter
% x= z' / d
% y= (p' / d)^2
% yy= p' / d

m= 1/2;
x= -m : 0.01: 0.0;
y=4.*m.*(m+x);
yy= sqrt(y);
plot(x, yy, 'k-' , x, -yy, 'k-' )
hold on

m= 1/3;
x= -m : 0.01: 0.0;
y=4.*m.*(m+x);
yy= sqrt(y);
```

```
plot(x, yy, 'k--' , x, -yy, 'k--' )
hold on

m= 1/4;
x= -m : 0.005: 0.0;
y=4.*m.*(m+x);
yy= sqrt(y);
plot(x, yy, 'k:' , x, -yy, 'k:' )
axis([-0.5 0 -0.5 0.5])
grid
```

It is observed, in Figure 8.2, that the axisymmetric; rotationally symmetric, paraboloid reflector is entirely defined by the two respective parameters: the diameter d and the focal length f. It is common that the parabola is specified in terms of d and the ratio f/d. When f/d approaches infinity, the reflector becomes flat. Figure 8.3 shows parabolic curves when f/d = 1/2, 1/3, and 1/4, respectively. The angle θ_0 from the feed (focal) point to the reflector's rim is related to f/d and expressed in (8.16). When f/d = 1/4, $\theta_0 = 90°$, and the focal point lies in the plane passing through the reflector's rim.

The focal distance f of a given reflector can be calculated after measuring its diameter d and its depth h_0, shown in Figure 8.2. Equation (8.18) can now be found by substituting $p¢ = d/2$ and $z¢ = f - h_0$ into (8.9).

$$f = \frac{d^2}{16h_0} \tag{8.18}$$

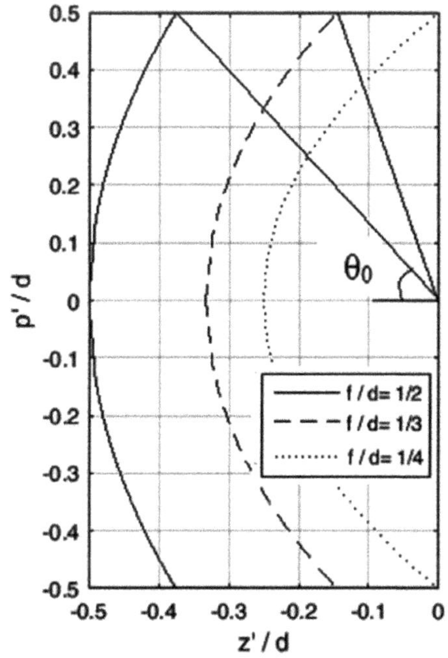

Figure 8.3 Variation of $p¢/d$ versus $z¢/d$ for various f/d of a parabolic reflector.

For example, if $f/d = 1/4$, then $h_0 = d/4$, from which $h_0 = f$. This means that the focal point is on the reflector's rim plane. If $f/d = 1/2$, $h_0 = d/8$, and $h_0 = f/4$.

The reflector design problem aims at matching the feed antenna pattern to the reflector. Usually, the feed pattern must be at about a −10-dB level in the direction of the rim.

The geometry of the paraboloid reflector has two valuable features. (1) All rays leaving the focal point O are collimated along the reflector's axis after reflection. (2) All overall ray path lengths (from the focal point to the reflector and on to the aperture plane) are the same and equal to $2F$.

The above properties are proven by the geometrical optics methods, and, hence, they are true only if the following conditions hold:

- The radius of the curvature of the reflector is large compared to the wavelength and the local region around each reflection point can be treated as planar.
- The radius of the curvature of the incoming wave from the feed is large and can be treated locally at the reflection point as a plane wave.
- The reflector is a perfect conductor.

We now show that all rays emanating from the focal point are reflected from the surface of the paraboloid in parallel to the z-axis, shown in Figure 8.2.

The parabolic curve equation can be easily derived from (8.5) as:

$$C_p = f - r'\cos^2\left(\frac{\theta}{2}\right) = 0 \tag{8.19}$$

The gradient of the parabolic curve, in spherical coordinates, can be obtained as follows:

$$\nabla C_p = -\boldsymbol{a}_r \cos^2\left(\frac{\theta}{2}\right) + \boldsymbol{a}_\theta \cos\left(\frac{\theta}{2}\right)\cdot\sin\left(\frac{\theta}{2}\right) \tag{8.20}$$

and

$$|\nabla C_p| = \cos\left(\frac{\theta}{2}\right) \tag{8.21}$$

therefore,

$$\boldsymbol{a}_n = \frac{\nabla C_p}{|\nabla C_p|} = -\boldsymbol{a}_r \cos\left(\frac{\theta}{2}\right) + \boldsymbol{a}_\theta \sin\left(\frac{\theta}{2}\right) \tag{8.22}$$

Thus, the angle α, between the incident ray and the normal unit vector $\boldsymbol{n}$, is:

$$\cos\alpha = -\boldsymbol{a}_r \cdot \boldsymbol{n} = \cos\left(\frac{\theta}{2}\right) \tag{8.23}$$

Thus, the angle β, between the reflected ray and the unit vector $\boldsymbol{a}_n$, is:

$$\begin{aligned}\cos\beta &= \boldsymbol{a}_z \cdot \boldsymbol{n} = \left(-\boldsymbol{a}_r \cos\theta + \boldsymbol{a}_\theta \sin\theta\right)\cdot \boldsymbol{n} \\ &= \cos\theta\cos\left(\frac{\theta}{2}\right) + \sin\theta\sin\left(\frac{\theta}{2}\right) = \cos\left(\frac{\theta}{2}\right)\end{aligned} \tag{8.24}$$

Therefore, the incident and reflected rays fulfill Snell's law and $\alpha = \beta$. In addition, $\theta = \alpha + \beta$, and the reflected ray emerges in parallel to the z-axis.

8.2 Aperture Distribution Method

There are two primary methodologies for assessing the radiation characteristics of reflectors. One of these is the current distribution method, a framework rooted in physical optics (PO) approximation. It operates on the premise of a known incident field from the feed, leading to the generation of surface currents on the reflector's surface—expressed as $\boldsymbol{J}_s = 2\boldsymbol{a}_n \times \boldsymbol{H}$. These surface currents are subsequently integrated to compute the far-zone field. The PO method presupposes a reflector with a perfectly conducting surface and employs image theory. Furthermore, it assumes that the incident wave emanating from the primary feed creates a locally plane far-zone field.

The aperture distribution method begins by analyzing the field on a plane perpendicular to the reflector's axis at its focal point, using geometrical optics; a ray-tracing technique. This method creates equivalent sources across this plane, assuming that they do not extend beyond the reflector's aperture. Understanding the field distribution at the reflector antenna's aperture is essential for calculating parameters such as the far-field pattern and directivity. Although all rays from the feed travel the same distance to the aperture, the phase of the aperture distribution is uniform while the amplitude varies due to the power density decrease of departing rays, following a $1/r'^2$ reduction. After reflection, there is minimal spreading loss, as the rays stay parallel, and the amplitude distribution of the aperture field changes in proportion to $1/r'$.

There are four types of parabolic reflector antennas. These are:

- Prime focus reflector antenna (Figure 8.1);
- Offset fed reflector (Figure 8.4);
- Cassegrain reflector, a dual reflector antenna composed of a secondary convex reflector at the focal point, which is fed by a primary feed (Figure 8.5);
- Gregorian reflector, a dual reflector antenna composed of a secondary concave reflector at the focal point, which is fed by a primary feed (Figure 8.5).

8.2.1 Establishing the Fields in the Aperture

Consider a feed antenna located at the focal point and a parabolic reflector located in the far field of the feed antenna. Assume that the far field radiated from the feed antenna as:

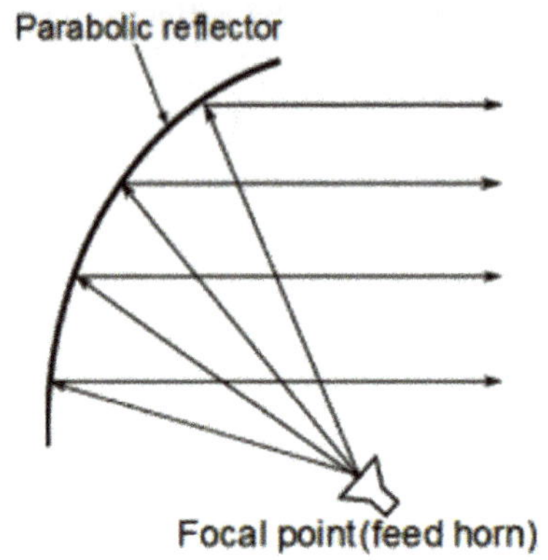

Figure 8.4 Parabolic reflector antenna with an offset feed at focal point.

$$E_f(r',\theta',\varphi') = E_0 F_f(\theta',\varphi')\frac{e^{-jkr'}}{r'} \tag{8.25}$$

where r',θ', φ' are coordinates on the reflector surface, which may be assumed as a source.

$F(\theta', \varphi')$ is the spatial distribution function of the feed antenna. It may be expressed as:

$$F_f(\theta',\varphi') = F_{\theta'}(\theta',\varphi')a_{\theta'} + F_{\varphi'}(\theta',\varphi')a_{\varphi'} \tag{8.26}$$

For a very large reflector dimension in which $D/\lambda > 100$, geometrical optics can be used to calculate the approximate fields of the aperture.

The reflection coefficients for the field when aligned and when perpendicular to the plane of incidence that contains the incident and reflected rays are $\Gamma_{\parallel} = 1$ and $\Gamma_{\perp} = -1$, respectively. It follows that:

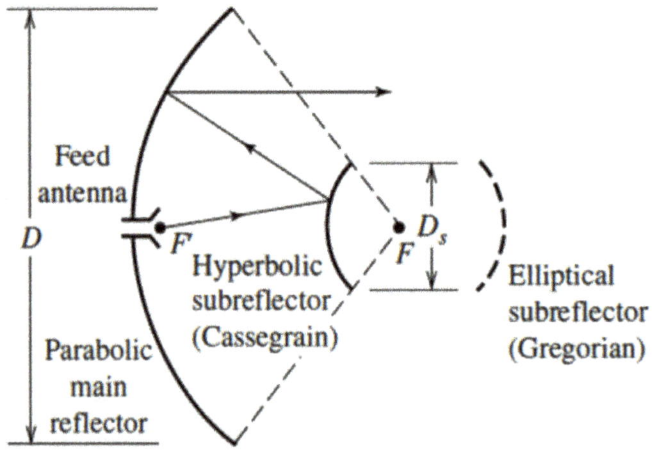

Figure 8.5 Dual-reflector parabolic antenna showing Cassegrain and Gregorian subreflectors.

Table 8.1 Advantages and Disadvantages of Different Types of Parabolic Antenna Reflectors

Type of Reflector	*Advantage*	*Disadvantage*
Prime focus	Direct feed: the prime focus feed is located at the focal point of the main reflector, providing a simple and direct feed arrangement	Blockage effect: the presence of the feed at the focal point can lead to blockage effects, impacting the overall aperture efficiency
	Unobstructed aperture: since there are no secondary reflectors, the aperture is unobstructed, resulting in high aperture efficiency	Higher spillover: prime focus feeds may exhibit higher spillover radiation compared to offset feeds
Offset feed	Simplicity: offset feeds are relatively simple and cost-effective compared to more complex feed systems	Slightly lower gain: offset feeds may exhibit slightly lower gain compared to Cassegrain and Gregorian systems
	Lower spillover: offset feeds can offer lower spillover radiation, reducing interference with nearby antennas	Feed placement: the off-center placement of the feed can result in distortion effects, impacting the overall antenna performance
Cassegrain subreflector	Compact design: Cassegrain antennas are known for their compact design, making them suitable for applications with limited space	Slightly more complex: the addition of a secondary reflector makes the Cassegrain system somewhat more complex than the offset feed system
	Versatility: Cassegrain systems can be designed for various frequency bands and applications	Potential blockage: the presence of the secondary reflector can lead to blockage effects, impacting the antenna's aperture efficiency
Gregorian subreflector	Aberration control: the concave secondary reflector in the Gregorian system allows for better control over aberrations, resulting in improved antenna performance	Complexity: Gregorian systems are more complex than offset feeds and Cassegrain systems, leading to higher manufacturing and maintenance costs
	High performance: Gregorian systems can achieve high antenna gain and low sidelobe levels	Mechanical precision: achieving optimal performance may require precise alignment of the primary and secondary reflectors

$$(\boldsymbol{E}_i + \boldsymbol{E}_r) \times \boldsymbol{n} = 0 \tag{8.27}$$

$$(\boldsymbol{E}_i - \boldsymbol{E}_r) \cdot \boldsymbol{n} = 0 \tag{8.28}$$

From (8.27), it is possible to conclude that:

$$\boldsymbol{n} \times (\boldsymbol{E}_i + \boldsymbol{E}_r) \times \boldsymbol{n} = 0 \tag{8.29}$$

Using the "bac-cab" rule, it follows that:

$$(\boldsymbol{E}_i + \boldsymbol{E}_r)(\boldsymbol{n} \cdot \boldsymbol{n}) - \boldsymbol{n}(\boldsymbol{n} \cdot (\boldsymbol{E}_i + \boldsymbol{E}_r) = 0 \tag{8.30}$$

Substituting $E_i = E_r$, from (8.28) into (8.30) yields:

$$(\boldsymbol{E}_i + \boldsymbol{E}_r) - \boldsymbol{n}(2\boldsymbol{n} \cdot \boldsymbol{E}_i) = 0 \tag{8.31}$$

$$\boldsymbol{E}_r = 2\boldsymbol{n}(\boldsymbol{n} \cdot \boldsymbol{E}_i) - \boldsymbol{E}_i \tag{8.32}$$

Hence, the reflected electric field is given in terms of the incident field. In the aperture field method, we are interested to find the reflected field at the aperture. There will therefore be a change of phase due to the movement of the field from the reflector to the aperture.

From Figure 8.2, PQ is related to r', θ', and f as follows:

$$PQ = r' \cos\theta = 2f - r' \tag{8.33}$$

The electric field at the aperture, hence, works out as:

$$\boldsymbol{E}_a = \left[2\boldsymbol{n}(\boldsymbol{n} \cdot \boldsymbol{E}_f) - \boldsymbol{E}_f\right] e^{-j\beta(2f - r')} \tag{8.34}$$

where E_f is the incident field on the reflector due to the feed antenna at O and β is the propagation constant.

The far-field magnetic field can be derived from the following expression:

$$\boldsymbol{H}_a = \frac{1}{\eta_0} \boldsymbol{a}_z \times \boldsymbol{E}_a \tag{8.35}$$

where η_0 is the intrinsic impedance in free space and equals 377Ω and $\boldsymbol{a}_z$ is a unit vector in the z-axis direction.

Now the unit vector can be expressed as:

$$\boldsymbol{n} = -\boldsymbol{a}_{r'} \cos\left(\frac{\theta'}{2}\right) + \boldsymbol{a}_{\theta'} \sin\left(\frac{\theta'}{2}\right) \tag{8.36}$$

Hence,

$$\begin{aligned} \boldsymbol{n} \cdot \boldsymbol{E}_f(r', \theta', \varphi') &= \left[-\boldsymbol{a}_{r'} \cos\left(\frac{\theta'}{2}\right) + \boldsymbol{a}_{\theta'} \sin\left(\frac{\theta'}{2}\right)\right] \\ &\quad \cdot E_0\left[F_{\theta'}(\theta', \varphi')\boldsymbol{a}_{\theta'} + F_{\varphi'}(\theta', \varphi')\boldsymbol{a}_{\varphi'}\right]\frac{e^{-jkr'}}{r'} \\ &= \left[\sin\left(\frac{\theta'}{2}\right)\right] \cdot E_0\left[F_{\theta'}(\theta', \varphi')\right]\frac{e^{-jkr'}}{r'} \end{aligned} \tag{8.37}$$

and

$$2\boldsymbol{n}\left[\boldsymbol{n} \cdot \boldsymbol{E}_f(r', \theta', \varphi')\right] = 2\left[\sin\frac{\theta'}{2} \cdot E_0\left[F_{\theta'}(\theta', \varphi')\right]\frac{e^{-jkr'}}{r'}\right]\left[-\boldsymbol{a}_{r'} \cos\left(\frac{\theta'}{2}\right) + \boldsymbol{a}_{\theta'} \sin\left(\frac{\theta'}{2}\right)\right] \tag{8.38}$$

Hence,

$$\begin{aligned}
&2\boldsymbol{n}\left[\boldsymbol{n}\cdot \boldsymbol{E}_f(r',\theta',\varphi')\right]-\boldsymbol{E}_f(r',\theta',\varphi')\\
&=2\left[\sin\frac{\theta'}{2}\cdot E_0\left[F_{\theta'}(\theta',\varphi')\right]\frac{e^{-jkr'}}{r'}\right]\left[-\boldsymbol{a}_{r'}\cos\left(\frac{\theta'}{2}\right)+\boldsymbol{a}_{\theta'}\sin\left(\frac{\theta'}{2}\right)\right]\\
&-E_0\left[F_{\theta'}(\theta',\varphi')\boldsymbol{a}_{\theta'}+F_{\varphi'}(\theta',\varphi')\boldsymbol{a}_{\varphi'}\right]\frac{e^{-jkr'}}{r'}\\
&=E_0\frac{e^{-jkr'}}{r'}\left[F_{\theta'}(\theta',\varphi')(\boldsymbol{a}_{r'}\sin\theta'+\boldsymbol{a}_{\theta'}\cos\theta')+F_{\varphi'}(\theta',\varphi')\boldsymbol{a}_{\varphi'}\right]
\end{aligned}\tag{8.39}$$

Substituting (8.39) into (8.34), we get:

$$\boldsymbol{E}_a=E_0\frac{e^{-jkr'}}{r'}\left[F_{\theta'}(\theta',\varphi')(\boldsymbol{a}_{r'}\sin\theta'+\boldsymbol{a}_{\theta'}\cos\theta')+F_{\varphi'}(\theta',\varphi')\boldsymbol{a}_{\varphi'}\right]e^{-j\beta(2f-r')}\tag{8.40}$$

This equation can be expressed in Cartesian coordinates by using the matrix equation:

$$\begin{pmatrix}E_{ax'}\\E_{ay'}\\E_{az'}\end{pmatrix}=\begin{pmatrix}\sin\theta'\cos\varphi' & \cos\theta'\cos\varphi' & -\sin\varphi'\\ \sin\theta'\sin\varphi' & \cos\theta'\sin\varphi' & \cos\varphi'\\ \cos\theta' & -\sin\theta' & 0\end{pmatrix}\begin{pmatrix}E_{ar'}\\E_{a\theta'}\\E_{a\varphi'}\end{pmatrix}\tag{8.41}$$

and bearing in mind to substitute from (8.33) for r' in terms of f and θ', we get:

$$E_{ax'}=E_0\frac{e^{-j\beta 2f}}{2f}(1+\cos\theta')\left[F_{\theta'}(\theta',\varphi')\cos\varphi'-F_{\varphi'}(\theta',\varphi')\sin\varphi'\right]\tag{8.42}$$

$$E_{ay'}=E_0\frac{e^{-j\beta 2f}}{2f}(1+\cos\theta')\left[F_{\theta'}(\theta',\varphi')\sin\varphi'+F_{\varphi'}(\theta',\varphi')\cos\varphi'\right]\tag{8.43}$$

8.2.2 Projecting the Effective Aperture at the Aperture Plane

As shown in Figure 8.6, we have a parabolic reflector, which is a 3D structure. So basically the reflector needs to be projected to a 2D effective aperture at the aperture plane $z = 0$.

We shall find the 2D Fourier transform of the aperture fields, E_{ax} and E_{ay}, over the aperture plane from the following expression:

$$\tilde{E}_{ax}=\int_0^{d/2}\int_0^{2\pi}E_{ax'}e^{j\beta\cdot r'}\rho\,d\rho\,d\varphi'\tag{8.44}$$

where $\boldsymbol{r}'$ lies in the aperture plane and then $\theta' = \pi/2$. The propagation wave vector is expressed as:

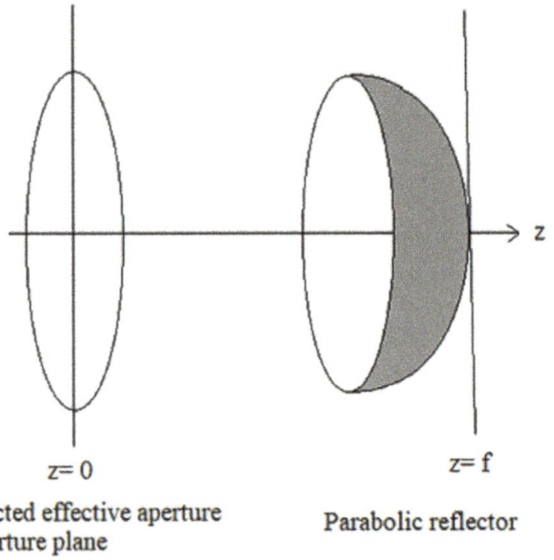

Figure 8.6 Projected effective aperture at aperture plane from parabolic reflector.

$$\boldsymbol{\beta} = \beta \boldsymbol{a}_r = \beta\left(\sin\theta\cos\varphi \boldsymbol{a}_x + \sin\theta\sin\varphi \boldsymbol{a}_y + \cos\theta \boldsymbol{a}_z\right) \tag{8.45}$$

Therefore,

$$\begin{aligned}\boldsymbol{\beta}\cdot\boldsymbol{r}' &= \beta\left(\sin\theta\cos\varphi \boldsymbol{a}_x + \sin\theta\sin\varphi \boldsymbol{a}_y + \cos\theta \boldsymbol{a}_z\right)\cdot\rho\left(\boldsymbol{a}_x\cos\varphi' + \boldsymbol{a}_y\sin\varphi'\right)\\ &= \beta\rho\left(\sin\theta\cos\varphi\cos\varphi' + \sin\theta\sin\varphi\sin\varphi'\right) = \beta\rho\left(\sin\theta\cos\left(\varphi-\varphi'\right)\right)\end{aligned} \tag{8.46}$$

Hence,

$$\tilde{E}_{ax} = \int_0^{d/2}\int_0^{2\pi} E_{ax'}e^{j\beta\rho(\sin\theta\cos(\varphi-\varphi'))}\rho\, d\rho\, d\varphi' \tag{8.47}$$

We may convert the above integral over the feed angles, θ' and φ'. In reference to Figure 8.7,

$$\rho' = r'\sin\theta' = 2r'\sin\left(\frac{\theta'}{2}\right)\cos\left(\frac{\theta'}{2}\right) \tag{8.48}$$

Substituting (8.5) into (8.47) yields:

$$\rho' = 2f\sec^2\left(\frac{\theta}{2}\right)\sin\left(\frac{\theta'}{2}\right)\cos\left(\frac{\theta'}{2}\right) = 2f\tan\left(\frac{\theta'}{2}\right) \tag{8.49}$$

Therefore,

$$\rho' d\rho' = 2f\tan\left(\frac{\theta'}{2}\right)\cdot f\sec^2\left(\frac{\theta'}{2}\right)d\theta' = 2f\tan\left(\frac{\theta'}{2}\right)\cdot r' d\theta' \tag{8.50}$$

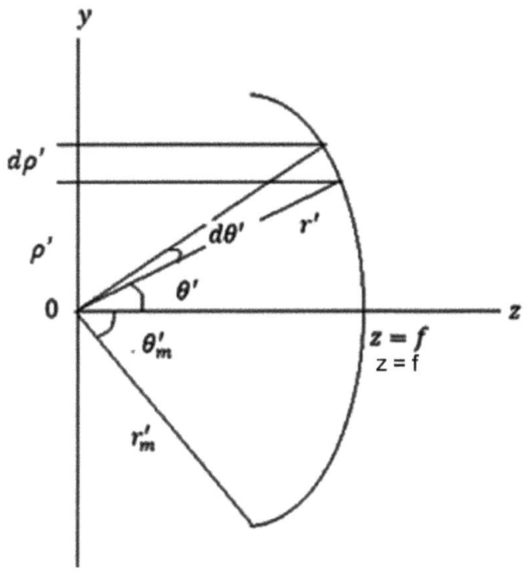

Figure 8.7 Cross-section of parabolic reflector used for converting (8.47) over the feed angles, θ' and φ'.

Hence, substituting (8.50) into (8.44) gives:

$$\tilde{E}_{ax} = 2f^2 \int_0^{\theta'_m} \int_0^{2\pi} E_{ax'} e^{j\beta 2f \tan\left(\frac{\theta'}{2}\right)(\sin\theta\cos(\varphi-\varphi'))} \cdot \tan\left(\frac{\theta'}{2}\right) \sec^2\left(\frac{\theta'}{2}\right) d\theta' d\varphi' \tag{8.51}$$

Similarly,

$$\tilde{E}_{ay} = 2f^2 \int_0^{\theta'_m} \int_0^{2\pi} E_{ay'} e^{j\beta 2f \tan\left(\frac{\theta'}{2}\right)(\sin\theta\cos(\varphi-\varphi'))} \cdot \tan\left(\frac{\theta'}{2}\right) \sec^2\left(\frac{\theta'}{2}\right) d\theta' d\varphi' \tag{8.52}$$

The above integrations can be performed numerically, if analytical integration is not possible.

8.2.3 Calculating the Electric Far Field

The process of finding the electric and magnetic far fields, $\boldsymbol{E}_{ff}$ and $\boldsymbol{H}_{ff}$, of an aperture antenna involves calculating the magnetic and electric vector potentials $\boldsymbol{A}$ and $\boldsymbol{F}$, respectively. $\boldsymbol{A}$ and $\boldsymbol{F}$, in turn, depend on finding the electric current and magnetic current densities $\boldsymbol{J}_S$ and $\boldsymbol{M}_S$, respectively, which can be derived from the fields in the aperture, $\boldsymbol{E}_a$ and $\boldsymbol{H}_a$, where:

$$\boldsymbol{J}_S = \boldsymbol{n} \times \boldsymbol{H}_a \tag{8.53}$$

$$\boldsymbol{M}_S = -\boldsymbol{n} \times \boldsymbol{E}_a \tag{8.54}$$

Considering (8.53) and substituting in our case for $\boldsymbol{n} = \boldsymbol{a}_z$ and for $\boldsymbol{H}_a = H_{ax}\boldsymbol{a}_x + H_{ay}\boldsymbol{a}_y$, gives:

$$J_S = \boldsymbol{a}_z \times \left(H_{ax}\boldsymbol{a}_x + H_{ay}\boldsymbol{a}_y\right) = -H_{ay}\boldsymbol{a}_x + H_{ax}\boldsymbol{a}_y \tag{8.55}$$

In general, the electric vector potential can be expressed as:

$$\boldsymbol{A} = \frac{\mu}{4\pi}\iint J_S \frac{e^{-j\beta|r-r'|}}{|r-r'|}\,dS' \tag{8.56}$$

After manipulating and performing far-field approximation, the phasor ratio term in (8.56) reduces to:

$$\frac{e^{-j\beta|r-r'|}}{|r-r'|} \cong \frac{e^{-j\beta\left(r-\frac{xx'}{r}-\frac{yy'}{r}\right)}}{r} = \frac{e^{\left(-j\beta r+j\beta_x x'+j\beta_y y'\right)}}{r} \tag{8.57}$$

where

$$\frac{x}{r} = \frac{\beta_x}{\beta} = \sin\theta\cos\varphi,\ \frac{y}{r} = \frac{\beta_y}{\beta} = \sin\theta\sin\varphi \tag{8.58}$$

Substituting (8.55) and (8.57) into (8.56) gives:

$$\begin{aligned}\boldsymbol{A} &= \frac{\mu e^{-j\beta r}}{4\pi r}\iint\left(-H_{ay}\boldsymbol{a}_x + H_{ax}\boldsymbol{a}_y\right)e^{\left(j\beta_x x'+j\beta_y y'\right)}\,dx'\,dy' \\ &= \frac{\mu e^{-j\beta r}}{4\neq r}\left(-\tilde{H}_{ay}\boldsymbol{a}_x + \tilde{H}_{ax}\boldsymbol{a}_y\right)\end{aligned} \tag{8.59}$$

where $\tilde{H}_{ay}$ and $\tilde{H}_{ax}$ are the 2D Fourier transform of H_{ay} and H_{ax}, respectively.

At this point, it is worth mentioning that Fourier transformation (FT) plays a significant role in finding the fields of aperture antennas such as parabolic reflectors.

Similarly, the electric vector potential can be derived as:

$$\boldsymbol{F} = -\frac{\varepsilon e^{-j\beta r}}{4\pi r}\left(-\tilde{E}_{ay}\boldsymbol{a}_x + \tilde{E}_{ax}\boldsymbol{a}_y\right) \tag{8.60}$$

where $\tilde{E}_{ay}$ and $\tilde{E}_{ax}$ are the 2D Fourier transform of E_{ay} and E_{ax}, respectively.

Now, bearing in mind that $A_r = F_r = 0$, in the far field, and substituting for the Cartesian unit vectors in terms of their equivalent ones in spherical coordinates in (8.59) and (8.60) yields:

$$A_\theta = \frac{\mu e^{-j\beta r}}{4\pi r}\left(-\tilde{H}_{ay}\cos\theta\cos\varphi + \tilde{H}_{ax}\cos\theta\sin\varphi\right) \tag{8.61}$$

$$A_\varphi = \frac{\mu e^{-j\beta r}}{4\pi r}\left(\tilde{H}_{ay}\sin\varphi + \tilde{H}_{ax}\cos\varphi\right) \tag{8.62}$$

Similarly,

$$F_\theta = -\frac{\varepsilon e^{-j\beta r}}{4\pi r}\left(-\tilde{E}_{ay}\cos\theta\cos\varphi + \tilde{E}_{ax}\cos\theta\sin\varphi\right) \tag{8.63}$$

$$F_\varphi = -\frac{\varepsilon e^{-j\beta r}}{4\pi r}\left(\tilde{E}_{ay}\sin\varphi + \tilde{E}_{ax}\cos\varphi\right) \tag{8.64}$$

The electric field in the far field will have contributions from both **A** and **F**. The spherical components of the far field are approximated as follows:

$$E_{ff\theta} \cong -j\omega A_\theta - j\omega\eta F_\varphi \tag{8.65}$$

$$E_{ff\varphi} \cong -j\omega A_\varphi + j\omega\eta F_\theta \tag{8.66}$$

Upon substitution of (8.61) and (8.64) into (8.65):

$$\begin{aligned} E_{ff\theta} \cong & -j\omega\frac{\mu e^{-j\beta r}}{4\pi r}\left(-\tilde{H}_{ay}\cos\theta\cos\varphi + \tilde{H}_{ax}\cos\theta\sin\varphi\right) \\ & + j\omega\eta\,\frac{\varepsilon e^{-j\beta r}}{4\pi r}\left(\tilde{E}_{ay}\sin\varphi + \tilde{E}_{ax}\cos\varphi\right) \end{aligned} \tag{8.67}$$

We can still eliminate ω, μ, and ε by using the relations $\beta = \omega\sqrt{\mu\varepsilon}$ and $\eta = \omega\mu/\beta$ in (8.67). Hence,

$$E_{ff\theta} \cong \frac{j\beta e^{-j\beta r}}{4\pi r}\left[\eta\left(\tilde{H}_{ay}\cos\theta\cos\varphi - \tilde{H}_{ax}\cos\theta\sin\varphi\right) + \left(\tilde{E}_{ay}\sin\varphi + \tilde{E}_{ax}\cos\varphi\right)\right] \tag{8.68}$$

Similarly, and upon substitution of (8.62) and (8.63) into (8.66):

$$E_{ff\varphi} \cong \frac{j\beta e^{-j\beta r}}{4\pi r}\left[-\eta\left(\tilde{H}_{ay}\sin\varphi + \tilde{H}_{ax}\cos\varphi\right) + \left(\tilde{E}_{ay}\cos\theta\cos\varphi - \tilde{E}_{ax}\cos\theta\sin\varphi\right)\right] \tag{8.69}$$

We can now substitute for $\tilde{H}_{ax}$ and $\tilde{H}_{ay}$ from the following equation, where

$$\tilde{\boldsymbol{H}}_a = \frac{1}{\eta}\boldsymbol{a}_z \times \left(\tilde{E}_{ax}\boldsymbol{a}_x + \tilde{E}_{ay}\boldsymbol{a}_y\right) = \frac{1}{\eta}\left(\tilde{E}_{ax}\boldsymbol{a}_y - \tilde{E}_{ay}\boldsymbol{a}_x\right)$$

where

$$\tilde{H}_{ax} = -\frac{1}{\eta}\tilde{E}_{ay} \quad , \quad \tilde{H}_{ay} = \frac{1}{\eta}\tilde{E}_{ax} \tag{8.70}$$

It follows that:

$$E_{ff\theta} \cong \frac{j\beta e^{-j\beta r}}{4\pi r}(1+\cos\theta)\left(\tilde{E}_{ax}\cos\varphi + \tilde{E}_{ay}\sin\varphi\right) \tag{8.71}$$

$$E_{ff\varphi} \cong \frac{j\beta e^{-j\beta r}}{4\pi r}(1+\cos\theta)\left(-\tilde{E}_{ax}\sin\varphi + \tilde{E}_{ay}\cos\varphi\right) \tag{8.72}$$

A MATLAB code follows that calculates and plots the electric far fields $E_{ff\theta}$ and $E_{ff\varphi}$ by first performing a numerical calculation for $\tilde{E}_{ax}$ and $\tilde{E}_{ay}$ as expressed in (8.51) and (8.52). $E_{ax'}$ and $E_{ay'}$ are assumed to be emanating from a half-wave dipole feed of spatial distribution functions depicted as $F_{\theta'}(\theta',\varphi') = \cos\theta'$ and $F_{\varphi'}(\theta',\varphi') = 1$. The diameter of the reflector was considered 100λ while the 3 cases of f/d = 1/2, 1/3, and 1/4 were considered, respectively.

MATLAB Program 10

```
% Matlab program to calculate the electric far fields Effθ
and Effφ
% Define the polar function f(theta,phi)
% Define the limits of radiation angle
% d is parabolic reflector's diameter (λ)
d= 10;
th_min = -5*pi/180;
th_max = 5*pi/180;

th = linspace(th_min,th_max,201);
ph= pi/2 ;
% FD is focus/diameter  ratio
FD= 1/2;

% Define the limits of integration
theta_min = eps(0);
theta_max = 2*pi;
phi_min = eps(0);
phi_max =  2*atan(1/(4*FD));
% Define the integration step size
delta_theta = 0.068;
delta_phi = 0.02;

for i= 1:201

fx= @(theta ,phi) (-1-cos(phi)).*(exp(j*2*pi*d/100*tan(theta/2).*
(sin(th(i)).*cos(ph-phi)))).*tan(theta/2).*(sec(theta/2)).^2;
fy= @(theta ,phi) (1+cos(phi)).*cos(phi).*(exp(j*2*pi*d/100*tan(t
heta/2).*(sin(th(i)).*cos(ph-phi)))).*tan(theta/2).*(sec(th
eta/2)).^2;

% Initialize the integral value
integral_value_x = 0;
integral_value_y = 0;

% Perform double integration using nested loops
for theta = theta_min:delta_theta:theta_max
```

```
  for phi = phi_min: delta_phi: phi_max
        integral_value_x = integral_value_x + (fx(theta,phi)  .*
delta_phi.* delta_theta);
        integral_value_y = integral_value_y + (fy(theta,phi)  .*
delta_phi.* delta_theta);
  end
end

% zero degree plane
E_ffth(i)= abs((1+cos(th(i))).*((integral_value_x) .*cos(ph)+
(integral_value_y).*sin(ph)));
% 90 degree plane
E_ffph(i)= abs((1+cos(th(i))).*(-(integral_value_x) .*sin(ph)+
(integral_value_y).*cos(ph)));

end

th=th*180/pi;
E_ffth=(E_ffth/max(E_ffth)).^1.5;
E_ffph=(E_ffph/max(E_ffph)).^1.5;
plot(th, 20*log10(E_ffth), 'k--', th, 20*log10(E_ffph), 'k-');
grid
```

It is worth mentioning that the edge angle in Figures 8.8 to 8.10 is the maximum angle between the focal point and the edge of the dish.

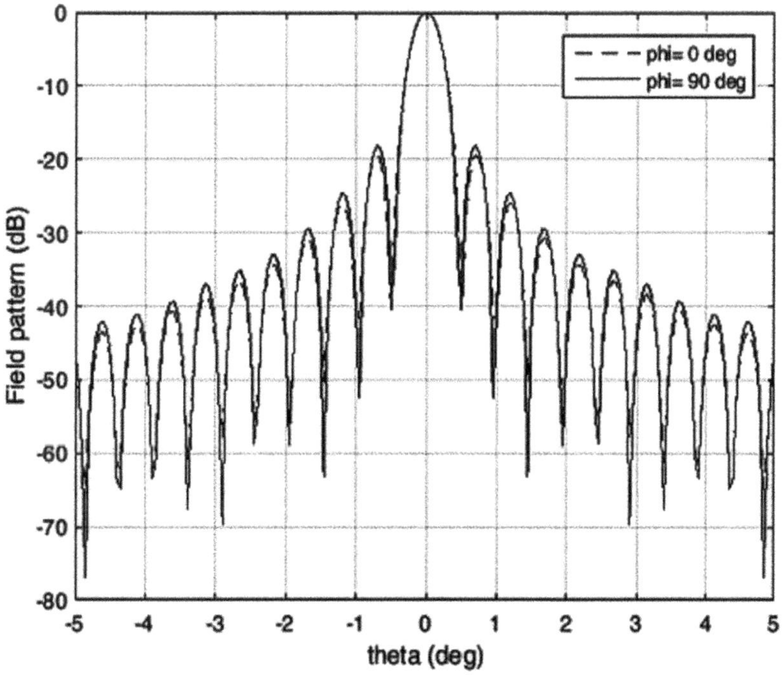

Figure 8.8 Radiation field pattern of a parabolic reflector antenna with diameter $D = 100\lambda$, $F/D = 1/2$, and an edge angle = 53° fed by a half-wave dipole located at the focus.

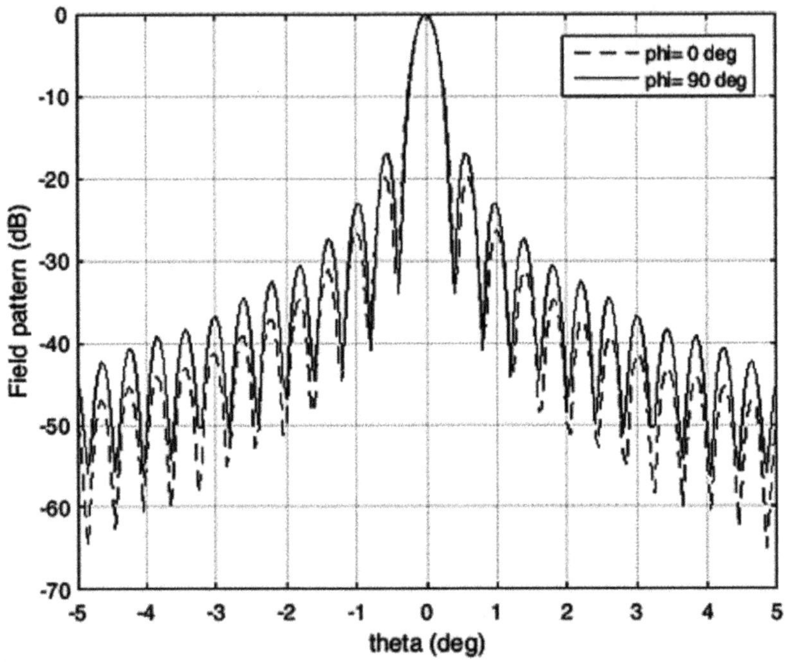

Figure 8.9 Radiation field pattern of a parabolic reflector antenna with diameter $D = 150\lambda$, $F/D = 1/3$, and an edge angle = 74° fed by a half-wave dipole located at the focus.

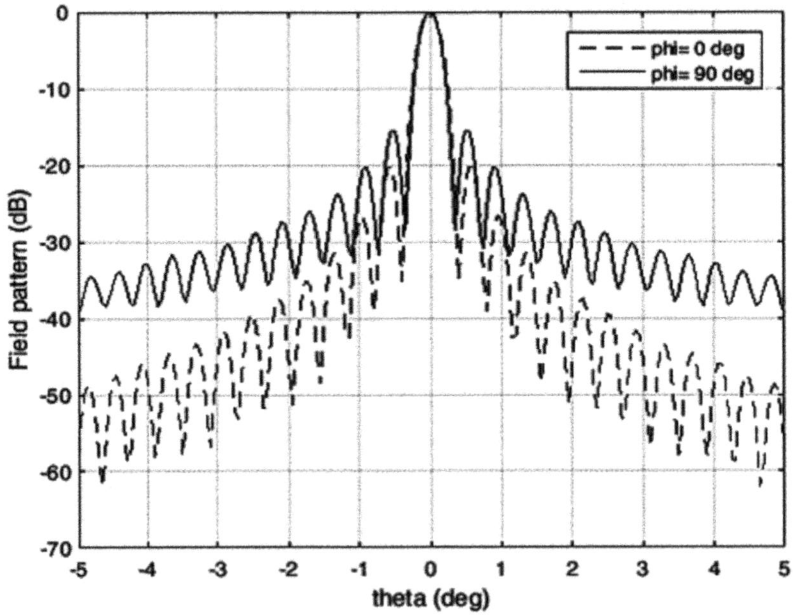

Figure 8.10 Radiation field pattern of a parabolic reflector antenna with diameter $D = 200\lambda$, $F/D = 1/4$, and an edge angle = 90° fed by a half-wave dipole located at the focus.

8.3 Gain of Reflector Antennas

The maximum achievable gain for an aperture antenna is

$$G_{\max} = D_u = \frac{4\pi}{\lambda^2} A_p \tag{8.73}$$

where D_u is the antenna directivity or maximum gain and $A_p = \pi(D/2)^2$ is the physical antenna aperture.

This gain is possible only if the following is true: uniform amplitude and phase distribution, no spillover, no ohmic losses. In practice, these conditions are not achievable, and the effective antenna aperture is less than its physical aperture:

$$G = \varepsilon_{ap} D = \frac{4\pi}{\lambda^2} \varepsilon_{ap} A_p = \varepsilon_{ap} \left(\frac{\pi D}{\lambda} \right)^2 \tag{8.74}$$

where $\varepsilon_{ap} \leq 1$ is the aperture efficiency. The aperture efficiency is expressed as a product of subefficiencies:

$$\varepsilon_{ap} = e_r \varepsilon_t \varepsilon_s \varepsilon_a \tag{8.75}$$

where e_r is the radiation efficiency (loss), ε_t is the aperture taper efficiency, ε_s is the spillover efficiency, and ε_a is the achievement efficiency.

The taper efficiency can be found using the following directivity expression for aperture antennas,

$$D_0 = \frac{4\pi}{\lambda^2} \frac{\left| \iint E_a \, ds' \right|^2}{\iint |E_a|^2 \, ds'} \tag{8.76}$$

$$A_{\text{eff}} = \frac{\left| \iint E_a \, ds' \right|^2}{\iint |E_a|^2 \, ds'} \tag{8.77}$$

$$\varepsilon_t = \frac{A_{\text{eff}}}{A_p} = \frac{1}{A_p} \frac{\left| \iint E_a \, ds' \right|^2}{\iint |E_a|^2 \, ds'} \tag{8.78}$$

For a circular aperture with radius a,

$$\varepsilon_t = \frac{1}{\pi a^2} \frac{\left| \iint E_a(\rho', \varphi') \, \rho' \, d\rho' \, d\varphi' \right|^2}{\iint |E_a(\rho', \varphi')|^2 \, \rho' \, d\rho' \, d\varphi'} \tag{8.79}$$

Now substituting for $\rho' d\rho'$ from (8.50) and for $E_a(\rho', \varphi')$ in terms of $F_f(\theta', \varphi')/r'$ gives:

$$\varepsilon_t = \frac{4F^2}{\pi a^2} \frac{\left| \int_0^{2\pi} \int_0^{\theta'_m} F_f(\theta',\varphi') \tan\left(\frac{\theta'}{2}\right) d\theta'\, d\varphi' \right|^2}{\int_0^{2\pi} \int_0^{\theta'_m} \left|F_f(\theta',\varphi')\right|^2 \sin\theta'\, d\theta'\, d\varphi'} \tag{8.80}$$

Once the feed pattern $F_f(\theta',\varphi')$ is known, it is possible to calculate ε_t.

If the feed pattern extends beyond the reflector's rim, a certain amount of power is not redirected by the reflector, which means some loss. This power loss is referred to as spillover. The spillover efficiency measures that portion of the feed pattern, which is intercepted by the reflector relative to the total feed power. Hence,

$$\varepsilon_s = \frac{\int_0^{2\pi} \int_0^{\theta'_m} \left|F_f(\theta',\varphi')\right|^2 \sin\theta'\, d\theta'\, d\varphi'}{\int_0^{2\pi} \int_0^{\pi} \left|F_f(\theta',\varphi')\right|^2 \sin\theta'\, d\theta'\, d\varphi'} \tag{8.81}$$

The reflector design problem includes a trade-off between aperture taper and spillover when the feed antenna is chosen. Taper and spillover efficiencies are combined to form the illumination efficiency, $\varepsilon_i = \varepsilon_t\varepsilon_s$. Multiplying (8.80) and (8.81) and substituting, from (8.49), for $\rho' = a = 2F\tan(\theta'_m/2)$, yield:

$$\varepsilon_i = \frac{D_f}{4\pi^2} \cot^2\left(\frac{\theta'_m}{2}\right) \left| \int_0^{2\pi} \int_0^{\theta'_m} F_f(\theta',\varphi') \tan\left(\frac{\theta'}{2}\right) d\theta'\, d\varphi' \right|^2 \tag{8.82}$$

where

$$D_f = \frac{4\pi}{\int_0^{2\pi} \int_0^{\pi} \left|F_f(\theta',\varphi')\right|^2 \sin\theta'\, d\theta'\, d\varphi'} \tag{8.83}$$

is the directivity of the feed antenna. An ideal feed antenna pattern would compensate for the spherical spreading loss by increasing the field strength as θ' of the feed increases and then would abruptly fall to zero in the direction of the reflector's rim in order to avoid spillover, as follows:

$$F_f(\theta',\varphi') = \begin{cases} \dfrac{\cos^2(\theta'_m/2)}{\cos^2(\theta'/2)} & , \theta' \le \theta'_m \\ 0 & , \theta' > \theta'_m \end{cases} \tag{8.84}$$

This ideal feed is not realizable. For practical purposes, the illumination efficiency has to be optimized with respect to the edge-illumination level.

The achievement efficiency, ε_a, is an integral factor including losses due to: random surface error, cross-polarization loss, aperture blockage, reflector phase error (profile accuracy), and feed phase error. A well-designed and well-made aperture antenna should have an overall aperture efficiency, $\varepsilon_{ap} \cong 0.65$.

The gain of a reflector antenna also depends on phase errors, which theoretically should not exist but are often present in practice. Any departure of the phase over the virtual aperture from the uniform distribution leads to a significant decrease of the directivity. For paraboloidal antennas, phase errors result from:

- Displacement of the feed phase center from the focal point;
- Deviation of the reflector surface from the paraboloidal shape, including surface roughness and other random deviations;
- Feed wavefronts not being exactly spherical.

The aperture phase deviation should be kept below $\pi/8$ if the gain is not to be affected much. This translates into a surface profile deviation from the ideal shape (e.g., paraboloid) of no more than $\lambda/16$.

8.4 Comparative Analysis of Reflector Antennas

Reflector antennas are highly regarded for their ability to achieve high gain and directivity, making them suitable for applications such as satellite communication, radio telescopes, and radar systems. Their design focuses on reflecting electromagnetic waves to a specific point, providing an enhanced signal strength in a desired direction. However, reflector antennas come with certain limitations, especially when compared to other antenna types such as dipole and horn antennas. Table 8.2 summarizes the key performance characteristics of each antenna type.

Reflector antennas offer high directivity, focusing signals in a specific direction with minimal dispersion, which allows for long-range communication. Their high gain further enhances their capability, making them essential for long-distance transmission. In contrast, dipole antennas, with their low directivity and gain, are more suited for short-range, omnidirectional communication applications, such as broadcast radio. Horn antennas fall between the two, offering medium directivity and gain, which makes them ideal for applications that require balanced performance, such as microwave communication and satellite feeds.

Table 8.2 Comparative Table of Performance Characteristics for Various Antennas

Antenna Type	*Directivity*	*Gain*	*Size and Portability*	*Complexity*	*Applications*
Reflector antenna	High	High	Large	Complex	Satellite communication, radio telescopes, radar
Dipole antenna	Low	Low	Small	Simple	Broadcast radio, wireless communication
Horn antenna	Medium	Medium	Medium	Moderate	Microwave communication, satellite feeds

When it comes to size and portability, reflector antennas are large and bulky, which can pose challenges in terms of transportation, installation, and maintenance. Dipole antennas are small and portable, making them easy to deploy and maintain. Horn antennas, although larger than dipoles, are more compact than reflectors, providing a middle ground in terms of size and ease of installation.

The complexity of reflector antennas is higher due to the precision required in their construction and alignment. Misalignment or surface distortions can lead to significant performance degradation. In contrast, dipole antennas are simple and easy to maintain, while horn antennas, although more complex than dipoles, are easier to handle than reflectors.

Reflector antennas are ideal for applications that require high performance and precision, such as satellite communication, radio telescopes, and radar systems. In contrast, dipole antennas are often used in applications where simplicity, low cost, and omnidirectional coverage are more important, such as in broadcast radio and wireless communication. Horn antennas serve applications that require moderate performance in terms of both directivity and gain, such as microwave communication and satellite feeds.

In conclusion, although reflector antennas excel in high-performance applications, their size, complexity, and installation challenges make them less suitable for situations requiring portability and ease of use. The choice of antenna depends largely on the specific application requirements, with dipole and horn antennas offering viable alternatives when different trade-offs are needed.

8.5 Real-World Challenges in Reflector Antennas

Although reflector antennas offer significant advantages, several practical challenges must be addressed to ensure their optimal performance in real-world applications. These challenges primarily stem from the need to maintain high precision, stability, and efficiency in various environments and operational conditions.

- *Feed alignment and positioning:* Reflector antennas rely on the precise placement of the feed at the focal point to ensure efficient signal reception and transmission. Any misalignment of the feed from the optimal position can cause signal loss, reduced efficiency, and beam distortion. This challenge becomes more critical in applications requiring high precision, such as satellite communication and radar systems. Even small deviations in alignment can lead to significant degradation in antenna performance, affecting the quality of communication or signal processing. Achieving and maintaining correct feed alignment require sophisticated tracking systems and continuous monitoring, especially in dynamic environments.
- *Surface distortions:* The performance of a reflector antenna depends heavily on the shape and smoothness of its reflective surface. The surface must be accurately shaped to focus incoming signals to the feed point. However, minor deformations due to mechanical stress, thermal expansion, or manufacturing defects can lead to surface distortions. These distortions can cause the antenna to lose its ability to focus waves effectively, leading to reduced

gain and directivity. In large-scale reflector antennas, such as those used in radio telescopes and satellite communication, maintaining the precise geometry of the reflector is challenging, especially over time and under varying environmental conditions. Advanced materials and technologies such as adaptive optics are being developed to address these issues by allowing real-time surface corrections.

- *Environmental effects:* Reflector antennas are subject to various environmental factors, such as wind loads, thermal expansion, and precipitation, which can affect their performance. For example, strong winds can introduce vibrations or deflections in the antenna structure, leading to misalignment or distortions. Thermal expansion can cause changes in the dimensions of the reflector, which may alter its focal properties. Precipitation, such as rain or snow, can cause signal attenuation or increase the mechanical load on the antenna. These effects are particularly pronounced for large antennas used in space applications, where the antenna must maintain high precision despite extreme conditions. To mitigate these challenges, advanced materials such as lightweight composites and structural reinforcements are increasingly used to enhance the durability and stability of the antenna.
- *Size and installation challenges:* One of the most significant challenges of reflector antennas is their large size, particularly in applications such as satellite communication, radar, and radio astronomy. Large parabolic reflectors require considerable space for installation, which can be a logistical challenge. The size also makes transportation and maintenance difficult, as specialized equipment is needed to move and install the components. Furthermore, these large antennas must be equipped with precise tracking systems to ensure proper alignment with the signal source or target. The complexity of installation and maintenance increases with the antenna's size and location, particularly when operating in remote or inaccessible areas, such as in space or on offshore platforms.
- *Cost and maintenance:* Reflector antennas, especially those designed for high-performance applications such as satellite communication or deep-space exploration, are expensive to fabricate and maintain. The production of large, high-precision reflectors requires advanced manufacturing techniques, which can drive up costs. Additionally, the maintenance of such systems can be challenging and costly, requiring periodic inspections, repairs, and realignment. The need for high-precision materials and components, as well as the potential for wear and tear due to environmental factors, can add to the operational costs. To reduce these expenses, some operators are exploring the use of adaptive optics, which can correct distortions in real time, and other cost-saving innovations, such as modular designs that allow easier maintenance and upgrades.

8.6 Applications of Reflector Antennas

Reflector antennas are used in a variety of applications due to their ability to effectively direct and amplify signals, making them crucial in several fields that require long-range, high-gain communication or detection capabilities. These applications leverage the inherent advantages of reflector antennas, such as high directivity, precision, and signal amplification.

- *Satellite communication:* Reflector antennas are widely used in satellite communication for both transmitting and receiving signals over long distances. In satellite communication systems, large parabolic reflectors are employed to ensure that signals can be focused accurately on the satellite's receiver or transmitted back to Earth. These antennas are essential for providing global communication services, such as television broadcasting, internet access, and military communications. The high gain and directivity of reflector antennas ensure that the transmitted signals can cover vast distances with minimal loss.

  Figure 8.11 shows Antenna array 2 of the Earth station Fuchsstadt, Germany. The Earth station with its around 50 parabolic antennas is part of one of the world's largest satellite communications systems, operated by Intelsat, a US company. Antenna array 2, one of a total of four, consists of one antenna with a diameter of 32m from 1985, four antennas with a diameter of 16.4m, and various smaller ones.
- *Radio astronomy:* Large reflector antennas are critical for radio astronomy, where they are used for deep-space observation and cosmic signal detection. Telescopes such as the Arecibo Observatory, shown in Figure 8.12, and the FAST (Five-hundred-meter Aperture Spherical Telescope) rely on parabolic reflectors to collect faint signals from distant celestial objects. These signals, which may originate from galaxies, stars, or other cosmic phenomena, are weak and require highly sensitive and large antenna systems to detect them accurately. The high precision and sensitivity of reflector antennas make them indispensable tools in the field of radio astronomy.
- *Radar systems:* Reflector antennas are employed in radar systems for air traffic control, weather monitoring, and military applications. Radar systems

Figure 8.11 Fuchsstadt Satellite Earth Station. (Source: Rainer Lippert. Wikimedia.)

Figure 8.12 Arecibo Observatory Aerial View. (Credit: Author H. Schweiker/WIYN and NOAO/AURA/NSF. Source: Wikimedia Commons.)

use these antennas to track objects, such as aircraft or weather patterns, over large distances. The high directivity and gain of reflector antennas allow radar systems to provide precise measurements of object location, speed, and direction. In military applications, radar antennas are used for surveillance, missile guidance, and reconnaissance, whereas, in civilian applications, they are used for weather forecasting and air traffic control.

Figure 8.13 shows a mobile radar site with set up Russian surveillance and radar systems. The vehicle consists of a P-40 radar antenna and an M109 Howitzer Chassis. The P-40 Radar ("Long Track") is a Soviet-era mobile air defense radar used for early warning and target tracking. It operated in the L-band with a range of up to 370 km and was mounted on a tracked vehicle for mobility. The M109 Howitzer Chassis is the base structure of the M109 self-propelled howitzer, an American tracked artillery vehicle designed for 155-mm artillery fire support. It provides mobility and protection for battlefield operations.

- *Deep space communications:* NASA's Deep Space Network (DSN) uses large parabolic reflector antennas to maintain communication with spacecraft traveling beyond Earth's orbit. These antennas are vital for sending commands to spacecraft and receiving data from interplanetary probes, such as those exploring Mars, Jupiter, or other distant celestial bodies. Reflector antennas are capable of transmitting signals across vast distances with minimal signal

Figure 8.13 P-40 radar system antenna and M109 chassis. (Source: Sgt. Nicole Snell. Wikimedia Commons.)

loss, making them ideal for deep-space communication where signal strength needs to be preserved over millions or even billions of kilometers.

Figure 8.14 shows the Canberra Deep Space Communications Complex with its 70-m antenna, in Tidbinbilla, 35 km from the city center.

- *Broadcasting and telecommunications:* Reflector antennas are also used in broadcasting and telecommunications, particularly for television signal transmission and high-frequency communication networks. In television broadcasting, large parabolic dishes are used to transmit signals from a central station to a wide geographic area. These antennas help to improve the quality of the broadcast by focusing the signal and minimizing interference. Similarly, reflector antennas are used in point-to-point communication systems, providing high-quality links between remote locations.

  Figure 8.15 shows radio masts and towers, which are typically tall structures designed to support antennas for telecommunications and broadcasting, including television. There are two main types: guyed (i.e., supported by tensioned cables that are anchored to the ground) and self-supporting structures (i.e., free-standing towers that do not need guy wires). Radomes can be seen covering the antennas for weather protection.
- *Earth observation:* Remote sensing satellites utilize reflector antennas to collect environmental and geographical data. These antennas enable the transmission of high-resolution images and data back to Earth, which are then used for weather forecasting, natural disaster monitoring, and environmental studies. The high gain and directivity of reflector antennas ensure that the data can

Figure 8.14 Canberra Deep Space Communications Complex. (Source: Canberra Space Center. Wikimedia Commons.)

Figure 8.15 Radio broadcasting and telecommunications towers. (Source: acopbc (Flickr). Wikimedia Commons.)

be transmitted over long distances without significant degradation, allowing for real-time monitoring of Earth's surface and atmosphere.

Figure 8.16 is an Indian satellite for earth observation. The TRISHNA (Thermal Infra-Red Imaging Satellite for High-resolution Natural Resource Assessment) Satellite indicates a collaborative mission between ISRO and CNES. The Indian Space Research Organisation (ISRO) and its French counterpart Centre National de Etudes Spatiales (CNES) have a rich history of cooperation and collaboration spanning more than six decades. The TRISHNA satellite will use a parabolic dish antenna or a horn antenna for X-band data transmission. The parabolic dish is common in Earth observation satellites due to its high directivity, efficient wave focusing, and strong ground station link. The satellite is scheduled for launch in 2026 aboard a Polar Satellite Launch Vehicle (PSLV) from the Satish Dhawan Space Centre. This joint mission between ISRO and CNES will focus on monitoring Earth's surface temperature and water management globally.

Solved Examples

Example 8.1 A parabolic reflector has a surface accuracy (RMS deviation) of 1 mm. Evaluate the impact of this deviation on the gain of the antenna at a frequency of 30 GHz.

Solution 8.1 The surface efficiency affects the efficiency of the antenna and this is described as $\eta = e^{(4\pi\sigma/\lambda)^2}$. Given $\sigma = 1$ mm = 0.001m, $\lambda = 0.01$m, η works out as 0.206. The efficiency hence drops significantly to 20.6%. Since gain and efficiency are linearly related ($G = \eta D$), this means a 20.6% drop in gain.

Example 8.2 Discuss the concepts of spillover and aperture efficiency in the context of parabolic reflector antennas. Provide numerical examples to illustrate how these factors affect the overall antenna efficiency.

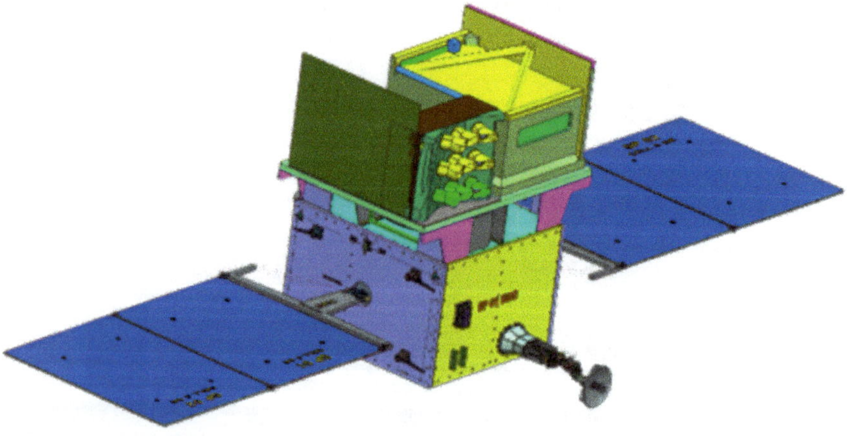

Figure 8.16 TRISHNA Earth Observation satellite in deployed configuration. (Source: Indian Space Research Organization. Wikimedia Commons.)

Solution 8.2 Spillover efficiency (η_s) is the fraction of the feed's power that is intercepted by the reflector. It can be less than 100% if some power radiates outside the reflector's edge.

Aperture efficiency (η_a) is the product of several efficiency factors, including spillover η_s, illumination η_i, phase efficiency η_p, and polarization efficiency η_o. Hence, $\eta_a = \eta_s \cdot \eta_i \cdot \eta_p \cdot \eta_o$. For example, $\eta_a = 0.9 \times 0.85 \times 0.95 \times 0.98 = 0.71$ where the numbers represent the individual efficiencies, respectively. This implies the overall efficiency is 0.71, reducing the gain by 29%.

Example 8.3 A parabolic reflector with $F/D = 1/2$, fed by a half-wave dipole centered at its focus, is designed to receive an electromagnetic signal at 10 GHz. Plot the azimuth and elevation radiation pattern if its diameter $D = 1$m, and measure the HPBW. Measure the first and second sidelobes' values and their angular locations.

Solution 8.3 The wavelength, $\lambda = 3$ cm at 10 GHz. Hence, the reflector's diameter, $D = 100$ cm/3 $= 33.3\lambda$.

Now the same MATLAB Program 10 can be used after making required changes: $F/D = 1/2$ and $D = 33.3\lambda$. Figure 8.17 shows the radiation field of parabolic reflector antenna (F/D= 1/2, D= 1m, frequency 10 GHz).

```
HPBW= 2*0.55= 1.1 degrees.
1st side lobe is -19.44 dB at 2.1 degrees on both sides.
2nd side lobe is -25.83 dB at 3.55 degrees on both sides.
```

Example 8.4 Design a parabolic reflector antenna for a satellite communication system operating at 14 GHz (uplink) and 11 GHz (downlink). Determine the required dimensions and performance parameters to achieve a gain of 45 dBi at both frequencies.

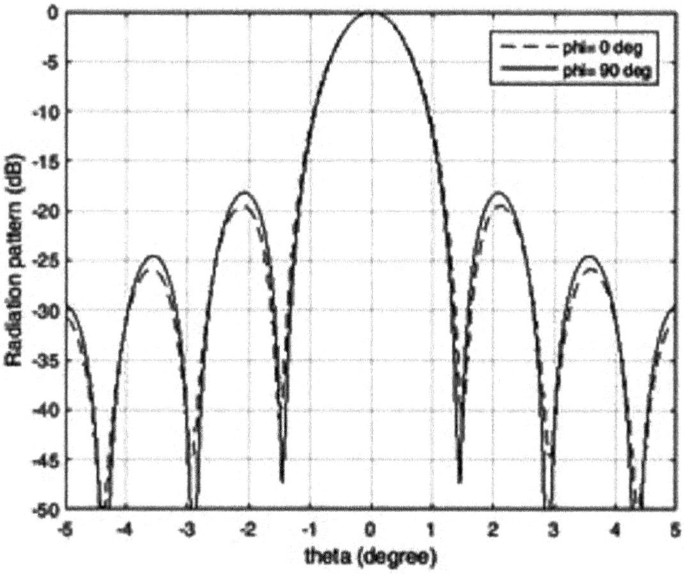

Figure 8.17 Radiation field of parabolic reflector antenna, $F/D = ½$, $D = 1$m, frequency 10 GHz.

Solution 8.4 The design specification for the required parabolic reflector are provided: $G = 45$ dBi, uplink frequency $f_u = 14$ GHz, and downlink frequency $f_d = 11$ GHz. Assume that $\eta = 0.7$.

$$\lambda_u = \frac{c}{f_u} = \frac{30}{14} = 2.14 \text{ cm}, \lambda_d = \frac{c}{f_d} = \frac{30}{11} = 2.73 \text{ cm}$$

$$G = 10^{4.5} = 31{,}622.8$$

D can now be calculated from $G = \eta(\pi D/\lambda)^2$.

$$D = \frac{\lambda_u}{\pi}\sqrt{\frac{G}{\eta}} = 144.8 \text{ cm}, D = \frac{\lambda_d}{\pi}\sqrt{\frac{G}{\eta}} = 184.7 \text{ cm}$$

To achieve a gain of 45 dBi at both 14 GHz and 11 GHz, the diameter of the parabolic reflector should be at least 184.7 cm to cover the uplink and downlink frequencies. Since the larger diameter will also satisfy the lower frequency requirement, the reflector should be designed with a diameter = 1.85m.

Problems for Students

8.1 Given a parabolic reflector with a diameter of 2m and a depth of 0.25m, calculate the focal length of the reflector.

8.2 A parabolic reflector antenna operates at a frequency of 10 GHz. The diameter of the reflector is 1.5m. Calculate the gain of the antenna, assuming an efficiency of 70%.

8.3 For a parabolic reflector antenna with a diameter of 2m operating at a frequency of 5 GHz, calculate the HPBW.

8.4 A parabolic reflector antenna has a focal length of 0.75m. If the reflector is 1.5m in diameter, determine the proper positioning of the feed to ensure optimal performance.

8.5 Explain the phenomenon of cross-polarization in parabolic reflector antennas. How does the feed design impact the cross-polarization performance? Provide a theoretical analysis.

8.6 Compare the performance of parabolic reflector antennas made from different materials (e.g., aluminum, mesh, carbon fiber). Discuss the trade-offs in terms of weight, cost, and performance.

Problems for Instructors

8.1 Basics of Parabolic Reflectors: Explain the principle of operation for a single parabolic reflector antenna. Calculate the focal length required for a parabolic reflector with a diameter of 2m operating at 12 GHz. Discuss the relationship between the focal length and beamwidth and how these parameters impact the directivity and efficiency of the reflector.

8.2 Aperture Distribution and Efficiency: Given a parabolic reflector antenna with a diameter of 1.5m operating at 10 GHz, assume a uniform aperture distribution across the surface. Calculate the effective aperture area and aperture efficiency. Discuss how different aperture distributions (e.g., uniform versus tapered) impact sidelobe levels and overall antenna performance.

8.3 Field Establishment in the Aperture: Using the method outlined in Section 8.2.1, establish the field distribution within the aperture of a parabolic reflector with an edge illumination taper of −10 dB. Assume that the operating frequency is 15 GHz and the reflector diameter is 3m. Calculate the on-axis gain and compare it to a reflector with uniform illumination. Discuss the trade-offs between edge taper and gain.

8.4 Projecting Effective Aperture: For a parabolic reflector with an effective aperture area of 1.2 square meters, calculate the projected aperture at the aperture plane when the antenna is operating at 8 GHz. Analyze how the effective aperture changes with frequency and explain its effect on the gain and beamwidth of the antenna.

8.5 Calculating the Far Field: Using Section 8.2.3, calculate the electric far-field pattern of a single parabolic reflector with a diameter of 2m at an operating frequency of 10 GHz. Assume a uniform aperture distribution and a feed positioned at the focus. Plot the far-field radiation pattern, and estimate the main lobe beamwidth and the first null position. Discuss the implications of these results for the antenna's application in satellite communications.

8.6 Gain Calculation of a Reflector Antenna: A parabolic reflector antenna operates at a frequency of 20 GHz with a diameter of 4m and an aperture efficiency of 70%. Calculate the theoretical gain of the antenna. Then determine the gain if the aperture efficiency is reduced to 50%. Discuss the impact of aperture efficiency on the overall performance of the antenna and identify strategies for optimizing it.

8.7 MATLAB Simulation of Far-Field Patterns: Using MATLAB, simulate the far-field pattern of a parabolic reflector with a diameter of 3m operating at 5 GHz. Assume a tapered illumination with an edge taper of −12 dB. The simulation should:
   (a) Plot the normalized radiation pattern.
   (b) Calculate the beamwidth and estimate the sidelobe level.
   (c) Determine the gain and compare it with theoretical expectations. Discuss any differences between the simulated and theoretical patterns and the practical considerations for designing and using reflector antennas in high-frequency applications.

Selected Bibliography

Akan, V., and E. Yazgan, "Antennas for Space Applications," *Advanced Radio Frequency Antennas for Modern Communication and Medical Systems*, 2020, p. 139.

Archer, J. S., "High-Performance Parabolic Antenna Reflectors," *Journal of Spacecraft and Rockets*, Vol. 17, No. 1, 1980, pp. 20–26.

Baars, J. W., and H. J. Kärcher, *Radio Telescope Reflectors*, Vol. 447, New York: Springer, 2018.

Balanis, C. A., *Antenna Theory: Analysis and Design*, Second Edition, New York: John Wiley & Sons, 1997.

Rahmat-Samii, Y., "Reflector Antennas," in *Antenna Handbook: Theory, Applications, and Design*, Boston, MA: Springer, 1988, pp. 949–1072.

Rao, S., L. Shafai, and S. K. Sharma, (eds.), *Handbook of Reflector Antennas and Feed Systems, Volume III: Applications of Reflectors*, Norwood, MA: Artech House, 2013.

Sharma, S. K., S. Rao, and L. Shafai, (eds.), *Handbook of Reflector Antennas and Feed Systems, Volume I: Theory and Design of Reflectors*, Norwood, MA: Artech House, 2013.

Stutzman, W. L., and G. A. Thiele, *Antenna Theory and Design*, New York: John Wiley & Sons, 2012.

Swenson, G., "The Paraboloidal Reflector Antenna in Radio Astronomy and Communication: Theory and Practice," *Physics Today*, Vol. 61, No. 7, July 2008, pp. 53–54.

CHAPTER 9

Sidelobe Suppression in E-Sectoral Horn Antennas

9.1 Modification of Aperture Field Distribution

Antenna designers are, in general, widely concerned with radiation beams of low sidelobe level. This is desired, in many applications, to reduce interference effects. A strong interfering signal incident on a receiving antenna in a direction corresponding to a sidelobe will interact with a weaker desired signal along the direction of the main lobe. Often, it is necessary to reduce the sidelobe level to some 30 dB or more below the main lobe.

The effect of phase error distribution across the radiation aperture of an electromagnetic horn is well known. The general effect of the phase error is to raise both the sidelobe level and the level of the minima. Although standard horns of sufficient axial lengths retain aperture fields with acceptable in-phase distribution, large apertures require impracticably long optimized horns. Large, radiating, open-ended waveguides would excite undesirable higher-order modes with a deteriorating effect on the resulting aperture field.

Classical techniques of phase-correcting the aperture field included the geometrical design of a dielectric lens or a metal plate lens, loaded at the aperture. Although both techniques enabled the reduction of the axial length to a comparable aperture size, the horns suffered aperture field reflections from boundary layers and reduction in bandwidth.

For most purposes, where a horn of high directivity is required, the relevant polar diagram can never be sharper than that of a uniformly illuminated aperture with a constant phase across it. While a rectangular aperture with a uniform field results in a sidelobe of only 13 dB below the main lobe level, a uniformly illuminated circular one produces a sidelobe level down by 17.6 dB, which is a modest improvement.

A tapered aperture distribution will generally result in a reduction of the sidelobe level. In arrays, the same technique could be used for reducing the sidelobe level, by tapering the element excitations toward the ends of the array. The penalty paid is an increase in the beamwidth and a reduction in the directivity brought about by a reduced effective aperture area. In a triangular aperture distribution, for example, the maximum field strength becomes one-half that of a uniformly illuminated aperture and the beamwidth between nulls would be doubled. The sidelobe level would be reduced from 13 dB to 26 dB, a pronounced tapering effect on the sidelobe level.

Several techniques have been used for this purpose and have resulted in the dielectrically loaded horn, the split-aperture horn, and the short-length, phase-corrected hyperbolic secant (HYSEC) horn, which will be demonstrated in Chapter 10. Dielectric loaded horns and split aperture horns are two sophisticated types of horn antennas extensively employed in microwave and RF communication systems. Dielectric loaded horns integrate dielectric materials within their structures to enhance performance parameters such bandwidth and efficiency. By altering the propagation characteristics of electromagnetic waves, dielectric loaded horns achieve improved impedance matching and reduced reflections, resulting in reduced sidelobes, broader bandwidth, and enhanced radiation properties. These antennas find widespread use in radar systems, telecommunications, and wireless communication networks where reliable and efficient transmission is critical.

Split aperture horns feature a segmented or split aperture, typically achieved by inserting metallic partitions within the horn structure. This design enables precise control over the antenna's radiation pattern and beam properties, such as beamwidth and directivity. Engineers can tailor the antenna's performance by adjusting the size and placement of the split aperture, allowing for optimization of characteristics such as sidelobe suppression and polarization. Split aperture horns are vital components in applications requiring accurate beam control, including radar systems, satellite communication, and radio astronomy. Their versatile design provides flexibility to adapt to various operational requirements, making them indispensable in modern communication and sensing technologies.

9.2 Computational Approaches for Horn Antenna Design

The investigation of electromagnetic field propagation within a horn necessitates the determination of fields that satisfy Maxwell's equations. Solving these equations for a horn constructed from perfectly conducting material, regardless of its shape, presents considerable challenges, with exact solutions feasible only through analytical separation of variable techniques for a limited range of scenarios, typically involving standard cross-sections such as rectangular, circular, or elliptical shapes. However, many practical applications demand exploration of more intricate geometries.

For such cases, modifications to fundamental shapes are often necessary to adjust parameters such as cutoff frequency, bandwidth, radiation pattern, and sidelobe levels. Analytical separation of variables becomes notably intricate when dealing with complex boundary conditions. Consequently, diverse alternative methods, including variational and field-matching techniques, as well as conformal transformations and point-matching methods, have been pursued to broaden the scope of solvable problems. Yet these approaches often face limitations concerning either achievable accuracy or the range of waveguide shapes amenable to solution.

The emergence of high-speed computing has increasingly favored numerical analysis for tackling such challenges. Finite difference schemes, in particular, have demonstrated versatility and robustness, notably in investigating semiconductor rib waveguides for integrated optics. This chapter employs finite-difference techniques to explore the effectiveness of corner channel inserts as sidelobe suppression mechanisms in an E-plane sectoral horn.

9.3 Finite-Difference Formulation of the Wave Equation

The scalar wave equation describing the nonzero longitudinal field component φ in an air-filled metallic waveguide is:

$$\left(\nabla_T^2 + K^2\right)\Phi = 0 \tag{9.1}$$

where

$$\nabla_T^2 = \frac{\partial^2}{\partial x^2} + \frac{\partial^2}{\partial y^2} \tag{9.2}$$

is the transverse operator and the cutoff wavenumber $K^2 = \gamma^2 + k^2$ where γ and k are the propagation constants in the waveguide and free space, respectively.

$\Phi = H_z$, for TE modes and $\Phi = E_z$, for TM modes. Solutions to (9.1) must satisfy the Neumann ($\partial\Phi/\partial n = 0$) and Dirichlet ($\Phi = 0$) boundary conditions, respectively. A solution for Φ determines all other field components and the corresponding cutoff frequency. By multiplying (9.1) by Φ and integrating over the waveguide cross-section, S, the Rayleigh quotient can be obtained by solving for K^2 in the form:

$$K^2 = -\frac{\iint \Phi \nabla_T^2 \Phi \, dS}{\iint \Phi^2 \, dS} \tag{9.3}$$

The quotient gives an upper bound to the value of K^2, which corresponds to the fundamental mode. This can be shown by virtue of orthogonality of normal modes.

In applying the finite-difference method to a waveguide cross-section, the area is first subdivided into a rectangular mesh of length h_x, in the x-direction and h_y in the y-direction. Discrete values of Φ at mesh point only are considered and the partial derivatives of (9.1) are replaced at the point ($h_x I$, $h_y J$) by appropriate finite-difference expressions. After rearrangement, one obtains:

$$\varphi(I,J) = \frac{\varphi(I+1,J) + \varphi(I-1,J) + R^2\left[\varphi(I,J+1) + \varphi(I,J-1)\right]}{2\left[1+R^2\right] - h_x^2 K^2} \tag{9.4}$$

where $R = h_x/h_y$.

When (9.4) is applied to all mesh points within the cross-section, a system of n linear equations emerges. To model the TE-mode boundary condition $d\varphi/dn = 0$, a fictitious point outside the waveguide cross-section is introduced, where φ is equated to the value of a corresponding interior point. Implementing the TM-mode boundary condition $\varphi = 0$ is straightforward. Symmetry considerations are fully leveraged. Initially, all $\varphi(I, J)$ are assumed to be unity, and an estimate for K is made. Equation (9.4) is then sequentially applied to each mesh point, updating the φ values accordingly. After each full scan, the latest φ values are utilized to numerically assess the Rayleigh quotient, (9.3), yielding a refined estimation of K. Subsequently, the field values and K are updated iteratively until convergence on the eigenvalue is achieved.

For a TE mode, lines of constant φ can be shown to correspond to electric field lines in the transverse plane. The magnitude of the transverse field E_T can be derived from its components as:

$$|E_T| = \sqrt{|E_x|^2 + |E_y|^2} = \frac{k}{K^2}\sqrt{\frac{\mu}{\varepsilon}}\sqrt{\left[\frac{\partial\varphi}{\partial x}\right]^2 + \left[\frac{\partial\varphi}{\partial y}\right]^2} \tag{9.5}$$

Equation (9.5) can be readily numerically evaluated at each mesh point.

9.4 Aperture Field Calculation of the Channel-Fitted Sectoral Horn

The geometric identification of the aperture profile in any sectoral horn equipped with channel sections can be achieved through the parameters A, B, a/A, and b/B, illustrated in Figure 9.1. The field distribution across the horn's mouth is dictated by both the excitation and the aperture's shape. Analyzing the aperture of the new horn involves treating it as a cross-section of an infinitely long waveguide with perfectly conducting walls, propagating the appropriate TE or TM mode. To ascertain the transverse field, the finite-difference method is employed, yielding φ. Corrections for phase errors stemming from the flare of the walls can be incorporated into the aperture field expression. This analysis disregards higher-mode fields within the aperture and current on the exterior walls, a valid assumption supported by observed radiation patterns from sectoral horns with rectangular cross-sections, particularly for small flare angles and E-plane apertures larger than a wavelength.

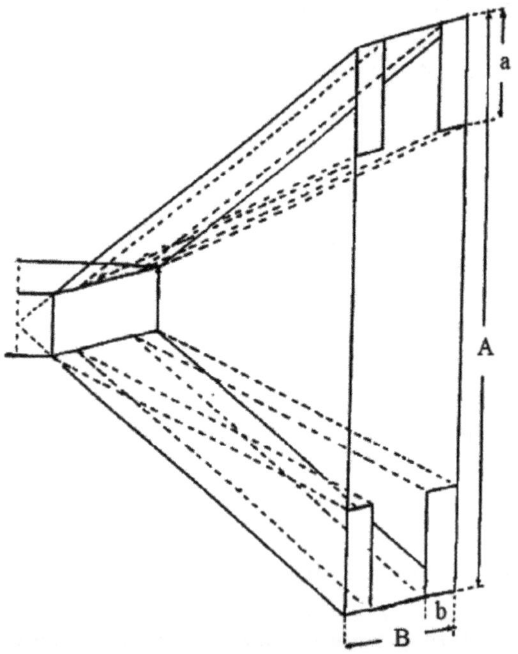

Figure 9.1 Schematic diagram of an E-sectoral horn with channel inserts.

Experimental results for a sectoral horn with a semi-flare angle of 18° further validate these assumptions.

The interaction of energy with higher-order modes in a waveguide-horn antenna configuration warrants further examination. In a straight lossless waveguide of uniform cross-section, the modes are orthogonal, ensuring no coupling between them. Typically, waveguides are engineered to propagate primarily the dominant TE_{10} mode within a specific frequency range, while other modes decay exponentially along the waveguide length if excited, such as at a cross-section discontinuity. Coupling between modes arises upon changes in the cross-section, occurring at discontinuities between the feed waveguide and the horn, as well as along the flared section of the horn itself. Notably, larger flare angles in horns exhibit stronger coupling. Despite this, at various positions along the horn's length corresponding to specific cross-sectional dimensions, many higher-order modes still decay evanescently at the operating frequency. For horns with small flare angles, the excitation and subsequent decay of higher-order modes create an attenuation region extending significantly from the apex. This attenuation region expands with higher-order waves, albeit with weak coupling for small flare angles, allowing the neglect of higher-order modes. Conversely, with increasing flare angle, the attenuation region contracts, predominantly confined to the throat, and with more pronounced coupling, necessitating consideration of higher-order modes in the analysis.

Solutions for φ were derived for multiple waveguide cross-sections resembling the configuration in Figure 9.2. By leveraging symmetry, the analysis was confined to a quarter of the cross-section to identify the dominant TE mode. Figure 9.3 displays the transverse electric-field lines for a sequence of quarter cross-sections with an aspect ratio of $A/B = 9.17$, alongside a set of a/A and b/B values. The results highlight a clear trend of increasing field distortion around the corners of the channel

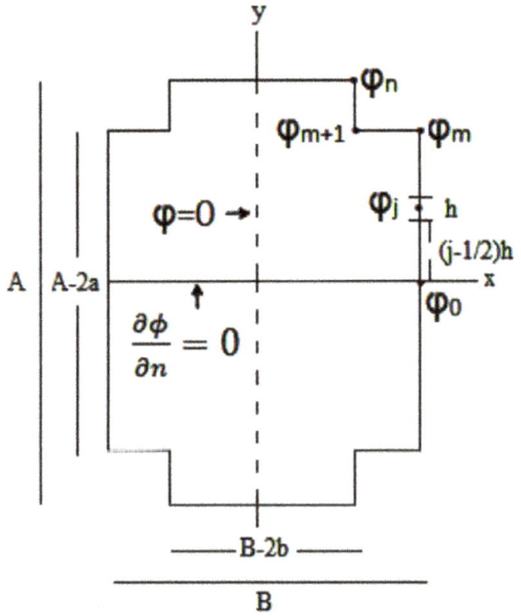

Figure 9.2 A channel-fitted sectoral horn that shows aperture cross-section. (Adapted from: [2].)

sections as the aspect ratios a/A and b/B increase. Notably, this distortion diminishes progressively as the channel ratios approach 0.5, where the waveguide transitions back to a rectangular shape.

The distortion in field patterns is closely linked to the amplitude tapering of the E_y and E_x components. Both components exhibit opposing variations in the x and y directions, with irregularities in the behavior of E_x becoming more pronounced at higher values of y within certain configurations. While the general shape of the field pattern remains consistent for varying values of A/B, a/A, and b/B, the degree of distortion around the corners intensifies with higher channel ratios, demonstrating a systematic trend across configurations. Further details can be found in [1].

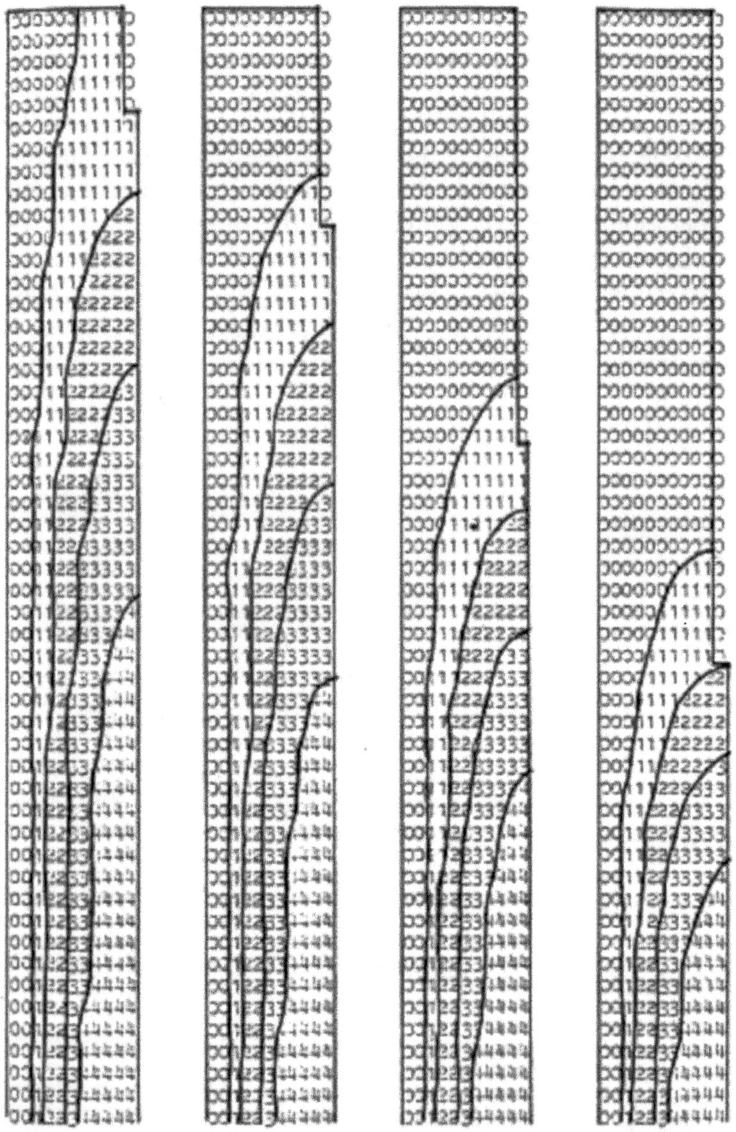

Figure 9.3 Transverse-electric field lines for $A/B = 9.17$. (Adapted from: [2].)

9.5 Radiation Pattern Calculation of the Channel-Fitted Sectoral Horn

An approximate calculation of the radiation pattern becomes possible, once the electric and magnetic field components across the horn aperture are known. Edge diffraction effects are neglected in this calculation, which is justified because the size of the horns studied is large compared to the wavelength operation in the E-plane of interest. The E-plane radiation field of any aperture, such as that of a channel-fitted horn, can be expressed as the θ-component of the electric field in the $\varphi = \pi/2$ plane, in a polar coordinate system centered on the aperture as:

$$E_\theta = \frac{j}{2\lambda R}\left(1+\frac{\lambda}{\lambda_g}\cos\theta\right)\exp\left(\frac{-j2\pi R}{\lambda}\right)\int\frac{\partial\varphi}{\partial x}\exp\left(\frac{j2\pi y}{\lambda}\sin\theta\right)dx\,dy \tag{9.6}$$

where λ is the free-space wavelength, λ_g is the guide wavelength, R is the radial distance for the aperture to the external point where the radiation field is calculated, E_y is the y-component of the aperture electric field, $dS = dxdy$ is an area element in the aperture plane, and $E_y = \partial\varphi/\partial x$ is the gradient of the longitudinal field.

Integrating with respect to x, the surface integral becomes:

$$\int_{-A/2}^{A/2}\exp\left(\frac{j2\pi y}{\lambda}\sin\theta\right)\int_{-L}^{L}\frac{\partial\varphi}{\partial x}dx = \int_{-A/2}^{A/2}\left[\varphi(L)-\varphi(-L)\right]\exp\left(\frac{j2\pi y}{\lambda}\sin\theta\right)dy \tag{9.7}$$

where

$$\pm L = \pm\frac{B}{2}\ \text{for}\ -\left(\frac{A}{2}-a\right)<y<\left(\frac{A}{2}-a\right)$$
$$= \pm\left(\frac{B}{2}-b\right)\ \text{for}\ \left(\frac{A}{2}-a\right)<y<-\left(\frac{A}{2}-a\right)$$

Hence, the surface integral reduces to a line integral and the calculation of the radiation pattern becomes possible, once the φ values on the boundary are known. Moreover, symmetry conditions about x and y-axes shown in Figure 9.2 give for the desired TE_{01} mode:

$$\varphi(x,y) = \varphi(x,-y) \tag{9.8}$$

$$\varphi(-x,y) = -\varphi(x,y) \tag{9.9}$$

Accordingly, $[\varphi(L) - \varphi(-L)]$ equals $2\varphi(L)$ in (9.7) and since there is even symmetry in the field about the x-axis, only the real part of the exponential term need to be considered and (9.7) becomes:

$$\begin{aligned}4\int_{0}^{A/2}\varphi\cos\left(\frac{2\pi}{\lambda}y\sin\theta\right)dy &= 4\int_{y_0}^{y_m}\varphi\cos\left(\frac{2\pi}{\lambda}y\sin\theta\right)dy\\ &+4\int_{y_{m+1}}^{y_n}\varphi\cos\left(\frac{2\pi}{\lambda}y\sin\theta\right)dy\end{aligned} \tag{9.10}$$

The integration in (9.10) must be performed numerically since φ is provided at discrete mesh points by a finite-difference solution. It can be assumed that the variation of φ, within a mesh element, is negligible for a sufficiently fine mesh and hence (9.10) can be replaced by a summation of contributions involving φ_J given as:

$$4\varphi_J \int_{\left(J-\frac{1}{2}\right)h_y}^{\left(J+\frac{1}{2}\right)h_y} \cos\left(\frac{2\pi}{\lambda} y \sin\theta\right) dy = 4h_y \varphi_J \frac{\sin\left(\frac{\pi h_y}{\lambda}\sin\theta\right)}{\left(\frac{\pi h_y}{\lambda}\sin\theta\right)} \cos\left(\frac{2\pi J h_y}{\lambda}\sin\theta\right) \quad (9.11)$$

The contribution from an end point such as φ_n in Figure 9.2 is only one-half that for a general point. The summation of contributions with the form of (9.11) can be substituted into (9.6) to give the θ-component of the radiated electric field as:

$$E_\theta = G\left(1 + \frac{\lambda}{\lambda_g}\cos\theta\right) \frac{\sin\left(\frac{\pi h_y}{\lambda}\sin\theta\right)}{\left(\frac{\pi h_y}{\lambda}\sin\theta\right)} x$$
$$\left\{ \begin{aligned} &\frac{1}{2}\varphi_0 + \frac{1}{2}(\varphi_m + \varphi_{m+1})\cos\left(\frac{2\pi m h_y}{\lambda}\sin\theta\right) + \frac{1}{2}\varphi_n \cos\left(\frac{2\pi n h_y}{\lambda}\sin\theta\right) \\ &+ \sum_{J=1}^{m-1} \varphi_J \cos\left(\frac{2\pi J h_y}{\lambda}\sin\theta\right) + \sum_{J=m+2}^{n-1} \varphi_J \cos\left(\frac{2\pi J h_y}{\lambda}\sin\theta\right) \end{aligned} \right\} \quad (9.12)$$

where G is a constant.

The special treatment of φ_m and φ_{m+1} in the above equation can be avoided if they are substituted by one field value: $\varphi_s = 1/2(\varphi_m + \varphi_{m+1})$. Furthermore, the value of λ_g can be determined from:

$$\lambda_g = \left\{ \left(\frac{1}{\lambda}\right)^2 - \left(\frac{1}{\lambda_c}\right)^2 \right\}^{-1/2} \quad (9.13)$$

where λ_c is the cutoff wavelength computed from the finite-difference solution of the field.

A close examination of (9.12) shows that the θ-component of the radiated field is predominantly determined by φ_J values along the rim of the aperture. The variation in these values therefore determines the shape and degree of tapering in the aperture field distribution and hence the shape of the main lobe and the sidelobe levels in the radiation pattern.

Figures 9.4 and 9.5 show the variation of φ_J along the rim for different values of a/A, for $A/B = 9.17$ and $b/B = 0.05$ and $b/B = 0.1$, respectively. It is seen that the tapering of φ along the rim first increases with a/A but then decreases as a/A increases further until $a/A = 0.5$ where φ_J becomes constant. For a smaller aspect ratio, $A/B = 2.5$, a similar behavior has also been noticed but with a smaller tapering magnitude in φ_J along the rim compared to that found for $A/B = 9.17$. It is desirable

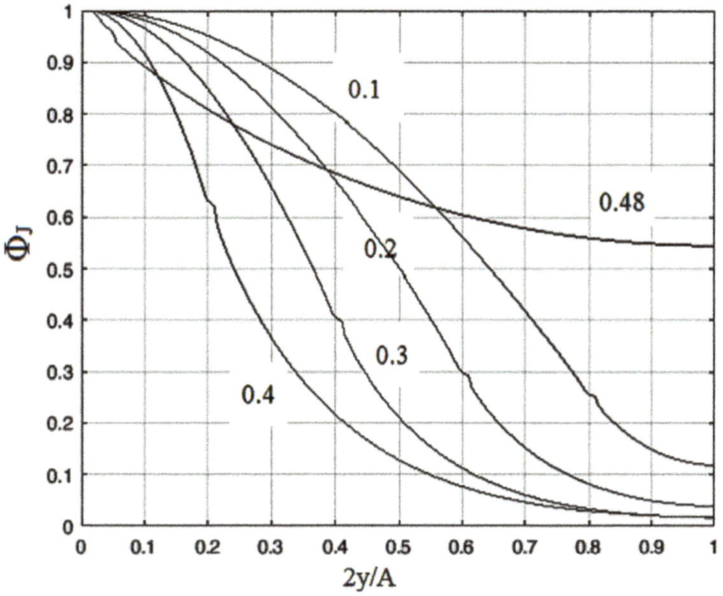

Figure 9.4 Variation of φ_J along the aperture rim for $A/B = 9.17$ and $b/B = 0.05$.

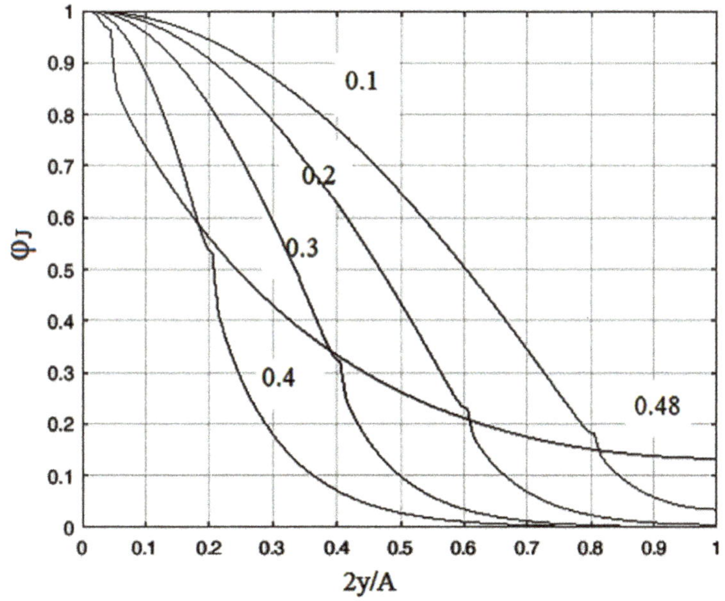

Figure 9.5 Variation of φ_J along the aperture rim for $A/B = 9.17$ and $b/B = 0.1$.

to find those parameters, which give φ_J variations along the rim, that provide the lowest sidelobe level in the radiation pattern.

MATLAB Program 11

```
% Matlab program to compute the transverse electric field of the
E-sectoral horn with channel inserts, considering only ¼ of the
```

```
aperture area because of symmetry about the x- and the
y- dimensions.

% a is ½ the x-dimension of the aperture
% b is ½ the y-dimension of the aperture
% m is the total number of meshes along 'a'
% m1 is the number of meshes along 'a' excluding the
insert channel
% n is the total number of meshes along 'b'
% n1 is the number of meshes along 'b' excluding the
insert channel
% k is an estimation of the wavenumber (assume an initial value
of 1)
% iter is number of iterations

a = input('Enter value for a: ');
b = input('Enter value for b: ');
m = input('Enter value for m: ');
m1 = input('Enter value for m1: ');
n = input('Enter value for n: ');
n1 = input('Enter value for n1: ');
ki = input('Enter value for ki: ');
iter = input('Enter value for iter: ');

dx = a / m;
dy = b / n;
k = ki;
e = ones(m+2, n+2);

for count = 1:iter
  for i = 2:m1+1
    for j = 2:n+1
      ea = e(i, j+1) + e(i, j-1);
      eb = e(i+1, j) + e(i-1, j);
      et = dx^2 * ea + dy^2 * eb;
      ebot = 2 * dx^2 + 2 * dy^2 - k^2 * dx^2 * dy^2;
      e(i, j) = et./ ebot;
    end
  end

  for i = m1+2:m+1
    for j = 2:n1+1
      ea = e(i, j+1) + e(i, j-1);
      eb = e(i+1, j) + e(i-1, j);
      et = dx^2 * ea + dy^2 * eb;
      ebot = 2 * dx^2 + 2 * dy^2 - k^2 * dx^2 * dy^2;
      e(i, j) = et./ ebot;
    end
  end
```

```
% Apply boundary conditions
for i= 1:m+1
  e(i,1)= e(i,3);
end

for j=1:n1+1
  e(m+2, j) = e(m, j);
end

for i=m1+1:m+1
  e(i, n1+2) = e(i, n1);
end

  e(m1+2, n1) = e(m1, n1+2);
  e(m1+2, n1+2) = e(m1+2, n1);

for j=n1+1:n+1
  e(m1+2,j) = e(m1, j);
end

for i=1:m1+1
  e(i, n+2) = e(i, n);
end

for j=1:n+1
  e(1,j) = 0;
end

  s = 0;
  s1 = 0;
  s2 = 0;

  for i = 2:m1+1
    for j = 2:n+1
      ec = e(i+1, j) + e(i-1, j) - 2 * e(i, j);
      ed = e(i, j+1) + e(i, j-1) - 2 * e(i, j);
      s1 = s1 + e(i, j).* dx * dy * (ec/dx^2 + ed/dy^2);
      s2 = s2 + e(i, j).^2 * dx * dy;
    end
  end

  for i = m1+2:m+1
    for j = 2:n1+1
      ec = e(i+1, j) + e(i-1, j) - 2 * e(i, j);
      ed = e(i, j+1) + e(i, j-1) - 2 * e(i, j);
      s1 = s1 + e(i, j).* dx * dy * (ec/dx^2 + ed/dy^2);
      s2 = s2 + e(i, j).^2 * dx * dy;
    end
  end
```

```
    k = sqrt(abs(s1/s2));

end

for i = 1:m1+1
  for j = 1:n+1
ee(i,j)= e(i,j);
  end
end

for i = m1+2:m+1
  for j = 1:n1+1
ee(i,j)= e(i,j);
  end
end

% y is the normalized dimension along the rim of the ¼ aperture
of the E-sectoral horn with channels
% erim is the electric field along the rim of the horn.
ee= ee';
ee(1:n1+1,m1+1)= ee(1:n1+1,m+1);
erim= ee(:,m1+1)./max(ee(:,m1+1));
for i= 1:n+1
y(i) = 1/n*i;
end

axis([0 1 0 1])
plot(y,erim, 'k-')
grid
```

The θ-component of the electric field in (9.12) relates to an aperture field distribution of constant phase. In order to account for the effect of aperture phase deterioration on the radiation field, a quadratic function of y is simplistically assumed, although this is normally used in calculations with ordinary rectangular aperture horns of small semi-flare angles not exceeding about 20°. Such a quadratic function takes the form, $q = y^2/2R_h$, under the condition $2R_h > \lambda$ where R_h represents the axial length of the horn and q is the radial difference between a point on the aperture and another one on a phase front, centered on the feed and just emerging from the aperture. Hence, the additional term to be considered in accounting for the phase deterioration across the aperture takes the form $\exp(-j2\pi q/\lambda_h)$ where λ_h is the horn wavelength. For a small finite difference mesh length h_y the exponential term can be considered as a constant across a given interval. The radiation expression for E_θ is modified by replacing φ_J values in (9.12) by $\exp(-j\pi J^2 h_y^2/R_h\lambda_h)$.

MATLAB Program 12

```
% Matlab program to compute the far-field radiation pattern of
the E-sectoral horn with channel inserts, assuming a quadratic
aperture phase.
```

```
% a is ½ the actual x-dimension of the aperture
% b is ½ the actual y-dimension of the aperture
% m is the total number of meshes along 'a'
% m1 is the number of meshes along 'a' excluding the
insert channel
% n is the total number of meshes along 'b'
% n1 is the number of meshes along 'b' excluding the
insert channel
% k is an initial estimate of the wavenumber (assume an initial
value of 1)
% iter is number of iterations

a = input('Enter value for a: ');
b = input('Enter value for b: ');
m = input('Enter value for m: ');
m1 = input('Enter value for m1: ');
n = input('Enter value for n: ');
n1 = input('Enter value for n1: ');
freq = input('Enter value for frequency in GHz: ');
ki = input('Enter value for ki: ');
iter = input('Enter value for iter: ');

dx = a / m;
dy = b / n;
k = ki;

e = ones(m+2, n+2);

for count = 1:iter
  for i = 2:m1+1
    for j = 2:n+1
      ea = e(i, j+1) + e(i, j-1);
      eb = e(i+1, j) + e(i-1, j);
      et = dx^2 * ea + dy^2 * eb;
      ebot = 2 * dx^2 + 2 * dy^2 - k^2 * dx^2 * dy^2;
      e(i, j) = et./ ebot;
    end
  end

  for i = m1+2:m+1
    for j = 2:n1+1
      ea = e(i, j+1) + e(i, j-1);
      eb = e(i+1, j) + e(i-1, j);
      et = dx^2 * ea + dy^2 * eb;
      ebot = 2 * dx^2 + 2 * dy^2 - k^2 * dx^2 * dy^2;
      e(i, j) = et./ ebot;
    end
  end

% Apply boundary conditions
```

```
for i= 1:m+1
  e(i,1)= e(i,3);
end

for j=1:n1+1
  e(m+2, j) = e(m, j);
end

for i=m1+1:m+1
  e(i, n1+2) = e(i, n1);
end

  e(m1+2, n1) = e(m1, n1+2);
  e(m1+2, n1+2) = e(m1+2, n1);

for j=n1+1:n+1
  e(m1+2,j) = e(m1, j);
end

for i=1:m1+1
  e(i, n+2) = e(i, n);
end

for j=1:n+1
  e(1,j) = 0;
end

  s = 0;
  s1 = 0;
  s2 = 0;

  for i = 2:m1+1
    for j = 2:n+1
      ec = e(i+1, j) + e(i-1, j) - 2 * e(i, j);
      ed = e(i, j+1) + e(i, j-1) - 2 * e(i, j);
      s1 = s1 + e(i, j).* dx * dy * (ec/dx^2 + ed/dy^2);
      s2 = s2 + e(i, j).^2 * dx * dy;
    end
  end

  for i = m1+2:m+1
    for j = 2:n1+1
      ec = e(i+1, j) + e(i-1, j) - 2 * e(i, j);
      ed = e(i, j+1) + e(i, j-1) - 2 * e(i, j);
      s1 = s1 + e(i, j).* dx * dy * (ec/dx^2 + ed/dy^2);
      s2 = s2 + e(i, j).^2 * dx * dy;
    end
  end

  k = sqrt(abs(s1/s2));
```

```
end

lc = 2 * pi / k;

for i = 1:m1+1
  for j = 1:n+1
EE(i,j)= e(i,j);
  end
end

for i = m1+2:m+1
  for j = 1:n1+1
EE(i,j)= e(i,j);
  end
end

A = zeros(41,401);
B = zeros(1,181);
C = zeros(1,181);

S1 = 0;
S2 = 0;
R1 = 0;
Ep = zeros(1,181);

% lmc is cutoff wavelength
lmc= lc;
% lm is free space wavelength
lm= 30/freq

% Calculate the phase term where Rh is the horn axial length
(semi flare angle is 18 degree)

Rh = b/tand(18);
g = lm*sqrt(1.0/(lm^2) - 1.0/(lmc^2));
lmh = lm/sqrt(1.0 - (lm/lmc)^2);

% quadratic aperture phase (otherwise w=0 if constant aperture
phase is desired)

% w = pi*(dy^2)/(lmh*Rh);
w= 0;
jj = complex(0.0, 1.0);

for I = 1:m1+1
  for J = 1:n+1
A(I,J)= EE(I,J);
  end
end
```

```
for I = m1+2:m+1
  for J = 1:n1+1
A(I,J)= EE(I,J);
  end
end

for p = 1:181
  S1 = complex(0.0, 0.0);
  S2 = complex(0.0, 0.0);
  Ep(p) = complex(0.0, 0.0);
  ta = (p-91)*pi/180.0;
  u = pi*dy*sin(ta)/lm;
  if abs(u) < 1.0E-12
     uu = 1.0;
  else
     uu = sin(u)/u;
  end
  for ji = 2:2:n1
    S1 = S1 + (A(m+1, ji)*cos(2*ji*u))*exp(-jj*w*ji^2);
  end
  for jk = n1+2:2:n
    S2 = S2 + (A(m1+1, jk)*cos(2*jk*u))*exp(-jj*w*jk^2);
  end
  R1 = 0.5*A(m+1,1) + 0.5*(A(m+1, n1+1)+A(m1+1,
n1+1))*cos(2*u*(n1+1))*exp(-jj*w*(n1+1)^2) + 0.5*A(m1+1,
n+1)*cos(2*u*(n+1))*exp(-jj*w*(n+1)^2);
   Ep(p) = (uu*(1 + g*cos(ta)))*(R1 + 1*S1 + 1*S2);
   B(p) = 20*log10(sqrt(real(Ep(p))^2 + imag(Ep(p))^2));
   C(p) = atan2(imag(Ep(p)), real(Ep(p)))*180.0/pi;
end

for K = 1:181
  th(K)= K-91;
end
B= B-max(B);
plot(th,B, '-k')
axis([-90 90 -50 0])
set(gca, 'xlim', [-90 90]);
set(gca, 'xtick', -90:30:90);
grid
```

Figures 9.6 to 9.8 show that, upon varying a/A between 0 and 0.5, the sidelobe levels first decrease and then increase. However, they are seen to be better for an A/B of 9.17 than for A/B of 2.5 shown in Figures 9.9 and 9.10. The effect of increasing b/B has been studied on apertures having an aspect ratio of 2.5 and constant a/A. For $a/A = 0.3$, for example, Figure 9.10 shows that lower sidelobes were obtained in the case where $b/B = 0.1$ than when $b/B = 0.2$.

Figures 9.11 to 9.13 show radiation patterns for the horn with a quadratic aperture phase $q_p = (-j\pi y^2/\lambda_h R_h)$ that is incorporated in the MATLAB program. Those

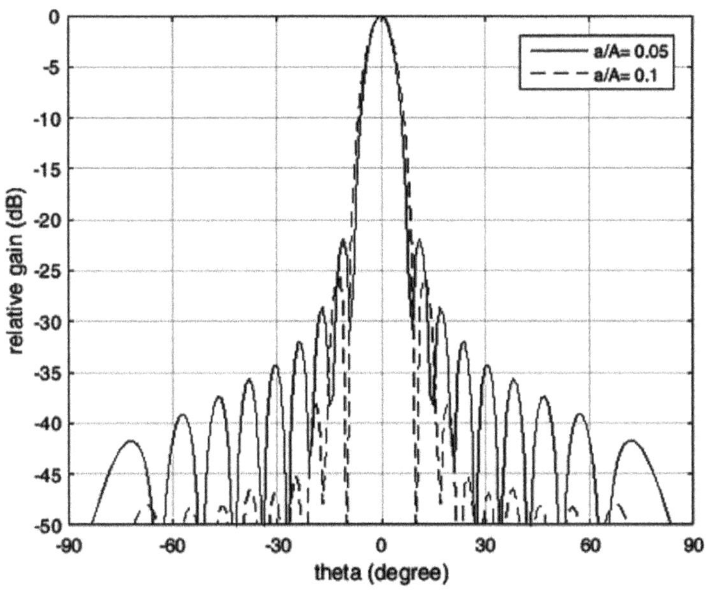

Figure 9.6 Theoretical radiation pattern of a channel fitted E-sectoral horn with $B = 2.286$ cm. $A/B = 9.17$ and $b/B = 0.05$, $a/A = 0.05$ and 0.1, with a constant aperture phase at $f = 12.77$ GHz.

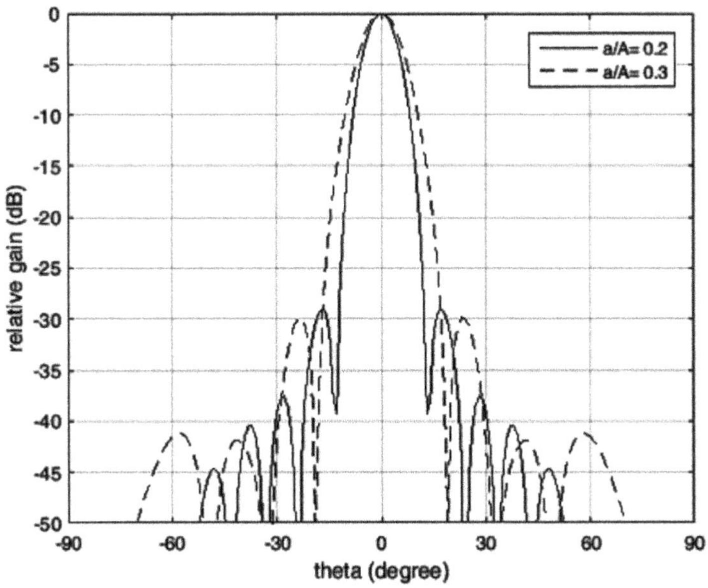

Figure 9.7 Theoretical radiation pattern of a channel fitted E-sectoral horn with $B = 2.286$ cm. $A/B = 9.17$ and $b/B = 0.05$, $a/A = 0.2$ and 0.3, with a constant aperture phase at $f = 12.77$ GHz.

were compared with patterns acquiring a constant aperture phase at frequencies of 7, 10, and 12.4 GHz respectively. These frequencies represent the lower, middle, and higher portions of the X-band frequency range. The general effect of introducing a quadratic phase error is to broaden the main lobe of the radiation pattern. Shoulders appear, also between approximately −25 dB and −40 dB, depending on

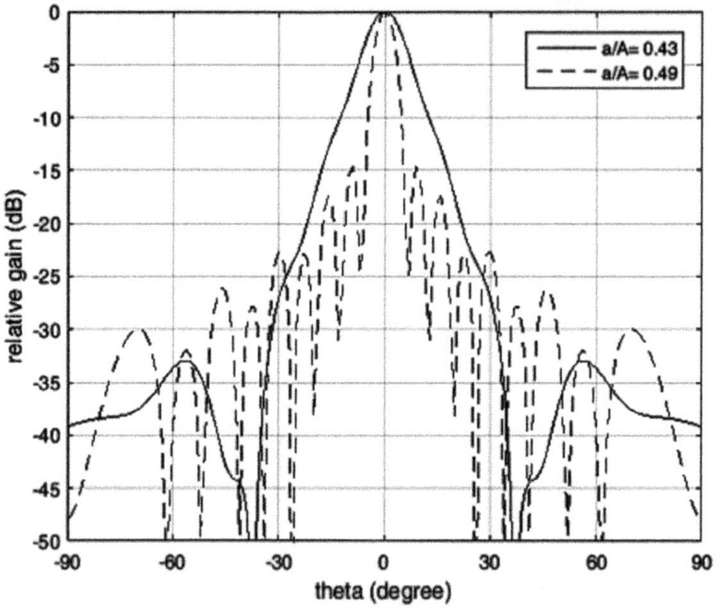

Figure 9.8 Theoretical radiation pattern of a channel fitted E-sectoral horn with $B = 2.286$ cm. $A/B = 9.17$ and $b/B = 0.05$, $a/A = 0.43$ and 0.49, with a constant aperture phase at $f = 12.77$ GHz.

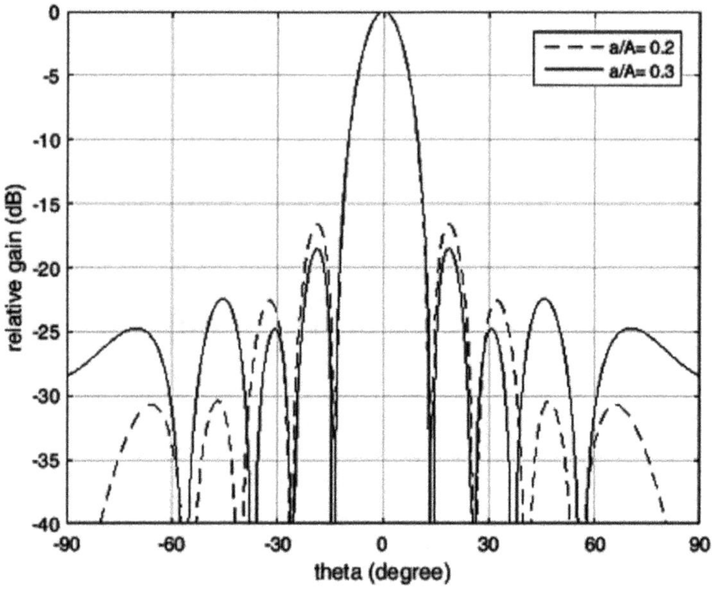

Figure 9.9 Theoretical radiation pattern of a channel fitted E-sectoral horn with $B = 2.286$ cm. $A/B = 2.5$ and $b/B = 0.05$, $a/A = 0.2$ and 0.3, with a constant aperture phase at $f = 12$ GHz.

the frequency. Furthermore, all sidelobes, with the quadratic aperture phase at the three frequencies, are suppressed below the −40-dB level.

A second horn with the same geometry but with no channel inserts was built for comparison purposes. The radiation patterns were compared at the same frequencies,

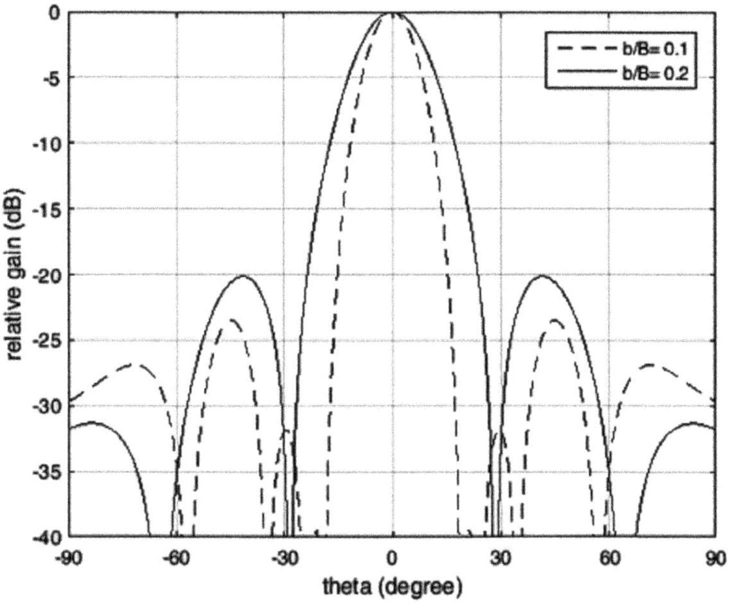

Figure 9.10 Theoretical radiation pattern of a channel fitted E-sectoral horn with $B = 2.286$ cm. $A/B = 2.5$ and $a/A = 0.3$, $b/B = 0.1$ and 0.2, with a constant aperture phase at $f = 12$ GHz.

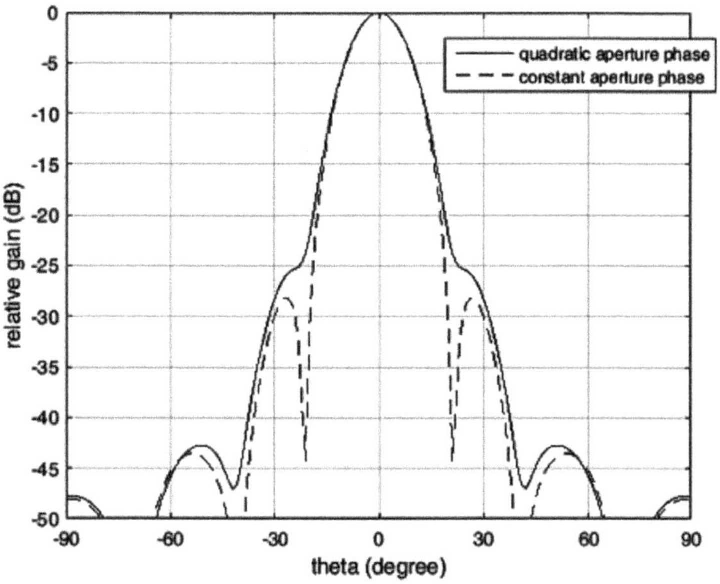

Figure 9.11 Radiation patterns of an E-sectoral horn with corner inserts with $B = 2.286$ cm. $A/B = 9.17$, $a/A = 0.15$, $b/B = 0.05$ and semi-flare angle 18° at $f - 7$ GHz.

7 GHz, 10 GHz, and 12.4 GHz, as shown in Figures 9.14 to 9.16, respectively. The addition of the corner inserts is expected to have two positive effects on the aperture field, namely amplitude tapering and phase correction. Phase compensation is expected because the narrowing of the horn width at the outer edges decreases the

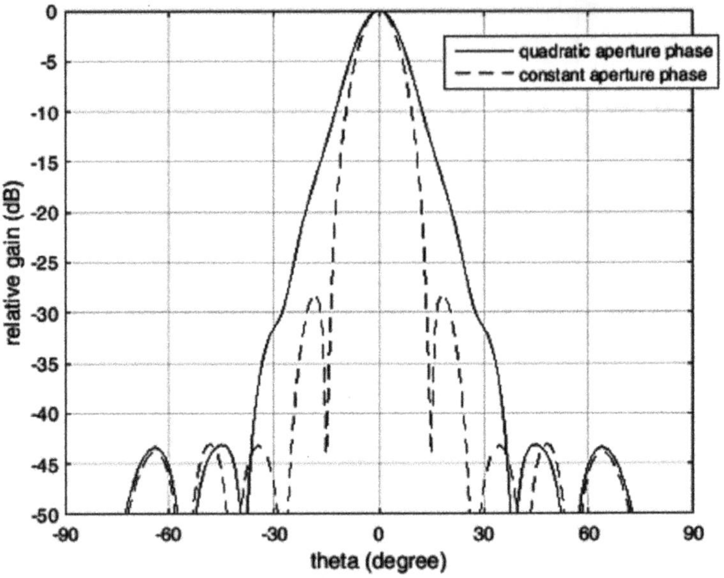

Figure 9.12 Radiation patterns of an E-sectoral horn with corner inserts with $B = 2.286$ cm. $A/B = 9.17$, $a/A = 0.15$, $b/B = 0.05$, and semi-flare angle 18° at $f = 10$ GHz.

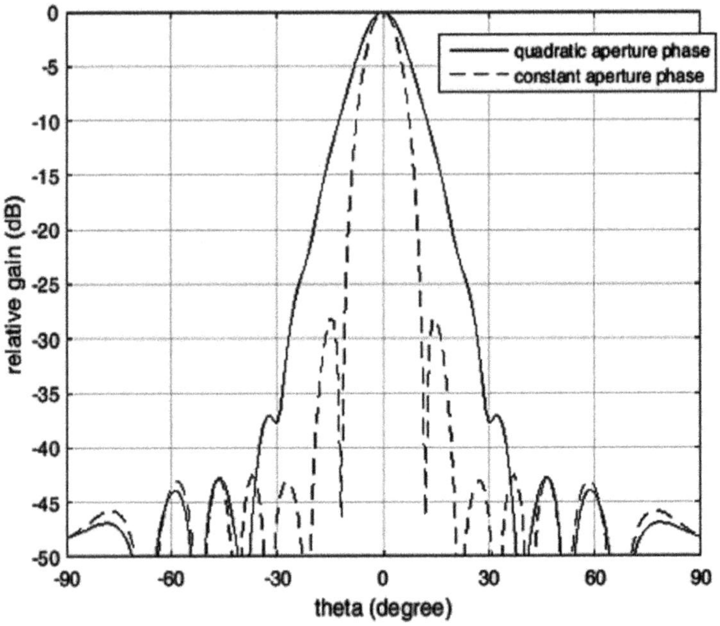

Figure 9.13 Radiation patterns of an E-sectoral horn with corner inserts with $B = 2.286$ cm. $A/B = 9.17$, $a/A = 0.15$, $b/B = 0.05$, and semi-flare angle 18° at $f = 12.4$ GHz.

local propagation constant in this region. In Figure 9.16, the aperture phase error has resulted in splitting and broadening of the main beam confirming that the aperture dimension chosen exceeds the optimum value for a horn of semi-flare angle 18°.

These factors imply that for the same semi-flare angle the modified horn can be expected to produce optimum performance (optimum power gain and lower sidelobe

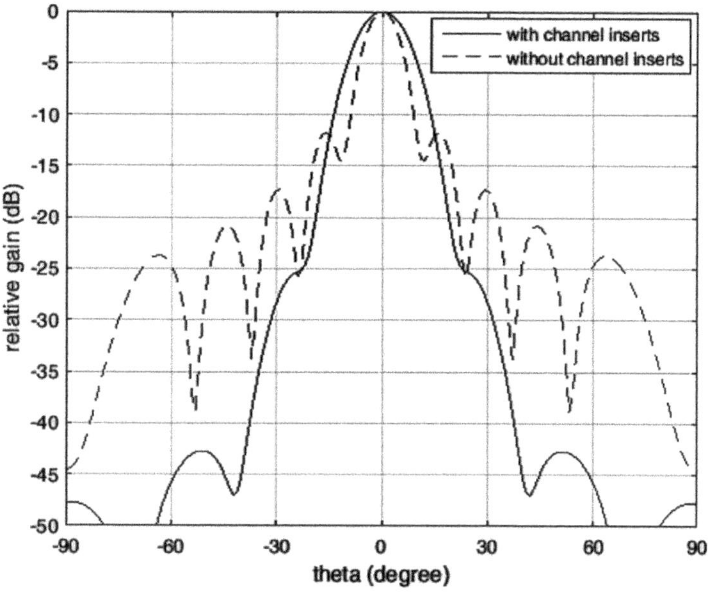

Figure 9.14 Comparison between radiation patterns of an E-sectoral horn with and without corner inserts. $B = 2.286$ cm, $A/B = 9.17$, $a/A = 0.15$, $b/B = 0.05$, and semi-flare angle 18° at $f = 7$ GHz.

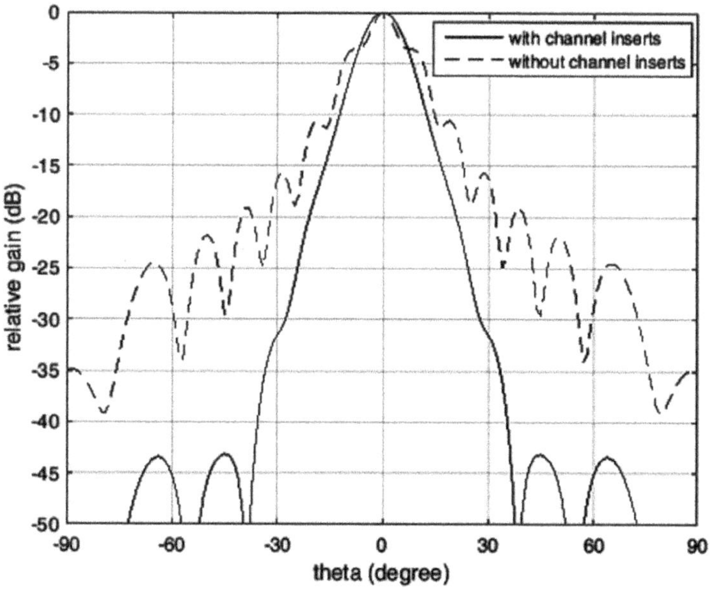

Figure 9.15 Comparison between radiation patterns of an E-sectoral horn with and without corner inserts. $B = 2.286$ cm, $A/B = 9.17$, $a/A = 0.15$, $b/B = 0.05$, and semi-flare angle 18° at $f = 10$ GHz.

level) for a larger value of A than the original horn. A slightly oversized horn was therefore studied to make the effect of the corner inserts clear.

Table 9.1 demonstrates the amount of relative reduction in sidelobe level, calculated at the three frequencies.

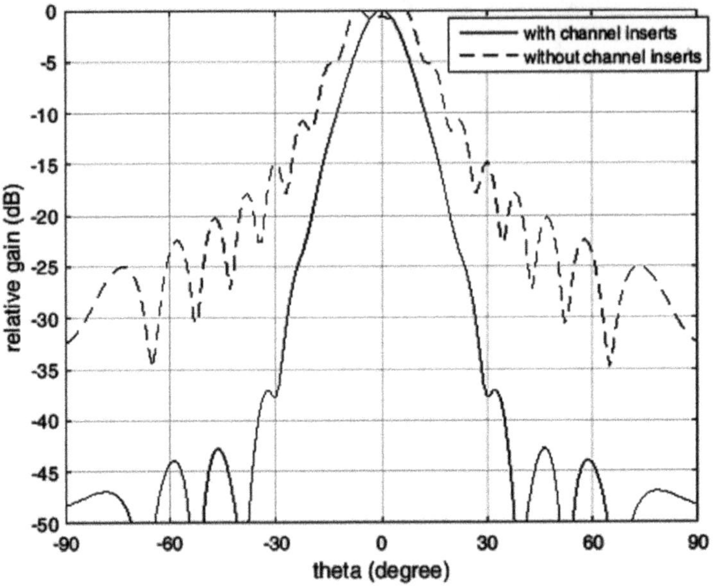

Figure 9.16 Comparison between radiation patterns of an E-sectoral horn with and without corner inserts. $B = 2.286$ cm, $A/B = 9.17$, $a/A = 0.15$, $b/B = 0.05$, and semi-flare angle 18° at $f =$ 12.4 GHz.

Figure 9.17 compares an experimental radiation pattern of an E-sectoral horn with channel inserts, at 10-GHz frequency, with three theoretical patterns, chosen with constant aperture phase, quadratic aperture phase, and partial quadratic aperture phase. The quadratic aperture phase was modeled as follows:

$$q_p^m = \frac{1}{2}\left[\frac{f_{op} - f_L}{f_H - f_L}\right] q_p \tag{9.14}$$

where q_p^m is the modified aperture phase and $f_L = 7$ GHz and $f_H = 12.4$ GHz are the lower and upper frequencies of our measurements. f_{op} is the operational frequency and the quadratic aperture phase $q_p = \exp(j\pi dy^2/R_h\lambda_h)$.

Table 9.1 Theoretical Results of the Sidelobe Level and HPBW for Two Horns, One with Channel Inserts and Another with a Similar Geometry but without Inserts

	$f = 7$ GHz	$f = 10$ GHz	$f = 12.4$ GHz
With channel inserts	HPBW = 16°	HPBW = 13°	HPBW = 12°
	SLL = −42.7 dB at 51°	SLL = −43.1 dB at 45°	SLL = −37 dB at 32°
	Shoulder = −25 dB at 23°	Shoulder = −31 dB at 29°	No shoulder
Without channel inserts	HPBW = 10°	HPBW = 10°	HPBW = 21°
	SLL= −11.7 dB at 16°	SLL = −10.6 dB at 19°	SLL= −10.7 dB at 22°
	No shoulder	Shoulder = −3.6 dB at 8°	Shoulder = −5.1 dB at 14°
SLL relative reduction	31 dB	32.5 dB	26.3 dB

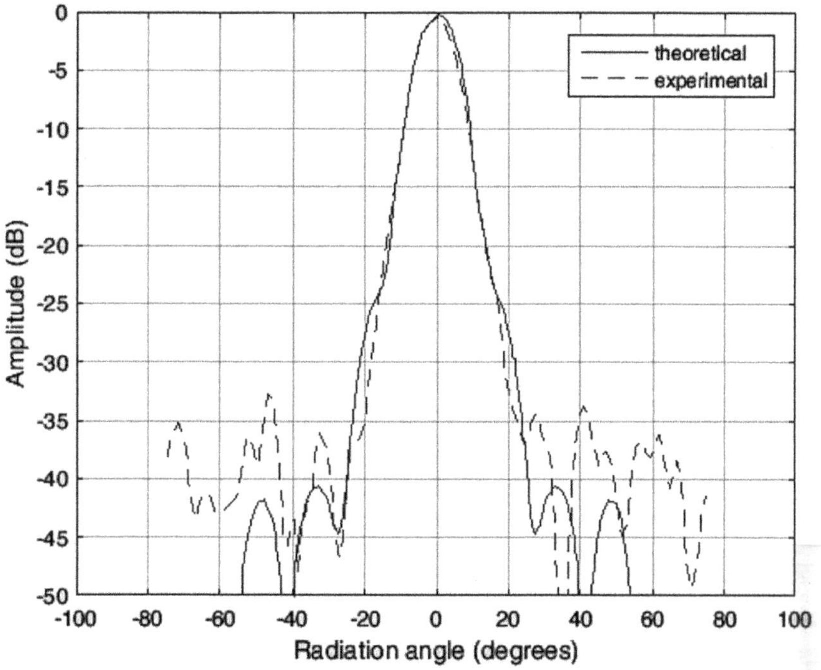

Figure 9.17 Radiation patterns of an E-sectoral horn with corner inserts with $A/B = 9.17$, $a/A = 0.15$, $b/B = 0.05$, and semi-flare angle 18° at $f_{op} = 10$ GHz. Solid line, theoretical with fraction quadratic aperture phase. Dashed line, experimental. (Adapted from: [2].)

In general, and as the theory predicts, an effective reduction in sidelobe level has been achieved by tapering the field at the edge of the aperture. Consequently, the beamwidth is found to be broader both theoretically and experimentally. Because the addition of corner inserts distorts the aperture field lines, one might expect the cross-polarization behavior of the horn to deteriorate, even though the net E_x-component is 0. The experimental measurement of cross-polarization for the case $\theta = 0$, at 10 GHz, reveals a deterioration from −34.9 dB up to −24.5 dB.

The radiation pattern in the H-plane of each horn was not measured experimentally but is expected to be broad, similar to that obtained for open-ended WG16 waveguide.

Problems for Students

9.1 An E-sectoral horn antenna is designed with a rectangular aperture of dimensions $a = 10$ cm and $b = 8$ cm. The aperture field is modified using a tapered distribution given by $E(x, y) = E_0(1 - x^2/a^2)(1 - y^2/b^2)$.
   (a) Plot the aperture field distribution over the aperture.
   (b) Calculate the effective aperture area.
   (c) Discuss how this tapering affects the sidelobe levels and beamwidth.

9.2 A sectoral horn antenna is designed with a conventional aperture field.
   (a) Explain the physical principle behind sidelobe suppression through aperture field modifications.

(b) If a cosine taper $E(x, y) = E_0\cos(\pi x/a)\cos(\pi y/b)$ is applied, derive the expression for the far-field radiation pattern.

9.3 Using the finite-difference formulation of the wave equation for an E-sectoral horn antenna, simulate the field distribution in the horn channel. Assume the following parameters: frequency = 10 GHz, horn channel length = 20 cm, and aperture dimensions a = 8 cm, b = 6 cm.
(a) Write the finite-difference equations for the wave equation.
(b) Solve the field distribution numerically using MATLAB or Python.

9.4 A channel-fitted horn antenna has its aperture field calculated based on the following distribution:

$$E(x,y) = E_0\exp\left(-\alpha x^2 - \beta y^2\right)$$

where $\alpha = 1/a^2$, $\beta = 1/b^2$. Assume that a = 12 cm, b = 10 cm.
(a) Calculate the directivity of the horn using this aperture field distribution.
(b) Discuss how the field variation parameters α and β influence sidelobe suppression.

9.5 For an E-sectoral horn with the parameters of aperture width a = 15 cm, b = 10 cm, and operating frequency = 12 GHz:
(a) Compute the far-field radiation pattern for a uniform aperture distribution.
(b) Compute the far-field radiation pattern for a cosine-apodized aperture distribution.
(c) Compare the two patterns in terms of sidelobe levels and beamwidth.

9.6 A horn antenna designer aims to achieve sidelobe suppression to below −25 dB using a tailored aperture field distribution.
(a) Propose a mathematical model for the aperture field distribution to achieve this goal.
(b) Using this model, estimate the beamwidth and directivity of the antenna.
(c) Discuss the trade-offs between sidelobe suppression and main lobe efficiency.

9.7 An E-sectoral horn antenna has an aperture with the following dimensions: width a = 20 cm, b = 15 cm. The aperture field is given by a Gaussian distribution:

$$E(x,y) = E_0\exp\left(-\frac{x^2}{2\sigma_x^2} - \frac{y^2}{2\sigma_y^2}\right)$$

where $\sigma_x = a/3$ and $\sigma_y = b/3$.
(a) Derive an expression for the power radiated through the aperture in terms of E_0, σ_x, and σ_y.
(b) Determine the effective aperture area and the antenna efficiency.
(c) Compare the efficiency of this Gaussian aperture distribution with that of a uniform aperture distribution of the same dimensions.
(d) Discuss the implications of Gaussian tapering on the sidelobe levels and beamwidth.

Problems for Instructors

9.1 Aperture Field Distribution Modification: Explain the role of aperture field distribution in sidelobe suppression for E-sectoral horn antennas. For an E-sectoral horn with a width of 0.4m operating at 8 GHz, modify the aperture field distribution to achieve a –20-dB sidelobe level. Describe the distribution you used and discuss how it affects the main lobe and sidelobe characteristics of the radiation pattern.

9.2 Designing with Computational Approaches: Using the computational approaches outlined in Section 9.2, design an E-sectoral horn antenna that achieves a sidelobe level of –25 dB at 10 GHz. Specify the key parameters, such as aperture size and flare angle, and explain the computational methods used to determine them. Discuss the effectiveness of these methods compared to traditional design techniques.

9.3 Finite-Difference Method for Wave Equation: Apply the finite-difference method to solve the wave equation for an E-sectoral horn antenna with an aperture of 0.3m at a frequency of 12 GHz. Describe the boundary conditions and grid discretization used, and plot the resulting electric field distribution at the aperture. Explain how this method aids in achieving optimal sidelobe suppression.

9.4 Aperture Field Calculation of a Channel-Fitted Sectoral Horn: Using Section 9.4, calculate the aperture field distribution for a channel-fitted E-sectoral horn with a tapered aperture profile. Assume a frequency of 9 GHz and an aperture height of 0.35m. Based on the field distribution, estimate the main lobe gain and discuss the effect of the channel fitting on sidelobe suppression and beamwidth.

9.5 Radiation Pattern Analysis of Channel-Fitted Horn: For an E-sectoral horn antenna designed with a fitted channel, calculate the far-field radiation pattern at 14 GHz. Plot the pattern, highlighting the main lobe, first sidelobe, and nulls. Compare the results with those of a conventional E-sectoral horn (without channel fitting) and analyze how channel fitting improves sidelobe suppression.

9.6 Numerical Optimization for Sidelobe Suppression: Design a numerical optimization algorithm that minimizes the sidelobe levels of an E-sectoral horn antenna at 10 GHz by adjusting the aperture tapering and flare angle. Describe the algorithm's steps and test it on an E-sectoral horn with an initial sidelobe level of –18 dB. Report the final sidelobe level achieved and discuss the trade-offs between main lobe gain and sidelobe suppression.

9.7 MATLAB Simulation of Aperture Field Distribution: Using MATLAB, simulate the aperture field distribution for an E-sectoral horn antenna with a width of 0.45m at 11 GHz. Apply a field tapering technique to achieve a sidelobe suppression of –22 dB. The simulation should:

(a) Display the aperture field distribution along the horn's width.

(b) Plot the far-field radiation pattern.

(c) Compare the simulated sidelobe level with theoretical values. Discuss the impact of aperture field tapering on the overall radiation pattern and sidelobe suppression.

References

[1] Ata, O. W., T. M. Benson, and A. Marincic, "An E-Plane Sectoral Horn Fitted with Side-Lobe Suppression Devices," *International Journal of Numerical Modelling: Electronic Networks, Devices and Fields*, Vol. 6, No. 2, May 1993, pp. 145–159.

[2] Ata, O. W., "Application of Ray Theory into Microwave Horn Antenna Design," PhD thesis, Cardiff University, UK, 1989.

Selected Bibliography

Benson, T. M., et al., "New Results for Rib Waveguide Propagation Constants," *IEE Proceedings Journal*, Vol. 136, 1989, pp. 97–102.

Binns, K. J., and P. J. Lawrenson, *Analysis and Computation of Electric and Magnetic Field Problems*, Oxford, UK: Pergamon Press, 1973.

Collins, J. H., and D. Daly, "Calculations for Guided Electromagnetic Waves Using Finite-Difference Methods," *J. Electronics and Control*, Vol. 14, 1963, pp. 361–380.

Davies, J. B., and C. A. Muilwyk, "Numerical Solution of Uniform Hollow Waveguides with Boundaries of Arbitrary Shape," *Proc. IEE*, Vol. 133, 1966, pp. 277–284.

Meinke, H. H., K. P. Lange, and J. F. Ruger, "TE- and TM-Waves in Waveguides of Very General Cross-Section," *Proc. IEEE*, Vol. 51, 1963, pp. 1436–1443.

Sander, K. F., and G. A. L. Reed, *Transmission and Propagation of Electromagnetic Waves*, Cambridge, U.K.: Cambridge University Press, 1978.

Yee, H. Y., and N. F. Audeh, "Uniform Waveguide with Arbitrary Cross-Section Considered by the Point-Matching Method," *IEEE Transactions on Microwave Theory and Techniques*, Vol. 13, 1965, pp. 847–851.

CHAPTER 10

Application of Ray Theory into the Design of a Short-Length, Phase-Corrected Horn Antenna

In the theoretical development of the channel fitted E-sectoral horn, the finite-difference method, based on Maxwell's equations, has been adopted for solving the relevant electromagnetic problem. Basically, the aperture and channel ratios should provide the necessary tapering controls in the E-plane transverse electric field, which would result in the required amplitude distribution across the double-regional aperture, for achieving low sidelobe levels in the radiation pattern. The channel-section type, whether continuously tapered or double-slope tapered, could also be used for adjusting the measured radiation pattern, and the final shape of the horn for the lowest sidelobe level would have to be determined experimentally in the broad frequency range.

In this chapter, an optical ray-tracing technique, based on Snell's law of refraction, is used to develop the theory and design of a microwave horn of short physical length. The short focus horn produces a plane phase front, as the wave rays emerge parallel to the central axis, from the aperture plane. The design, construction, and evaluation of such phase corrected primary feed horns will be described in a fair amount of detail.

10.1 Theory and Design

It is known that a graded refractive-index medium of a hyperbolic secant type allows all meridional rays to travel with periodic focusing. Meridional rays are those rays confined to a surface for which one coordinate of the system has a constant value. Planes that contain the axis of symmetry in a cylindrical system, for example, are meridional. At a distance of a quarter period length apart from the focal point, all meridional rays become parallel to the stratification of the medium that is parallel with the axis of propagation.

10.1.1 The Ray Trajectory Equation

Consider a medium with a refractive index medium of the form:

$$\eta(y) = \eta_{co}\,\text{sech}\left(\frac{\pi y}{2L}\right) = \frac{\eta_{co}}{\cosh\left(\frac{\pi y}{2L}\right)} \tag{10.1}$$

where η_{co} is the refractive index along the central axis, $2L$ is one-half the focus index period, and L is the horn length.

The restriction of the refractive index variation to the y-dimension only would reasonably simplify the scale of the mathematics in the present problem. Hence, an expression can be found for the trajectory, in the stratified medium, taken by any ray leaving a point axial source 0 at an angle θ with respect to the z-axis, perpendicular to the horn aperture plane, shown in Figure 10.1(a). Figure 10.1(b) demonstrates the refractive index profile of the horn.

The derivation of the ray trajectory equation starts by defining an infinitesimal elemental distance, ds, along the ray as:

$$ds^2 = dy^2 + dz^2 \tag{10.2}$$

Along the curve, Snell's law of refraction holds and hence:

$$\eta_{co}\cos\theta = \eta(y)\cos\theta_y = h_\theta \tag{10.3}$$

where h_θ is the ray constant, referred to the angle of emergence θ and

$$\cos\theta_y = \frac{dz}{ds} \tag{10.4}$$

hence

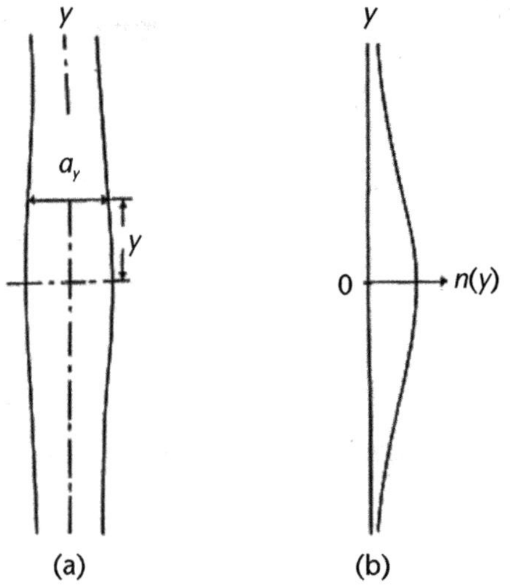

Figure 10.1 Horn aperture with variable refractive index profile. (a) Variable width cross-section. (b) Refractive index profile.

$$\eta(y)\frac{dz}{ds} = h_\theta \tag{10.5}$$

Pending modification of (10.5):

$$ds^2 = \frac{\eta(y)^2\, dz^2}{h_\theta^2} \tag{10.6}$$

Substitution for ds^2 from (10.2) into (10.6) yields:

$$dz = \frac{h_\theta \cdot dy}{\sqrt{\eta^2(y) - h_\theta^2}} \tag{10.7}$$

Integrating both sides:

$$z = \int_0^y \frac{h_\theta \cdot dy}{\sqrt{\eta^2(y) - h_\theta^2}} \tag{10.8}$$

The integral of the above equation can be solved by substituting for $\eta(y)$ from (10.1). Hence, let

$$u = \sinh\left(\frac{\pi y}{2L}\right) \tag{10.9}$$

then

$$du = \frac{\pi}{2L}\cosh\left(\frac{\pi y}{2L}\right)dy \tag{10.10}$$

The substitution for dy from (10.10) into (10.8) gives:

$$\begin{aligned} z &= \frac{2L}{\pi}\int_0^u \frac{h_\theta \cdot \sqrt{1+\sinh^2\left(\frac{\pi y}{2L}\right)}}{\cosh\left(\frac{\pi y}{2L}\right)\sqrt{\eta_{co}^2 - h_\theta^2 - h_\theta^2\sinh^2\left(\frac{\pi y}{2L}\right)}}du \\ &= \frac{2L}{\pi}\int_0^u \frac{du}{\sqrt{\left[\left(\eta_{co}^2 - h_\theta^2\right)/h_\theta^2\right] - u^2}}du \end{aligned} \tag{10.11}$$

Let

$$a = \sqrt{\frac{\left(\eta_{co}{}^2 - h_\theta^2\right)}{h_\theta^2}} \tag{10.12}$$

then

$$z = \frac{2L}{\pi}\int_0^u \frac{du}{\sqrt{a^2 - u^2}}du = \frac{2L}{\pi}\sin^{-1}\left(\frac{u}{a}\right) \tag{10.13}$$

where $|u| < a$.

The substitution for h_θ in (10.12) from (10.3) gives:

$$a = \sqrt{\frac{(1-\cos^2\theta)}{\cos^2\theta}} = \tan\theta \tag{10.14}$$

Consequently, upon substitution for u and a in (10.13) from (10.9) and (10.14), respectively, the following ray trajectory equation results as:

$$\sinh\left(\frac{\pi y}{2L}\right) = \tan\theta \cdot \sin\left(\frac{\pi z}{2L}\right) \tag{10.15}$$

The ray paths, which obviously obey (10.15), are periodic with respect to z with a distance period of $4L$. At $z = L$, y takes a maximum value, irrespective of θ, and for this z all rays travel parallel to the z-axis as shown in Figure 10.1(b).

10.1.2 The Optical Path Equation

The optical path length of a ray traveling in a refractive index medium can be thought of as the equivalent distance that it would have traveled with a relatively different velocity in free space.

Accordingly, the following expression holds

$$dP = \eta ds = \frac{c}{v}ds = cdt \tag{10.16}$$

where c is the velocity that a ray travels in free space, v is the velocity that a ray travels in the refractive medium, and η is the refractive index of the medium.

Integrating the first two sides of (10.16) and substituting (10.1) and (10.6):

$$P = \int \eta ds = \frac{\eta_{co}^2}{h_\theta}\int_0^z \operatorname{sech}^2\left(\frac{\pi y}{2L}\right)dz = \frac{L\eta_{co}}{\cos\theta}\int_0^{z/L} \frac{d(z/L)}{1+\sinh^2\left(\frac{\pi y}{2L}\right)} \tag{10.17}$$

Substituting (10.15) into (10.17) gives:

$$P = \frac{L\eta_{co}}{\cos\theta}\int_0^{z/L} \frac{d(z/L)}{1+\tan^2\theta\sin^2\left(\frac{\pi z}{2L}\right)} \tag{10.18}$$

A phase front consists of the end points of the ray curves (surface really) that are of equal optical distance from the origin. These end points related to a certain optical path length can be worked out by employing a computer library routine that solves the integral, at any required angle θ, numerically.

Alternatively, an analytical expression can be found using the solution of the following standard integral:

$$\int \frac{dx}{a + b\sin^2 x} = \frac{sign\ a}{\sqrt{a(a+b)}} \tan^{-1}\left(\sqrt{\frac{a+b}{a}} \tan x\right) \tag{10.19}$$

where a, b are constant and $b/a > -1$.

Let $x = \pi z/2L$ in (10.18); hence:

$$\begin{aligned} P &= \frac{2L\eta_{co}}{\pi\cos\theta} \int_0^x \frac{dx}{1 + \tan^2\theta\sin^2 x} \\ &= \frac{2L\eta_{co}}{\pi\cos\theta}\left[\frac{1}{\sqrt{1+\tan^2\theta}} \tan^{-1}\left(\sqrt{1+\tan^2\theta}\tan x\right)\right] \end{aligned} \tag{10.20}$$

Since $1 + \tan^2\theta = \sec^2\theta$, (10.20) is modified into:

$$P = \frac{2L\eta_{co}}{\pi} \tan^{-1}(\sec\theta\tan x) \tag{10.21}$$

$$\tan\left(\frac{\pi P}{2L\eta_{co}}\right) = \sec\theta\tan x = C \tag{10.22}$$

and

$$C^2 = \sec^2\theta\tan^2 x = \left(1 + \tan^2\theta\right) \cdot \frac{\sin^2 x}{\cos^2 x} \tag{10.23}$$

Substituting for $\tan\theta$ from (10.15) gives:

$$C^2 = \tan^2\left(\frac{\pi z}{2L}\right) + \frac{\sinh^2\left(\frac{\pi y}{2L}\right)}{\cos^2\left(\frac{\pi z}{2L}\right)} \tag{10.24}$$

It follows that:

$$C^2\cos^2\left(\frac{\pi z}{2L}\right) = \sin^2\left(\frac{\pi z}{2L}\right) + \sinh^2\left(\frac{\pi y}{2L}\right) = \cosh^2\left(\frac{\pi y}{2L}\right) - \cos^2\left(\frac{\pi z}{2L}\right) \tag{10.25}$$

therefore

$$\cos\left(\frac{\pi z}{2L}\right) = \rho \cdot \cosh\left(\frac{\pi y}{2L}\right) \tag{10.26}$$

where

$$\rho = \sqrt{\frac{1}{C^2+1}} = \sqrt{\frac{1}{\tan^2\left(\frac{\pi P}{2L\eta_{co}}\right)+1}} \qquad (10.27)$$

Figure 10.1(b) shows phase fronts at equal optical path lengths. At $z = L$, all rays are parallel to the z-axis and all optical paths from 0 are equal at this z. Hence, it is evident that the plane $z = L$ is an equiphase surface.

10.1.3 The Design Equation

A graded refractive index profile with the form of (10.1) is obtained by varying the plate separation a_y of a planar waveguide supporting a TE_{01} mode as shown in Figure 10.1(a).

By employing the dispersion relation for a TE_{01} mode, an effective refractive index $\eta(y)$ can be defined through:

$$\beta^2 = \omega^2\mu_0\varepsilon_0\varepsilon_{eff}(y) = \omega^2\mu_0\varepsilon_0 - \left(\frac{\pi}{a_y}\right)^2 \qquad (10.28)$$

where β is the propagation wave number and $\varepsilon_{eff}(y)$ is the variable effective permittivity,

Since $c = 1/\sqrt{\mu_0\varepsilon_0}$ and $\eta(y) = \sqrt{\varepsilon_{eff}(y)}$, then

$$\eta(y) = \eta_{co}\text{sech}\left(\frac{\pi y}{2L}\right) = \left[1-\left(\frac{\pi c}{a_y\omega_0}\right)^2\right]^{1/2} \qquad (10.29)$$

It is necessary to calculate η_{co} in (10.28) by selecting a_y at $y = 0$, the horn axial length, L, and the design frequency. These selected values are provided in the discussion of the horn construction in Section 10.2.

10.1.4 The Ray Trajectory Equation with Modified Refractive Index Variation

For angular frequencies (ω_1) above and below the design frequency (ω_0), the fixed plate separation give rise to a modified refractive index variation which can be expressed analytically.

Equation (10.29) can be employed to find η_{co} at a_{y0} when $y = 0$ and $\omega = \omega_0$ as:

$$\eta_{co}^2 = 1-\left(\frac{\pi c}{a_{y0}\omega_0}\right)^2 \qquad (10.30)$$

Substitution into (10.1) gives:

$$\left(\frac{\pi c}{a_y \omega_0}\right)^2 = 1 - \frac{1-\left(\dfrac{\pi c}{a_{y0}\omega_0}\right)^2}{\cosh^2\left(\dfrac{\pi y}{2L}\right)} \tag{10.31}$$

At the angular frequency ω_1,

$$\left(\frac{\pi c}{a_y \omega_1}\right)^2 = 1 - \overline{\eta}^2(y) \tag{10.32}$$

where $\overline{\eta}(y)$ represents the modified refractive index function.

Dividing (10.31) by (10.32) gives, after some modification:

$$\overline{\eta}^2(y) = p + q\,\mathrm{sech}^2\left(\frac{\pi y}{2L}\right)$$

where

$$p = 1 - \left(\frac{\omega_0}{\omega_1}\right)^2$$
$$q = \left(\frac{\omega_0}{\omega_1}\right)^2 \left\{1 - \left(\frac{\pi c}{a_{y0}\omega_1}\right)^2\right\} \tag{10.33}$$

Furthermore,

$$p + q = 1 - \left(\frac{\pi c}{a_{y0}\omega_1}\right)^2 = \overline{\eta}_{co}^2 \tag{10.34}$$

and if $\omega_1 = \omega_0$, then $p = 0$ and $q = \eta_{co}^2$. Figure 10.2 shows the variation of $\overline{\eta}^2(y)$ with y/L for 9.5, 10, and 10.5 GHz, as a result of the following MATLAB program.

MATLAB Program 13

```
% Matlab program to calculate modified refractive index variation
with y/L
f1= input('enter the value of angular frequency of operation w1
in GHz: ')
w1= 2*pi*f1*10^9;

% w0 is the design angular frequency (10 GHz)
w0 = 2*pi*10*10^9;
```

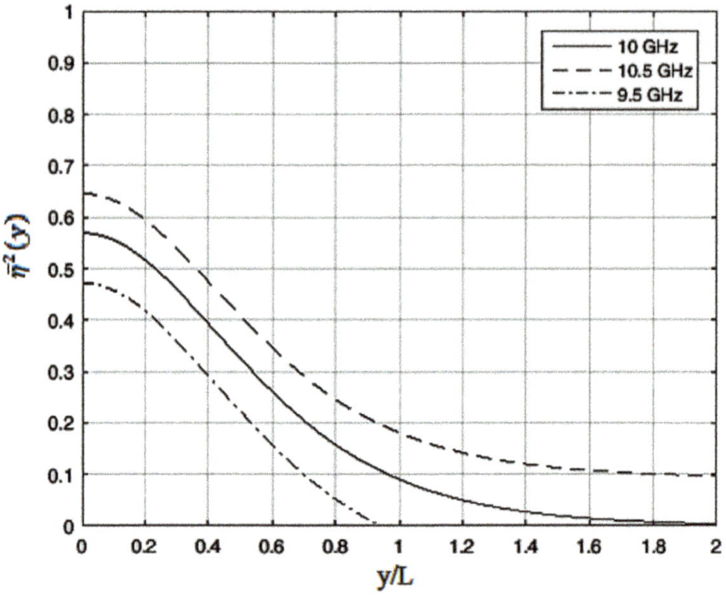

Figure 10.2 Variation of the modified refractive index versus normalized height *y/L* of the short length phase corrected horn.

```
% c is speed of electromagnetic signal in free space (m/s)
c= 3*10^8;
% ay is the width of a WG16 horn feed (m)
ay= 2.286*10^-2;

p= 1- (w0/w1)^2;
q= (w0/w1)^2*(1-(pi*c/(ay*w1))^2);

% yl is equivalent to the normalized aperture height y/L
% ny is the modified refractive index of the horn

for i= 1:101
yl(i) = (i-1)*0.02;
ny(i)= p+q*(sech(pi/2*yl(i)))^2;
end
plot(yl, ny, 'k-')
axis([0 2 0 1])
grid
```

In order to derive an expression for the new ray trajectory, substitution for $\overline{\eta}^2(y)$ from (10.33) is employed in (10.8) as:

$$z = \int_0^y \frac{\overline{h}_\theta \cdot dy}{\sqrt{p + q\,\mathrm{sech}^2\left(\dfrac{\pi y}{2L}\right) - \overline{h}_\theta^2}}$$

where

$$\bar{h}_\theta = \bar{\eta}_{co}\cos\theta = \sqrt{p+q}\cos\theta \tag{10.35}$$

Using the same mechanism as used for solving (10.8), the integral in (10.34) takes a standard form of a known solution as:

$$z = \frac{2L}{\pi}\sqrt{\frac{\bar{h}_\theta^2}{\bar{h}_\theta^2 - p}}\int_0^u \frac{du}{\sqrt{a^2-u^2}} = \frac{2L}{\pi}\sqrt{\frac{\bar{h}_\theta^2}{\bar{h}_\theta^2 - p}}\sin^{-1}\left(\frac{u}{a}\right)$$

where

$$u = \sinh\left(\frac{\pi y}{2L}\right) \quad \text{and} \quad \sqrt{\frac{q+p-\bar{h}_\theta^2}{\bar{h}_\theta^2 - p}} \tag{10.36}$$

Hence,

$$\sinh\left(\frac{\pi y}{2L}\right) = \sqrt{\frac{q+p-\bar{h}_\theta^2}{\bar{h}_\theta^2 - p}}\sin\left(\frac{\pi z}{2L}\sqrt{\frac{\bar{h}_\theta^2 - p}{\bar{h}_\theta^2}}\right) \tag{10.37}$$

Subsequently, the ray trajectory equation with modified refractive index variation is given as:

$$\sinh\left(\frac{\pi y}{2L}\right) = \frac{\sin\theta}{\sqrt{\cos^2\theta - A}}\sin\left(\frac{\pi z}{2L}\frac{\sqrt{\cos^2\theta - A}}{\cos\theta}\right)$$

where

$$A = \frac{p}{(p+a)} \tag{10.38}$$

The case might arise when $\cos^2\theta - A \leq 0$, for the higher portion of θ angles. That could only happen, with reference to (10.33), at $\omega_1 > \omega_0$. Equation (10.38) is then modified to:

$$\sinh\left(\frac{\pi y}{2L}\right) = \frac{\sin\theta}{\sqrt{A - \cos^2\theta}}\sinh\left(\frac{\pi z}{2L}\frac{\sqrt{A - \cos^2\theta}}{\cos\theta}\right) \tag{10.39}$$

10.1.5 The Optical Path Equation with Modified Refractive Index Variation

An optical-path expression for a ray traveling in a modified refractive index region can be found using (10.16) and substituting for ds from (10.6). The two sides of the equation can then be integrated as follows:

$$P = \int \bar{\eta}(y)\,ds = \frac{L}{h_\theta}\int_0^{z/L} \bar{\eta}^2(y)\,d\left(\frac{z}{L}\right) \tag{10.40}$$

Consequently,

$$\begin{aligned} P &= \int_0^{z/L} \frac{L}{\left(\sqrt{p+q}\right)\cos\theta}\left[p + q\,\text{sech}^2\left(\frac{\pi y}{2L}\right)\right] d\left(\frac{z}{L}\right) \\ &= \frac{Lp}{\left(\sqrt{p+q}\right)\cos\theta}\cdot\frac{z}{L} + \frac{Lq}{\left(\sqrt{p+q}\right)\cos\theta}\int_0^{z/L}\frac{d(z/L)}{1+B^2\sin^2\left(\frac{\pi z}{2L}D\right)} \end{aligned}$$

where

$$\begin{aligned} B &= \frac{\sin\theta}{\sqrt{\cos^2\theta - A}} \\ D &= \frac{\sqrt{\cos^2\theta - A}}{\cos\theta} \end{aligned} \tag{10.41}$$

It is noticed that at the design frequency, f_0, $p = 0$, and $q = \eta^2_{co}$. Hence, $A = 0$, $B = \tan\theta$, and (10.41) reduces back to (10.18).

New ray trajectories and curves of equal optical length are determined by submitting the relevant expression to numerical computation. Equiphase surfaces and predicted trajectories for frequencies at 9.5, 10, and 10.5 GHz are shown in Figures 10.3, 10.4, and 10.5, respectively. It can be seen that at the lower frequency 9.5 GHz, the horn overcompensates with respect to phase correction at $z/L = 1$, whereas at the higher frequency 10.5 GHz, it undercompensates. As expected, at the design frequency, an equiphase front is observed at $z/L = 1$.

The following MATLAB program has been developed towards plotting the mentioned figures. The program is divided into three parts and can be utilized at whatever operational frequency 9.5 to 10.5 GHz. The first part calculates a matrix that stores all z/L ranges at a fixed set of optical path values. The second part calculates y/L for each z/L of a particular range and then plots those phase fronts. The third part calculates the modified refractive index trajectories and plots them too.

MATLAB Program 14

```
% Matlab program to compute trajectories and phase fronts in a
modified refractive index horn region.
% Input freq 9.5 GHz to test for over-compensated phase front.
Input freq= 10.5 to test for under-compensated phase front.

% Compute the z/L range matrix associated with the phase fronts

% Constants and parameters
% omeg_0 is the angular design frequency
```

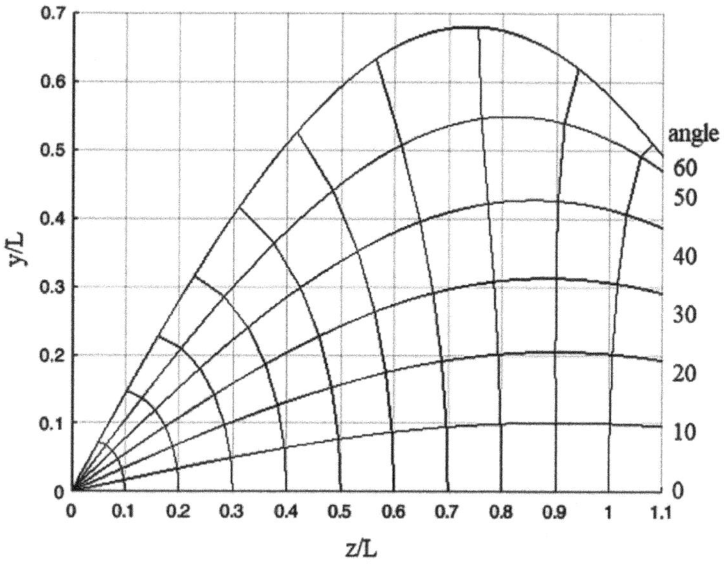

Figure 10.3 Modified ray trajectories at various angles and phase fronts calculated at a constant optical path length at 9.5-GHz operational frequency (an overcompensated phase front can be observed at $z/L = 1$).

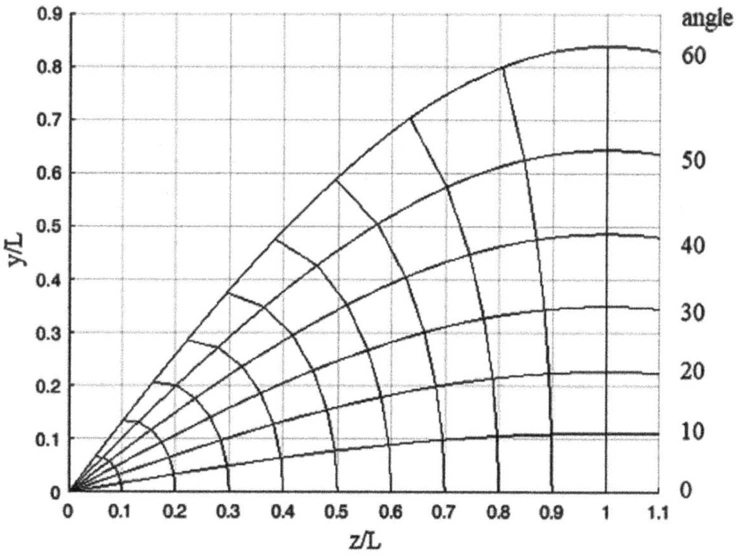

Figure 10.4 Ray trajectories at various angles and phase fronts calculated at a constant optical path length at the design frequency 10 GHz.

```
omeg_0 = 2*pi*10*10^9;
freq = input('Enter value for operational frequency in GHz: ');
omeg_1 = 2*pi*freq*10^9;
c = 3*10^8;
ay = 2.286*10^-2;
```

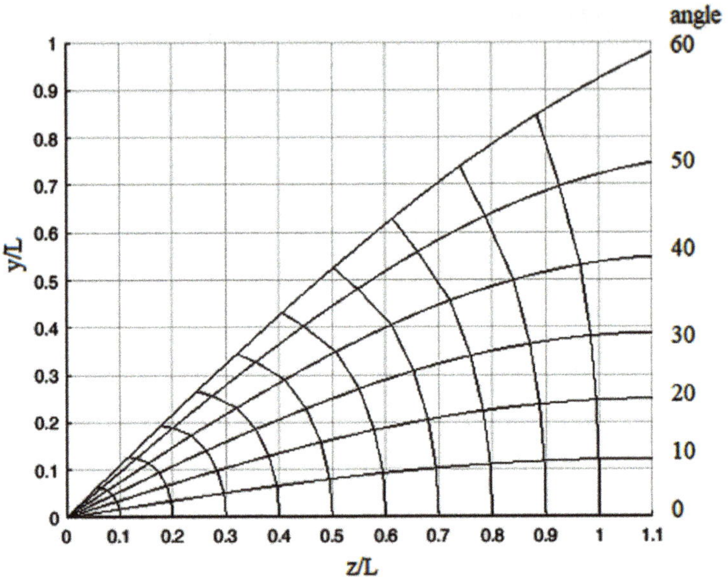

Figure 10.5 Modified ray trajectories at various angles and phase fronts calculated at constant optical path length at 10.5 GHz operational frequency (undercompensated phase front is observed at $z/L = 1$).

```
L = 5.72*10^-2;
p = 1 - ((omeg_0)/(omeg_1))^2;
q = ((omeg_0)/(omeg_1))^2*(1-((pi*c)/(ay*omeg_0))^2);
A = p./(p+q);

% Define the fixed value of OP
OP_values = [1, 0.9, 0.8, 0.7, 0.6, 0.5, 0.4, 0.3, 0.2, 0.1];
Z1 = zeros(numel(OP_values), 7);

for j = 1:numel(OP_values)
  OP = L * sqrt(p + q) * OP_values(j);

  for i = 1:7
    thet = 10*(i-1);
    thet_rad = thet * pi / 180; % angle in radians
    B = sin(thet_rad)/sqrt((cos(thet_rad))^2 - A);
    D = sqrt((cos(thet_rad))^2 - A)/cos(thet_rad);

    % Define the integrand function
    integrand = @(z1) 1 ./ (1 + B^2 * (sin((pi*D/2)*z1)).^2);

    % Define the search range for z1
    z1_range = 0:0.0001:1.1; % double floating range for z1

    % Initialize variables to store the closest value of z1 and
the corresponding OP
```

```
    closest_zl = [];
    closest_OP = inf;
    % Perform the search
    for zl = zl_range
      % Evaluate the integral
      integral_value = integral(integrand, 0, zl);
      % Calculate the value of OP for this zl
      current_OP = (L*p)/((sqrt(p+q)*cos(thet_rad)))*zl+(L*q)/
((sqrt(p+q)*cos(thet_rad)))*integral_value;
      % Check if this OP is closer to the fixed value
      if abs(current_OP - OP) < abs(closest_OP - OP)
        closest_zl = zl;
        closest_OP = current_OP;
      end
    end
    Z1(j, i) = closest_zl;
  end
  %disp([Z1(j, :)' repmat(OP_values(j), 7, 1)]);
end
zl_matrix= Z1;
if omeg_1 > omeg_0
zl_matrix(:,7) = 0.5*(zl_matrix(:,7) +zl_matrix(:,6));
end
zl_matrix

% Calculate y/L associated with z/L values in the matrix and plot
the phase fronts

% Define phi range in degrees
phi_deg = [0 10 20 30 40 50 60];
phi_rad = deg2rad(phi_deg);

% Initialize arrays to store yl values
yl = zeros(length(phi_deg), size(zl_matrix, 2));
yl1 = zeros(size(zl_matrix, 1), size(yl, 2));

for j = 1:size(zl_matrix, 1)

zl = zl_matrix(j, :);

  % Calculate yl for each phi and zl row of the above matrix
  for i = 1:length(phi_deg)
    p = 1 - ((omeg_0)/(omeg_1)).^2;
    q = ((omeg_0)/(omeg_1)).^2*(1-((pi*c)/(ay*omeg_0)).^2);
    A = p./(p+q);

    yl(i, :) = 2/pi * asinh(sin(phi_rad(i))./sqrt(abs(cos(phi_
rad(i)).^2 - A))) .* ...
      sin(pi/2 * zl .* sqrt(abs(cos(phi_rad(i)).^2 - A))./
cos(phi_rad(i)));
```

```
    end
    % Store only the desired values from yl
    yl1(j,:) = diag(yl);

    % Plot yl versus zl for each phi
    figure(1)
    hold on
    set(gca, 'xlim', [0 1.1]);
    set(gca, 'xtick', 0:0.1:1.1);
    plot(zl, yl1(j,:), 'k-')
  end

  % Calculate the ray trajectories of the modified refractive index
  region and plot the curves.
  figure(1)
  hold on
  % Angle range
  phi_deg = 0:10:60; % degrees
  phi_rad = deg2rad(phi_deg);

  % zl range
  zl = 0:0.001:1.1;

  % Initialize arrays to store results
  yl = zeros(length(phi_rad), length(zl));

  % Compute yl for each angle and zl
  for i = 1:length(phi_rad)
    p = 1 - ((omeg_0)./(omeg_1)).^2;
    q = ((omeg_0)/(omeg_1)).^2*(1-((pi*c)./(ay*omeg_0)).^2);
    A = p./(p+q);

    yl(i, :) = 2/pi.* asinh((sin(phi_rad(i)))./sqrt(abs(cos(phi_
  rad(i)).^2-A))) .* sin(pi/2.*zl.*(sqrt(abs(cos(phi_
  rad(i)).^2-A))./cos(phi_rad(i))));
  end
  hold on
  % Plot yl vs zl for each angle
  for i = 1:length(phi_rad)
    plot(zl, yl(i, :), 'k-')
  end
  grid
```

10.1.6 Calculation of the Aperture Amplitude Distribution at the Design Frequency

It has already been shown that a ray leaving a point source, at an angle θ with the z-axis of propagation, in a hyperbolic refractive index medium, reaches its maximum point y when $z = L$ (the aperture plane), at the design frequency.

At $z = L$, (10.15) becomes:

$$\tan\theta = \sinh\left(\frac{\pi y_L}{2L}\right) \tag{10.42}$$

The power radiated by the feed between the angle θ and $\theta + d\theta$ would therefore leave the aperture between y and $y + dy$.

Differentiating (10.42) yields:

$$\frac{d\theta}{dy} = \frac{\pi}{2L}\frac{\cos^2\theta}{\operatorname{sech}\left(\frac{\pi y_L}{2L}\right)} \tag{10.43}$$

Employing (10.3) gives:

$$\cos\theta = \frac{\eta(y_L)}{\eta_{co}} = \operatorname{sech}\left(\frac{\pi y_L}{2L}\right) \tag{10.44}$$

Hence,

$$\frac{d\theta}{dy} = \frac{\pi}{2L}\operatorname{sech}\left(\frac{\pi y_L}{2L}\right) \tag{10.45}$$

which is independent of the angle θ.

If $P(\theta)$ is the polar power distribution that the primary feed radiates, then the power distribution $P(y)$ across the aperture is given as:

$$P(y) = P(\theta)\frac{d\theta}{dy} \tag{10.46}$$

A further modification is required in the above expression, to account for the reflections occurring at the aperture. While the refractive index across the aperture is a function of position, it is unity outside the same aperture. The power transmission coefficient $4\eta(y)/(1 + \eta(y))^2$, derived in the Appendix, must be included and the actual power distribution in the aperture becomes:

$$P(y) = P(\theta)\frac{d\theta}{dy}\frac{4\eta(y_L)}{(1+\eta(y_L))^2} \tag{10.47}$$

If the polar distribution, radiated by the primary feed, is assumed to be of the $\cos\theta$ type, then

$$P(\theta) = \cos^2\theta = \operatorname{sech}^2\left(\frac{\pi y_L}{2L}\right) \tag{10.48}$$

and upon substituting for $\eta(y_L)$ from (10.44) in (10.47)

$$P(y) = \frac{2\pi}{L}\eta_{co}\frac{\text{sech}^4\left(\frac{\pi y_L}{2L}\right)}{\left[1+\eta_{co}\,\text{sech}\left(\frac{\pi y_L}{2L}\right)\right]^2} \tag{10.49}$$

Hence, the amplitude distribution $A(y)$ takes the form:

$$A(y) = \frac{\text{sech}^2\left(\frac{\pi y_L}{2L}\right)}{1+\eta_{co}\,\text{sech}\left(\frac{\pi y_L}{2L}\right)} \tag{10.50}$$

MATLAB Program 15 computes and plots the amplitude distribution A(y/L) versus y/L at the design frequency 10 GHz, as shown in Figure 10.6.

MATLAB Program 15

```
% Matlab program to compute the amplitude distribution A(y/L)
versus y/L at the design frequency 10 GHz

% speed of electromagnetic wave in meter unit
c= 3*10^8;
% separation between the plates at the center in meter unit
ay0= 2.286*10^-2;
% angular design frequency in Hz
w0= 2*pi*10*10^9;
% length of horn in cm
L= 5.72;
```

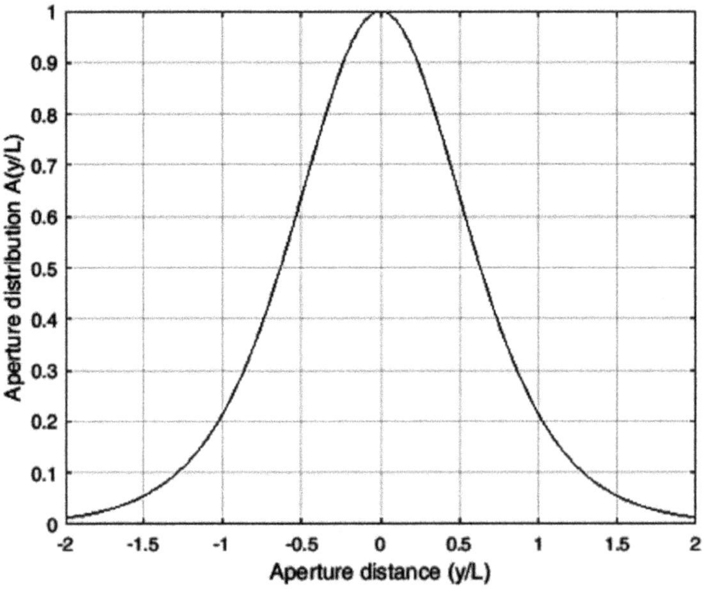

Figure 10.6 Normalized amplitude distribution $A(y/L)$ versus y/L at the design frequency 10 GHz.

```
% normalized height dimension of horn yl= y/L
yl=linspace(-2,2);

% calculation of refractive index at the center of the horn
nco =sqrt(1-(pi*c/(ay0*w0))^2);

% amplitude distribution calculation
A=(sech(pi*yl/2)).^2./(1+nco*sech(pi*yl/2));
A= A./max(A);
plot(yl,A, 'k-')
axis([-2 2 0 1])
grid on
```

10.1.7 Calculation of the Aperture Amplitude Distribution at a Modified Frequency

The ray trajectory (10.38) and (10.39) with $z = L$ can be differentiated to find an expression for $d\theta/dy$, which is necessary for the calculation of the aperture field distribution. Hence,

$$\sinh\left(\frac{\pi y_L}{2L}\right) = B \cdot \sin\left(\frac{\pi}{2}D\right) \tag{10.51a}$$

$$\sinh\left(\frac{\pi y_L}{2L}\right) = B \cdot \sinh\left(\frac{\pi}{2}D\right) \tag{10.51b}$$

where

$$B = \frac{\sin\theta}{\sqrt{|\cos^2\theta - A|}}, \quad D = \frac{\sqrt{|\cos^2\theta - A|}}{\cos\theta}$$

Differentiating B and D with respect to θ, provided $\cos^2\theta - A > 0$, gives:

$$B' = \frac{dB}{d\theta} = \frac{\cos\theta \cdot (1 - A)}{\left(|\cos^2\theta - A|\right)^{3/2}} \tag{10.52a}$$

$$D' = \frac{dD}{d\theta} = \frac{-A \cdot \sin\theta}{\cos^2\theta \cdot \left(|\cos^2\theta - A|\right)^{1/2}} \tag{10.52b}$$

Differentiating B and D with respect to θ, provided $\cos^2\theta - A < 0$, gives:

$$B' = \frac{dB}{d\theta} = \frac{\cos\theta \cdot (A - 1)}{\left(|\cos^2\theta - A|\right)^{3/2}} \tag{10.53a}$$

$$D' = \frac{dD}{d\theta} = \frac{A \cdot \sin\theta}{\cos^2\theta \cdot \left(\left|\cos^2\theta - A\right|\right)^{1/2}} \tag{10.53b}$$

It follows upon differentiating (10.51a) when $\cos^2\theta - A > 0$, as well as (10.51b) when $\cos^2\theta - A \leq 0$, respectively, that:

$$\left|\frac{d\theta}{dy_L}\right| = \frac{\pi}{2L} \cdot \frac{\cosh\left(\frac{\pi y_L}{2L}\right)}{\left|\left(\frac{\pi}{2}BD'\right)\cos\left(\frac{\pi}{2}D\right) + B'\sin\left(\frac{\pi}{2}D\right)\right|} \tag{10.54a}$$

$$\left|\frac{d\theta}{dy_L}\right| = \frac{\pi}{2L} \cdot \frac{\cosh\left(\frac{\pi y_L}{2L}\right)}{\left|\left(\frac{\pi}{2}BD'\right)\cosh\left(\frac{\pi}{2}D\right) + B'\sinh\left(\frac{\pi}{2}D\right)\right|} \tag{10.54b}$$

A modified power transmission coefficient to the one stated in (10.46) is used in the calculation of the power distribution $P(y)$ because rays leaving the aperture plane, at a frequency other than that of the design, are not parallel to the axis of propagation.

The modified power transmission coefficient is:

$$\bar{T} = \frac{4\bar{\eta}(y_L)\cos\theta_i^2}{\left[\bar{\eta}(y_L)\cos\theta_t + \eta_2\cos\theta_i\right]^2} = \frac{4\bar{\eta}(y_L)}{\left[\bar{\eta}(y_L)\left(\frac{\cos\theta_t}{\cos\theta_i}\right) + \eta_2\right]^2} \tag{10.55}$$

where θ_t is the angle that a ray leaving the aperture makes with the direction of stratification, θ_i is the angle that an incident ray on the aperture makes with the direction of stratification, and $\eta_2 = 1$ (the free space refractive index).

From Snell's law:

$$\bar{\eta}(y)\sin\theta_i = \sin\theta_t \tag{10.56}$$

hence

$$\cos\theta_t = \left(1 - \bar{\eta}^2(y)\sin^2\theta_i\right)^{1/2} \tag{10.57}$$

Substituting for $\cos\theta_t$ in (10.55) gives:

$$\bar{T} = \frac{4\bar{\eta}(y_L)\cos\theta_i^2}{\left[\bar{\eta}(y_L)\left(1 - \bar{\eta}^2(y)\sin^2\theta_i\right)^{1/2} + \cos\theta_i\right]^2} \tag{10.58}$$

If $P(\theta) = \cos^2\theta$, the amplitude distribution can be worked out from (10.54) and (10.58) as:

$$A(y) = \left[P(\theta) \cdot \left| \frac{d\theta}{dy_L} \right| \cdot \bar{T} \right]^{1/2} \tag{10.59}$$

The following MATLAB program computes and plots the aperture distribution (in amplitude and phase), as shown in Figure 10.7 and 10.8, respectively at the design frequency 10 GHz and modified frequencies 9.5 GHz and 10.5 GHz.

MATLAB Program 16

```
% Matlab program for computing aperture amplitude distribution of
HYSEC horn at various frequencies

% Input frequency of operation
freq_GHz = input('Enter the frequency in GHz: ');
lam = 0.3 / freq_GHz; % Wavelength in meter

% Parameters
c = 3e8; % speed of light in meters
ay0= 2.286*10^-2; % width of WG16 feed in meter
w0 = 2*pi*10*10^9;
w1 = 2*pi*freq_GHz*10^9;
```

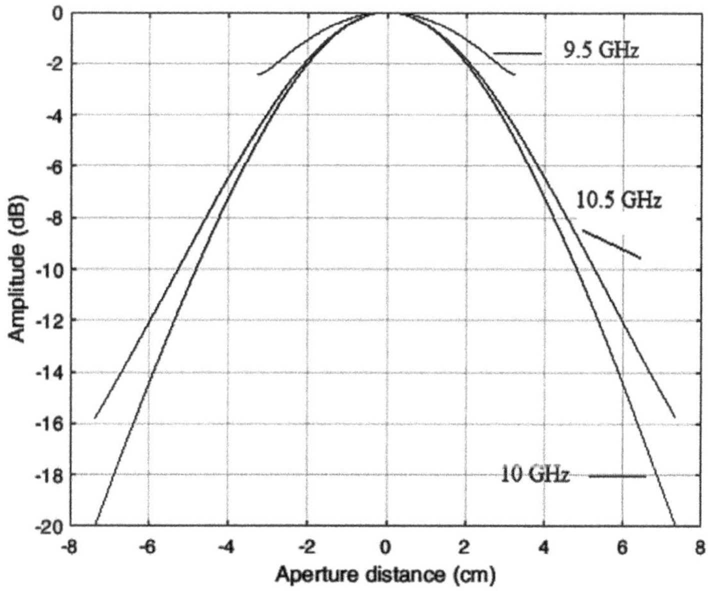

Figure 10.7 Amplitude distribution (dB) versus aperture distance (cm) at various frequencies.

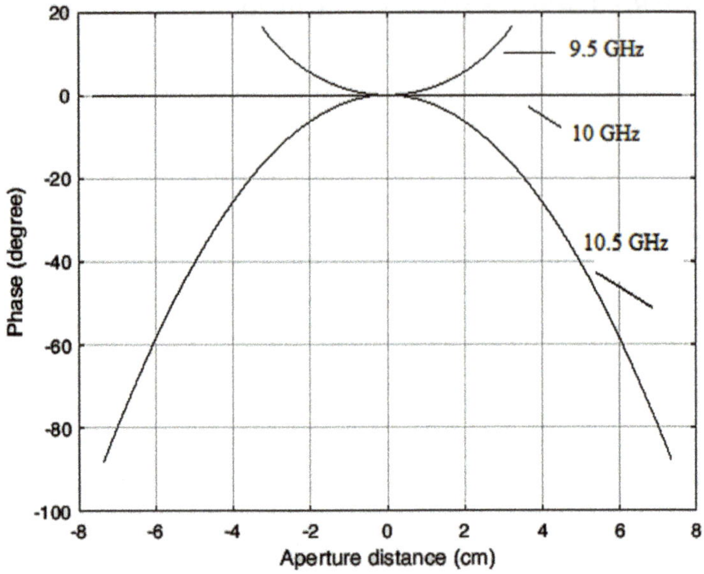

Figure 10.8 Phase distribution (degree) versus aperture distance (cm) at various frequencies.

```
ko = 2*pi/(100*lam);
L = 5.72; % length of horn in centimeter

% Define the range of thet. At the design frequency, this is 76
degree for max y/L.
thet = linspace(-1.326, 1.326, 1000); % Range of thet in radians

% Compute parameters of equation
p = 1 - (w0/w1)^2;
q = (w0/w1)^2 * (1 - (pi*c/(ay0*w0))^2);
A = p / (p + q);
n_co = sqrt(p + q);
B = sin(thet)./ sqrt(abs((cos(thet)).^2 - A));
D = sqrt(abs((cos(thet)).^2 - A))./ cos(thet);

% Pre-allocate arrays for Bd, Dd, d_thet, n_ym, and Tn
Bd = zeros(size(thet));
Dd = zeros(size(thet));
d_thet = zeros(size(thet));
n_ym = zeros(size(thet)); % n_ym is variable refractive index
Tn = zeros(size(thet)); % power transmission factor at
antenna aperture

% Calculate Bd and Dd based on conditions
for i = 1:length(thet)
  if (cos(thet(i))).^2 >= A
    Bd(i) = cos(thet(i)) * (1 - A) / (abs((cos(thet(i))).^2
- A)).^1.5;
```

```
    Dd(i) = -A * sin(thet(i)) / ((cos(thet(i))).^2 *
sqrt(abs((cos(thet(i))).^2 - A)));
  elseif (cos(thet(i))).^2 < A
    Bd(i) = cos(thet(i)) * (A - 1) / (abs((cos(thet(i))).^2
- A)).^1.5;
    Dd(i) = A * sin(thet(i)) / ((cos(thet(i))).^2 *
sqrt(abs((cos(thet(i))).^2 - A)));
  end
end

% Solve the equation for y at L that correspond to values of
angle thet
for i = 1:length(thet)
frac1(i)= sqrt(abs((cos(thet(i)))^2) - A);
if (cos(thet(i))).^2 >= A
  y_solution(i) = asinh(sin(thet(i))/frac1(i) * sin(pi/2 *
frac1(i)/cos(thet(i)))) * 2 * L / pi;
  elseif (cos(thet(i))).^2 < A
  y_solution(i) = asinh(sin(thet(i))/frac1(i) * sinh(pi/2 *
frac1(i)/cos(thet(i)))) * 2 * L / pi;
end

% Calculate d_thet based on conditions
  if (cos(thet(i))).^2 >= A
    d_thet(i) = cosh(pi*y_solution(i)/(2*L)) / abs((pi*B(i)*Dd(i)
/2)*cos(pi*D(i)/2) + Bd(i)*sin(pi*D(i)/2));
  elseif (cos(thet(i))).^2 < A
    d_thet(i) = cosh(pi*y_solution(i)/(2*L)) / abs((pi*B(i)*Dd(i)
/2)*cosh(pi*D(i)/2) + Bd(i)*sinh(pi*D(i)/2));
  end
end

% Calculate n_ym and Tn
for i = 1:length(thet)
n_ym(i) = sqrt(p + q*(sech(pi*y_solution(i)/(2*L)).^2));
thet_inc(i)= (n_co.*cos(thet(i))./n_ym(i)).^2;
Tn(i) = n_ym(i).* thet_inc(i) ./ (n_ym(i) .* sqrt(1 - n_
ym(i).^2.*(1 - thet_inc(i))) + sqrt(thet_inc(i))).^2;
end

% Power radiation from point source
P_thet = (cos(thet)).^2;

% Calculate OPO; optical ray length at thet= zero
theta_zero = 0;
B_zero = sin(theta_zero) / sqrt(abs((cos(theta_zero))^2 - A));
D_zero = sqrt(abs((cos(theta_zero))^2 - A)) / cos(theta_zero);
OPO_integrand = @(z1) 1./(1+B_zero^2.*(sin(pi*z1.*D_zero/2)).^2);
OPO = L*p./(sqrt(p+q).*cos(0)) + L*q./(sqrt(p+q).*cos(0)).*integr
al(OPO_integrand, 0, 1);
```

```
% Calculate Delta; optical length difference between ray
considered and referenced ray at theta_zero
OP = zeros(size(thet));
for i = 1:length(thet)
  OP(i) = L*p./(sqrt(p+q).*cos(thet(i))) + L*q./(sqrt(p+q).*cos(t
het(i))).*integral(@(zl) 1./(1+B(i)^2.*(sin(pi*zl.*D(i)/2)).^2),
0, 1);
end
Delta = OP - OP0;
% Calculate Amplitude in dB and phase in degree
Amp = sqrt(P_thet.*d_thet.*Tn) .* exp(-j*ko*Delta);
Amp_ab = abs(Amp)./max(abs(Amp));
Amp_ab = 20*log10(Amp_ab);
Phase= atan2(imag(Amp), real(Amp)).*180/pi;

% Plot amplitude and phase

figure(1)
plot(y_solution, Amp_ab, 'k')
xlabel('Aperture distance (cm)');
ylabel('Amplitude (dB');
axis([-8 8 -20 0])
grid on
hold on

figure(2)
plot(y_solution, Phase, 'k')
xlabel('Apertre distance (cm) ');
ylabel('Phase (degree)');
axis([-8 8 -100 20])
grid on
hold on
```

The computation of the aperture distribution at a frequency below that of the design needs a special treatment. At 9.5 GHz, the variation of y_L/L with θ is wildly unstable above around $\theta = 80°$. Hence, the aperture distribution contributions from rays in an such unstable region are ignored. For convenience, the effective aperture size is approximated at this frequency between $-0.61 < y_L/L < +0.61$ and if rays in the negative angle direction are also considered, then contributions from three different rays sum up at every known point within the aperture region considered. At a point near $y_L/L = 0.61$, the aperture field computation breaks down since $d\theta/dy_L$ tends to be infinity, which cannot be practically realizable.

10.1.8 Refocusing from an Offset Source in the Hyperbolic Secant Refractive Medium

It is already obvious that ray paths, emerging from a point source in a hyperbolic secant law refractive medium, refocus periodically at successive half-distance periods of $2L$ along the z-axis. It may be interesting at this point to see the effect of an off-axis displace source on the refocusing of the ray trajectories.

The ray trajectory (10.8) is used with some modification with respect to the integral bounds and the ray constant.

$$z = \int_d^y \frac{h_d \cdot dy}{\sqrt{\eta^2(y) - h_d^2}} \tag{10.60}$$

where d is the offset displacement of the source, $h_d = \eta_{cd}\cos\theta$, and $\eta_{cd} = \eta_{c0}\,\mathrm{sec}h(\pi d/2L)$.

Substituting for $\eta(y)$ in (10.60) and setting $u = \sin h(\pi y/2L)$ gives

$$z = \int_{u_d}^{u} \frac{du}{\sqrt{a^2 - u^2}} \tag{10.61}$$

where

$$a = \left(\frac{\eta_{co}^2 - h_d^2}{h_d^2}\right)^{1/2} = \left(\frac{\cosh^2\left(\frac{\pi d}{2L}\right) - \cos^2\theta}{\cos^2\theta}\right)^{1/2} \tag{10.62}$$

Solving (10.61) yields:

$$\frac{\pi z}{2L} = \sin^{-1}\left(\frac{u}{a}\right) - \sin^{-1}\left(\frac{u_d}{a}\right) \tag{10.63}$$

Consequently,

$$\sinh\left(\frac{\pi y}{2L}\right) = a * \sin\left(\sin^{-1}\left(\frac{\sinh\left(\frac{\pi d}{2L}\right)}{a}\right) + \frac{\pi z}{2L}\right) \tag{10.64}$$

If $y < d$, then (10.64) transforms to:

$$\sinh\left(\frac{\pi y}{2L}\right) = a * \sin\left(\sin^{-1}\left(\frac{\sinh\left(\frac{\pi d}{2L}\right)}{a}\right) - \frac{\pi z}{2L}\right) \tag{10.65}$$

MATLAB Program 17 calculates and plots the trajectories emanating from a focal feed point, placed off-axis by various values of d, as shown in Figure 10.9, Figure 10.10, and Figure 10.11.

MATLAB Program 17

```
% Matlab program to calculate the trajectories emanating from a
focal feed point placed off-axis by d.
```

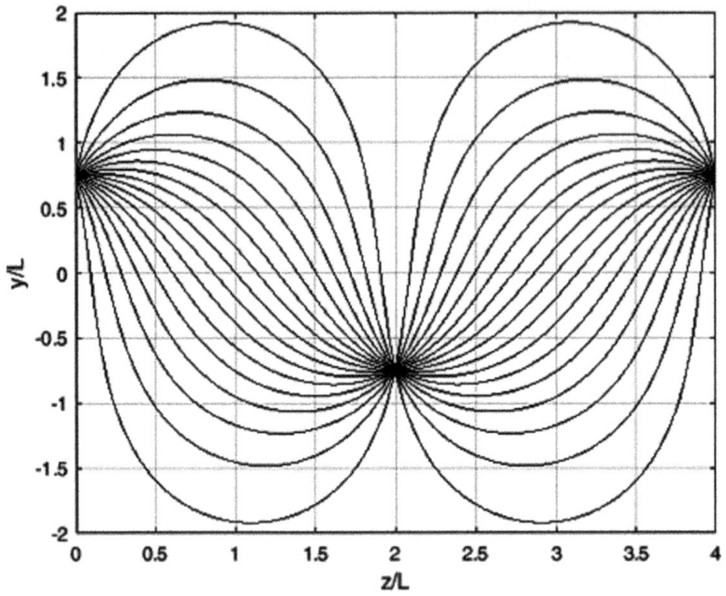

Figure 10.9 Ray trajectories from a positive off-axis displaced source ($d/L = 0.75$) at the design frequency.

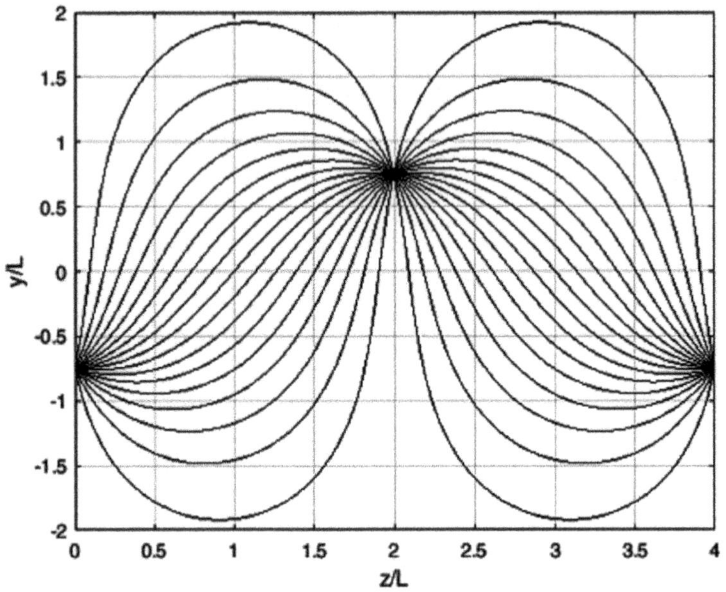

Figure 10.10 Ray trajectories from a negative off-axis displaced source ($d/L = -0.75$) at the design frequency.

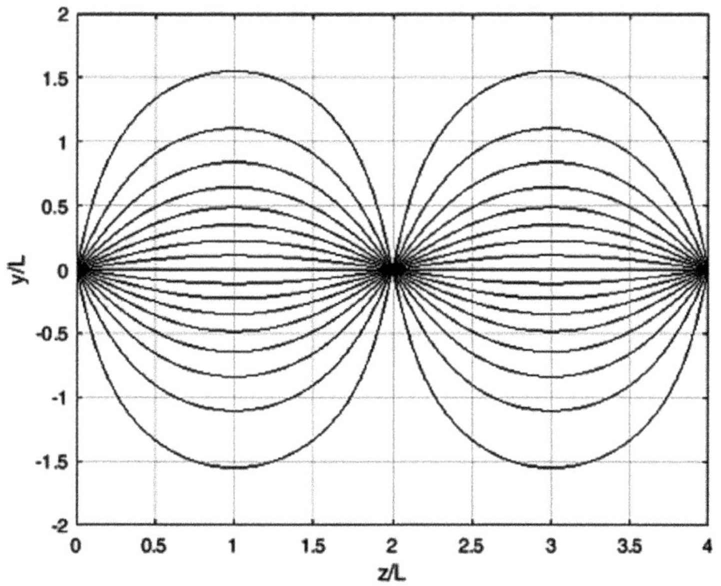

Figure 10.11 Ray trajectories from a central undisplaced source at the design frequency.

```
% Length of horn in cm
L = 5.72;
% off-axis deviation of focal feed point in cm (assumed as 75%
*L)
d = 4.29; % it can be positive, negative or zero

% Calculate theta angles
j = 1:9;
thet = 10 * pi / 180 * (j - 1);

% Calculate 'an' values
an = sqrt(((cosh(pi / 2 * d / L))^2 - (cos(thet)).^2) ./
(cos(thet)).^2);

% Initialize zl
i = 1:101;
zl = 0.04 * (i - 1);

% Calculate yl
zl_matrix = repmat(zl, 9, 1);
an_matrix = repmat(an', 1, 101);
zl_matrix_positive = zl_matrix;
zl_matrix_positive(zl_matrix_positive < 0) = 0;

% Calculate yl for positive and negative regions
yl_positive = 2 / pi * asinh(an_matrix .* sin(asin(sinh(pi / 2 *
d / L) ./ an_matrix) + pi / 2 * zl_matrix_positive));
yl_negative = 2 / pi * asinh(an_matrix .* sin(asin(sinh(pi / 2 *
d / L) ./ an_matrix) - pi / 2 * zl_matrix_positive));
```

```
% Plot the mirror images in the region zl = 0 to 2
yl_positive_mirror = yl_positive;
yl_negative_mirror = yl_negative;

% Plot the mirror images in the region zl = 2 to 4 above xl = 2
zl_matrix_mirror = 4 - zl_matrix;
yl_positive_mirror(zl_matrix >= 2) = yl_positive(zl_matrix >= 2);
yl_negative_mirror(zl_matrix >= 2) = yl_negative(zl_matrix >= 2);

yl_positive_mirror_flipped = yl_positive_mirror(end:-1:1,:);
yl_negative_mirror_flipped = yl_negative_mirror(end:-1:1,:);

% Plot the trajectory curves
plot(zl_matrix_mirror(end:-1:1,:)', yl_positive_mirror_flipped',
'k-', zl_matrix_mirror(end:-1:1,:)', yl_negative_mirror_flipped',
'k-')
xlabel('z/L')
ylabel('y/L')
axis([0 4 -2 2])
grid on
```

It is obvious that the point source displacement along the y-axis does not affect the refocusing of rays to an image, in an off-axis position.

10.1.9 Calculation of the Optical Path Length from an Extended Source

As the sec*h*-law refractive medium, of concern, is 1D and because a point source cannot be realized in the physical sense, it is practically more appropriate to consider an extended source to see its effect on the calculation of the optical ray-path length.

Consider an extended source of length $2y_0$, centered along the y-axis as shown in Figure 10.12. A ray from the center and one from each end of the extended source are assumed to travel in the sec*h*-law refractive medium and meet at a single point on the aperture plane at $z = L$.

The upper ray, u, which can be visualized as emerging from a point on the z-axis, has the following trajectory:

$$\sin\left(\frac{\pi}{2}\cdot\frac{z+z_0}{L}\right) = \cot\theta'\sinh\left(\frac{\pi}{2}\cdot\frac{y}{L}\right) \tag{10.66}$$

At $z = 0$, $y = y_0$ and

$$\sin\left(\frac{\pi}{2}\frac{z_0}{L}\right) = \cot\theta'\sinh\left(\frac{\pi}{2}\frac{y_0}{L}\right) \tag{10.67}$$

At $z = L$, $y = y_L$ and

$$\sin\left(\frac{\pi}{2}\left(1+\frac{z_0}{L}\right)\right) = \cos\left(\frac{\pi}{2}\frac{z_0}{L}\right) = \cot\theta'\sinh\left(\frac{\pi}{2}\frac{y_L}{L}\right) \tag{10.68}$$

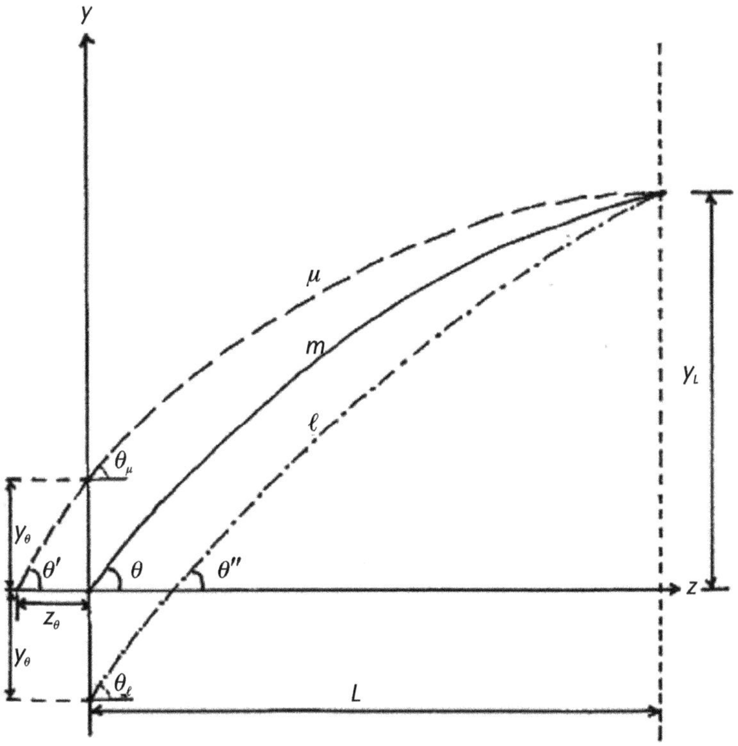

Figure 10.12 Rays emerging from an extended source feed and meeting at a point on the aperture plane.

Similarly, for the lower ray, l, with z_0, θ', being replaced by z'_0, θ'', respectively:

$$\sin\left(\frac{\pi}{2}\frac{z'_0}{L}\right) = \cot\theta'' \sinh\left(\frac{\pi}{2}\frac{y_0}{L}\right) \tag{10.69}$$

$$\cos\left(\frac{\pi}{2}\frac{z'_0}{L}\right) = \cot\theta'' \sinh\left(\frac{\pi}{2}\frac{y_L}{L}\right) \tag{10.70}$$

Upon squaring and summing (10.67) and (10.68), as well as (10.69) and (10.70) separately, the following new equations result:

$$1 = \cot^2\theta' \left(\sinh^2\left(\frac{\pi y_0}{2L}\right) + \sinh^2\left(\frac{\pi y_L}{2L}\right)\right) \tag{10.71a}$$

$$1 = \cot^2\theta'' \left(\sinh^2\left(\frac{\pi y_0}{2L}\right) + \sinh^2\left(\frac{\pi y_L}{2L}\right)\right) \tag{10.71b}$$

It is obvious that $\theta' = \theta''$. Furthermore, substituting from (10.42) into (10.71):

$$\tan^2\theta' = \sinh^2\left(\frac{\pi y_0}{2L}\right) + \tan^2\theta \tag{10.72}$$

Because $1 + \tan^2\theta = \sec^2\theta$, this also gives:

$$\sec^2\theta' = \sinh^2\left(\frac{\pi y_0}{2L}\right) + \sec^2\theta \tag{10.73}$$

The optical path length, expressed in (10.18), holds for the ray, m. For the ray, u, emerging from the extended source, the expression needs to be modified at the design frequency and, hence,

$$P = \frac{L\eta_{co}}{\cos\theta'}\int_0^{z/L}\frac{d(z/L)}{1+\tan^2\theta'\sin^2\left(\frac{\pi(z+z_0)}{2L}\right)} \tag{10.74}$$

Now

$$\sin\left(\frac{\pi(z+z_0)}{2L}\right) = \sin\left(\frac{\pi z}{2L}\right)\cos\left(\frac{\pi z_0}{2L}\right) + \cos\left(\frac{\pi z}{2L}\right)\sin\left(\frac{\pi z_0}{2L}\right) \tag{10.75}$$

Substituting for the terms that involve z_0 from (10.67), (10.68), and (10.42):

$$\tan'\theta\sin\left(\frac{\pi(z+z_0)}{2L}\right) = \sin\left(\frac{\pi z}{2L}\right)\tan\theta + \cos\left(\frac{\pi z}{2L}\right)\sinh\left(\frac{\pi y_0}{2L}\right) \tag{10.76}$$

Substituting back into (10.74) from (10.73) and (10.76) gives:

$$P = L\eta_{co}\left(\sinh^2\left(\frac{\pi y_0}{2L}\right) + \sec^2\theta\right)^{1/2}\int_0^{z/L}\frac{d(z/L)}{1+\left(\sin\left(\frac{\pi z}{2L}\right)\tan\theta + \cos\left(\frac{\pi z}{2L}\right)\sinh\left(\frac{\pi y_0}{2L}\right)\right)^2} \tag{10.77}$$

The integral can be easily worked out numerically, using MATLAB software.

Before finalizing this section, we show in Figure 10.13 ray trajectories emanating from the end points of an extended source in a sec*h*-law medium. All rays, as expected, refocus at two points, half cycle away at $z/L = 2$. If the medium to the right of $z/L = 1$ is free space, all meeting rays would emerge parallel to the z-axis. MATLAB Program 15 was utilized to draw the plot, where d was made equal to L and another time as $-L$.

10.1.10 Calculation of the Aperture Amplitude Distribution from an Extended Source

The calculation of the aperture amplitude distribution from an extended source is enabled by summing numerically, in amplitude and phase, all power flow contributions of rays, at every selected point on the aperture plane.

The polar power distribution of the extended source acting as a primary feed, is assumed to be of the $\cos^2\theta$ type. For a short-sized feed in the order of a wavelength,

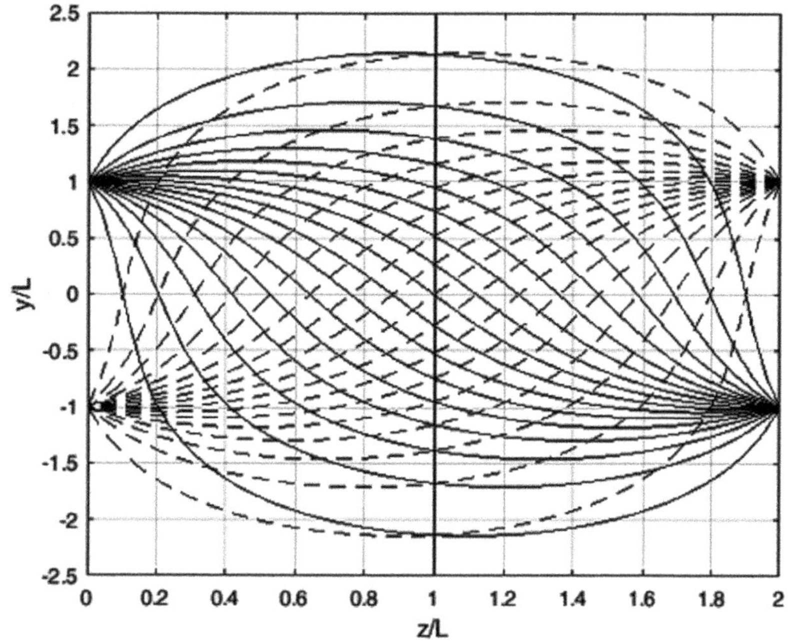

Figure 10.13 Ray trajectories emanating from the end points of an extended source in a sech-law medium.

this assumption is practically convenient because it provides a unidirectional polar power distribution and improves upon the directivity of the circular pattern predicted from a $\cos\theta$ type source.

In order to calculate the power flow contribution of any ray emerging from the extended source, the expression $P(\theta)(d\theta_n/d_{yL})T_n$ can be exployed for working out the amplitude part.

Consider the ray, u, which emerges from a point displaced by y_0 from the center of the extended feed. The following equation may be employed after rearranging (10.63):

$$\frac{u}{a} = \sin\left(\frac{\pi z}{2L} + c\right) \tag{10.78}$$

where

$$c = \sin^{-1}\left(\frac{u_{y_0}}{a}\right)$$

and a is given in (10.62).

At $z = z_0$, $y = y_0$ and $u = u_{y0}$; hence,

$$\frac{u_{y_0}}{a} = \sin c \tag{10.79}$$

At $z = z_L$, $y = y_L$ and $u = u_{yL}$; hence,

$$\frac{u_{y_L}}{a} = \cos c \tag{10.80}$$

Squaring and summing (10.79) and (10.80) and after substituting for u

$$\sinh^2\left(\frac{\pi y_0}{2L}\right) + \sinh^2\left(\frac{\pi y_L}{2L}\right) = a^2 \tag{10.81}$$

Squaring and rearranging (10.62):

$$\cosh^2\left(\frac{\pi y_0}{2L}\right) + \cosh^2\left(\frac{\pi y_0}{2L}\right) \cdot \tan^2\theta_u - 1 = a^2 \tag{10.82}$$

Consequently, equating (10.81) and (10.82) yields:

$$\sinh\left(\frac{\pi y_L}{2L}\right) = \tan\theta_u \cdot \cosh\left(\frac{\pi y_0}{2L}\right) \tag{10.83}$$

Differentiating (10.83) with respect to y_L gives:

$$\frac{d\theta_u}{dy_L} = \frac{\pi}{2L} \cdot \frac{\cosh\left(\frac{\pi y_L}{2L}\right)}{\sec^2\theta_u \cdot \cosh\left(\frac{\pi y_0}{2L}\right)} \tag{10.84}$$

while squaring and rearranging (10.83) produces:

$$\sec^2\theta_u = \frac{\sinh^2\left(\frac{\pi y_L}{2L}\right)}{\cosh^2\left(\frac{\pi y_0}{2L}\right)} + 1 \tag{10.85}$$

Substituting for $\sec^2\theta_u$ in (10.84) gives:

$$\frac{d\theta_u}{dy_L} = \frac{\pi}{2L} \cdot \frac{\cosh\left(\frac{\pi y_L}{2L}\right) \cdot \cosh\left(\frac{\pi y_0}{2L}\right)}{\sinh^2\left(\frac{\pi y_L}{2L}\right) \cdot \cosh^2\left(\frac{\pi y_0}{2L}\right)} \tag{10.86}$$

By including the power transmission coefficient, T_n, the amplitude part of the power flow contribution from a ray can be found. In addition, a phase error term must be included with the contribution from each ray. Such a phase term is expressed as $\exp(-jk_0\Delta P)$ where ΔP indicates the optical path difference between the ray considered and the middle one that meets at the same point on the aperture plane and k_0 is the free-space wave number. The effect of an extended source on the aperture amplitude distribution is shown utilizing the following developed MATLAB program.

MATLAB Program 18

```
% Matlab program to compute the amplitude distribution of a HYSEC
horn, at the 10 GHz design frequency, with extended feed sources
of various sizes

c = 3 * 10^8;
ay0 = 2.286 * 10^-2;
w0 = 2 * pi * 10 * 10^9;
L = 5.72; % Length of horn in cm
nco = sqrt(1 - (pi * c / (ay0 * w0))^2);
k0 = 2 * pi / (3);
y0 = input('Enter size of extended source at the design frequency
in λ: ')
y0= 0.5*(y0*3); % this is half of extended source in cm
% Define yl range
yl_range = linspace(-8, 8, 200); % Adjust the number of points
as needed

% Initialize arrays for amplitude magnitudes
A_mag = zeros(size(yl_range));

% Loop through each yl value
for idx = 1:numel(yl_range)
yl = yl_range(idx);

% Numerical computation of ray optical paths from extended source
(edge rays upper Pu1 and lower Pl1) and mid rays (upper Pu3 and
lower Pl3)
  Pu1 = L * nco * sqrt((sinh(pi * y0 / (2 * L))).^2 + (cosh(pi *
yl / (2 * L))).^2)...
    .*integral(@(z) 1./(1+(sin(pi*z/(2)).*sinh(pi*yl/
(2*L))+cos(pi*z/(2)).*sinh(pi*y0/(2*L))).^2), 0, 1);
  Pl1 = L * nco * sqrt((sinh(-pi * y0 / (2 * L))).^2 + (cosh(pi *
yl / (2 * L))).^2)...
    .*integral(@(z) 1./(1+(sin(pi*z/(2)).*sinh(pi*yl/
(2*L))+cos(pi*z/(2)).*sinh(-pi*y0/(2*L))).^2), 0, 1);
  Pu3 = L * nco * sqrt((sinh(pi * y0 / (4 * L))).^2 + (cosh(pi *
yl / (2 * L))).^2)...
    .*integral(@(z) 1./(1+(sin(pi*z/(2)).*sinh(pi*yl/
(2*L))+cos(pi*z/(2)).*sinh(pi*y0/(4*L))).^2), 0, 1);
  Pl3 = L * nco * sqrt((sinh(-pi * y0 / (4* L))).^2 + (cosh(pi *
yl / (2 * L))).^2)...
    .*integral(@(z) 1./(1+(sin(pi*z/(2)).*sinh(pi*yl/
(2*L))+cos(pi*z/(2)).*sinh(-pi*y0/(4*L))).^2), 0, 1);

% Optical path of ray emerging from center of extended source
  P0 = L * nco * sqrt((cosh(pi * yl / (2 * L))).^2)...
```

```
    .* integral(@(z) 1./(1+(sin(pi*z/(2)).*sinh(pi*yl/
(2*L))).^2), 0, 1);

% Optical path difference between considered ray and central ray
  delta_pul = Pul - P0;
  delta_pl1 = Pl1 - P0;
  delta_pu3 = Pu3 - P0;
  delta_pl3 = Pl3 - P0;

% Amplitude distribution calculation
  p_thet = (sech(pi * yl / (2 * L))).^2;
  dthet_dyl = cosh(pi * yl / (2 * L)) * cosh(pi * y0 / (2 *
L))...
    /((sinh(pi * yl / (2 * L))).^2 + (cosh(pi * y0 / (2 *
L))).^2);
  dthet_dy0 = cosh(pi * yl / (2 * L)) / ((sinh(pi * yl / (2 *
L))).^2+1);
  t_n = 4 * nco * sech(pi * yl / (2 * L)) ./ (1 + nco * sech(pi *
yl / (2 * L))).^2;

% Amplitude contributions from ray trajectories
  A_yul = sqrt(p_thet .* dthet_dyl .* t_n) .* exp(-1j * k0 *
delta_pul);
  A_yl1 = sqrt(p_thet .* dthet_dyl .* t_n) .* exp(-1j * k0 *
delta_pl1);
  A_yu3 = sqrt(p_thet .* dthet_dyl .* t_n) .* exp(-1j * k0 *
delta_pu3);
  A_yl3 = sqrt(p_thet .* dthet_dyl .* t_n) .* exp(-1j * k0 *
delta_pl3);

  A_y0 = sqrt(p_thet .* dthet_dy0 .* t_n);
  A_y = A_yul + A_yl1 + A_yu3 + A_yl3 +A_y0;

% Calculate and store amplitude magnitude
  A_mag(idx) = abs(A_y);
end

% Plot amplitude magnitude
plot(yl_range, 20*log10(A_mag./max(A_mag)), 'k-');
xlabel('Aperture distance (cm)');
ylabel('Amplitude (dB)');
axis([-8 8 -20 0])
grid on
```

Figure 10.14 demonstrates the amplitude distribution (dB) versus aperture distance (cm) from extended sources of various sizes. The plot shows that further tapering in the amplitude distribution is introduced by extending the source, although an extended source of 0.347λ (corresponding to the experimental WG16 feed) produces a distribution resembling that from a point source feed.

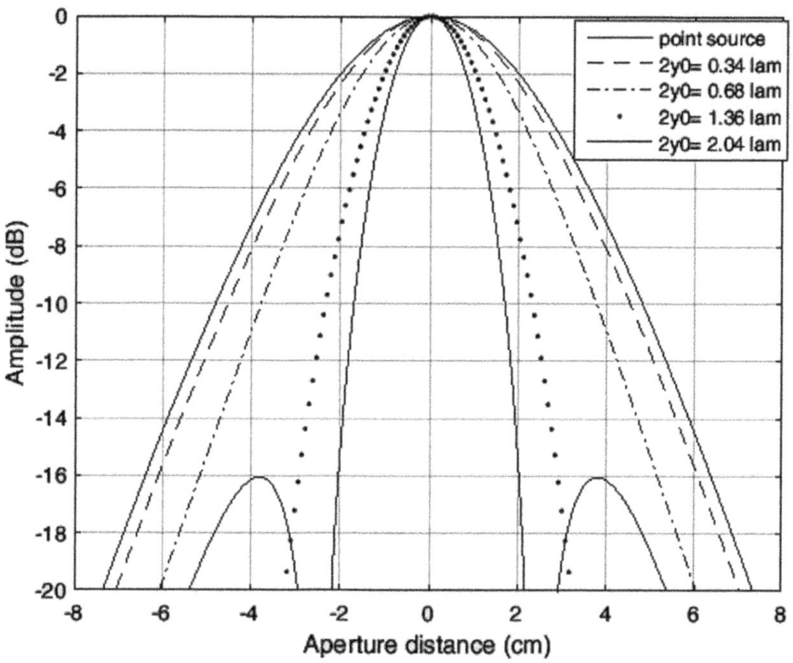

Figure 10.14 Amplitude distribution (dB) versus aperture distance (cm) from extended sources of various sizes.

10.2 The Horn Construction and Design Considerations

It is evident from the amplitude distribution expressed in (10.50) that the aperture dimension, y/L, can theoretically be infinite. The nature of such amplitude distribution reasonably approximates to a truncated near-Gaussian function. In a practical situation, the taper ratio of a predicted Gaussian amplitude distribution, which is dictated by the level of amplitude excitation at the edge relative to that at the center, determines the sidelobe level in the radiation pattern. It has been deduced that when $y/L = 1.5$, in the amplitude distribution expressed in (10.50), the taper ratio is about 0.05. Consequently, the radiation pattern of a near-Gaussian aperture distribution would show sidelobes approximately 30 dB below the main beam. For $y/L = 2$, the lower taper ration would result in radiation sidelobes of around 10 dB lower than the former value.

A short focus horn was manufactured by carefully milling and smoothly filing the required profile, determined from the design (10.28) into two blocks of brass which were then secured at the correct separation with plates at ±7.62 cm. The choice of a_y for $y = 0$, a_{yo}, which merely determined η_{co} in the same equation, at the design frequency 10 GHz, was chosen to be 2.29 cm, the width of the X-band waveguide (WG16), which fed the back plate attached to the horn. The axial length, L, of the horn was chosen to be 5.72 cm, as this was the maximum length that could be milled. Figure 10.15(a, b) show front and side views of the constructed horn with its waveguide feed.

(a)

(b)

Figure 10.15 Constructed short focus HYSEC horn: (a) side view shows wave guide feed, and (b) front view shows aperture profile.

The tolerance required on the plate spacing is another consideration to be observed. While such tolerance can be fairly generous at the center, it becomes progressively tighter with increasing values of y/L and can be considerable at $y/L > 1.5$.

For the horn dimensions considered in the construction of the horn, y_{max}/L works out to be 1.33 with a predicted taper ratio of 0.1.

10.3 Prediction of the Far-Field Radiation Pattern

It is possible to calculate the radiation pattern from any bounded plane aperture, once the field distribution across it is known. At the design frequency, the phase corrected aperture distribution of the HYSEC horn is expressed in (10.50), which is obviously dependent on the y-variable. Kirchhoff's integral expressed in (9.6) of Chapter 9 enables the calculation of the E-plane radiation field from the horn aperture. The integral is based on Huygens' principle, which regards the field in the radiation distribution as being the superposition of all contributions from the infinite number of the elementary sources across the radiating aperture. The factors that only involve the radial distance R will not affect the shape of the radiation pattern and hence can be ignored. In addition, for a fairly large aperture size in terms of free-space wavelengths, λ/λ_g approaches unity for a main beam largely confined to small values of θ, the factor $(1 + (\lambda/\lambda_g)\cos\theta)$ in (9.6) varies very slowly and is assumed to be a constant that can be ignored.

Because the aperture distribution, $A(y)$, is 1D, the radiation pattern $G(\sin\theta)$ is restricted to be that of an E-plane and the radiation expression reduces to:

$$G(\sin\theta) = \int_{-y}^{y} A(y)\exp\left(\frac{j2\pi y}{\lambda}\sin\theta\right)dy \tag{10.87}$$

The integral in (10.87) can only be solved by numerical techniques. This will be demonstrated in MATLAB Program 19, which computes the far field from the aperture (amplitude and phase) distribution data. Figure 10.16 shows the predicted far-filed amplitude distribution, computed from (10.87).

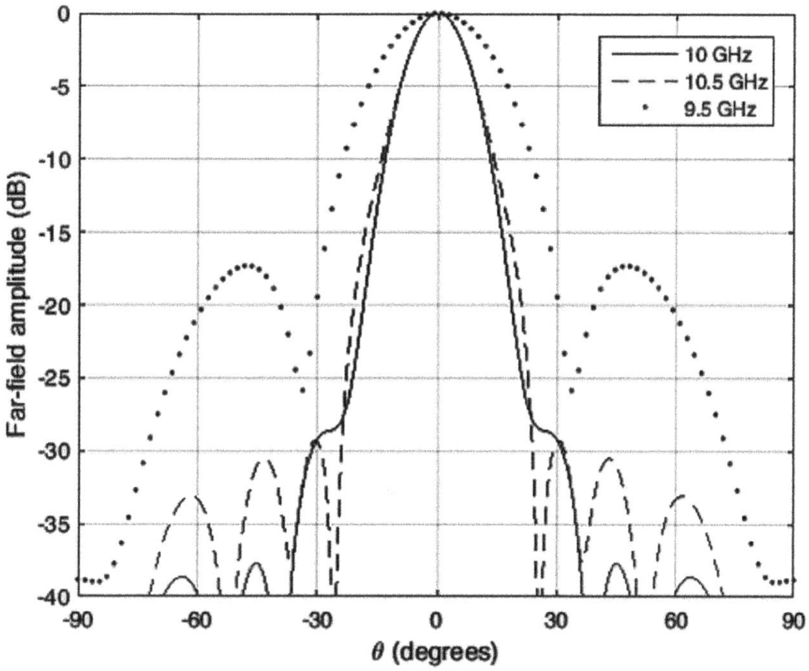

Figure 10.16 Far-field amplitude (dB) versus radiation angle (degree) of HYSEC horn at various frequencies.

MATLAB Program 19

```
% Matlab program to compute the far field radiation pattern of
the HYSEC horn aperture at various frequencies. Before running
this program, you would need to run Matlab program 15, which
computes the aperture amplitude and phase distribution.

% Far field calculation from aperture distribution Amp(j)
lam = lam*100; % Wavelength in centimeters
theta_deg = linspace(-90, 90, 1000); % Angle in degrees, covering
full 360 degrees
theta_rad = deg2rad(theta_deg); % Convert degrees to radians
E_far = zeros(size(theta_rad));
ii = complex(0,1);

for i = 1:length(theta_rad)
  integral_result = zeros(size(y_solution));
  for j = 1:length(y_solution)
    % integrand = Amp(j) * exp((ii*2*pi*y_solution(j)/
lam)*sin(theta_rad(i)));
  integrand = Amp(j) * exp((ii*2*pi*y_solution(j)/
lam)*sin(theta_rad(i)));
integral_result(j) = integrand;
end

% Perform numerical integration
E_far(i) = trapz(y_solution, integral_result); end
% Convert far field amplitude to dB
E_far = abs(E_far)/max(abs(E_far));
E_far = 20*log10(E_far);
figure (3);
plot(theta_deg, E_far, 'k.');
xlabel('\theta (degrees)');
ylabel('Far-field amplitude');
axis([-90 90 -40 0])
set(gca, 'xlim', [-90 90]);
set(gca, 'xtick', -90:30:90);
grid on
hold on
```

It is noticed that, at the design frequency, shoulders show up below −27 dB, while sidelobes are below −37 dB. At 10.5 GHz, the sidelobe level rises up to below −28 dB. The radiation pattern at 9.5 dB shows the highest sidelobe level at around −17 dB. At 10 GHz and 10.5 GHz, the radiation patterns have relatively narrower beamwidth, with an HPBW of approximately 14°, while, at 9.5 GHz, the radiation pattern is relatively the broadest, with an HPBW of around 27°. This is expected because the effective aperture size at 9.5 GHz is reduced to $2y_0 = 2 \times 5.72 \times 0.61 = 6.97$ cm from that of the physical aperture $2 \times 5.72 \times 1.33 = 15.21$ cm, which is relatively 46% of the original size.

The far-field radiation pattern can be analyzed from an extended source of different sizes that range from a point source through the WG16 feed ($2y_0 = 0.34\lambda$) to $2y_0 = 2.04\lambda$. MATLAB Program 20 is developed for such a purpose. Figure 10.17 shows the far-field radiation pattern curves, resulting from the simulation. It is noticed that as the size of the extended source increases, the radiation pattern broadens more while the radiation sidelobes disappear below the –40-dB normalized level. The pattern resulting from the horn with the WG16 feed, compared to a point source, shows relatively smoother but raised shoulders and notably a lower sidelobe level. As the size of the waveguide feed doubles and triples ($2y_0 = 0.68\lambda$, 1.36λ, and 2.04λ), the radiation beam becomes increasingly broader with a dip at the boresight angle, as the feed size becomes equal to 2.04λ.

MATLAB Program 20

```
% Matlab program to compute the far field pattern from the
amplitude distribution of a HYSEC horn, at 10 GHz, with an
extended feed source, 2y0= 0.34λ, 0.68λ, 1.36λ and 2.04λ.

c = 3 * 10^8;
ay0 = 2.286 * 10^-2;
w0 = 2 * pi * 10 * 10^9;
L = 5.72; % Length of horn in cm
nco = sqrt(1 - (pi * c / (ay0 * w0))^2);
k0 = 2 * pi / 3;

y0 = input('Enter the size of the extended source at the design
frequency in lambda: ')
```

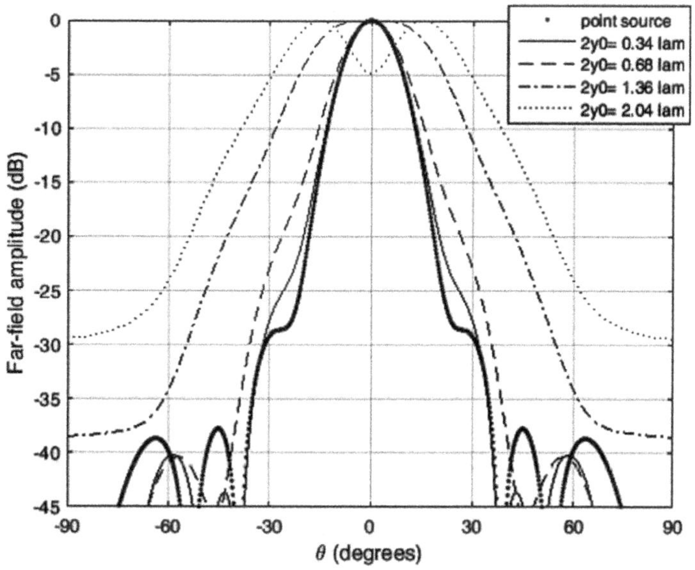

Figure 10.17 Far-field radiation pattern of a HYSEC horn, at the design frequency, from extended sources with various sizes.

```
y0= 0.5*(y0*3); % this is half of extended source in cm
% Define yl range
yl_range = linspace(-8, 8, 200); % Adjust the number of points
as needed

% Initialize arrays for amplitude magnitudes
A_mag = zeros(size(yl_range));

% Loop through each yl value
for idx = 1:numel(yl_range)
  yl = yl_range(idx);

% Numerical computation of ray optical paths from extended source
(edge rays upper Pu1 and lower Pl1) and mid rays (upper Pu3 and
lower Pl3)
  Pu1 = L * nco * sqrt((sinh(pi * y0 / (2 * L))).^2 + (cosh(pi *
yl / (2 * L))).^2)...
    .*integral(@(z) 1./(1+(sin(pi*z/(2)).*sinh(pi*yl/
(2*L))+cos(pi*z/(2)).*sinh(pi*y0/(2*L)))).^2), 0, 1);
  Pl1 = L * nco * sqrt((sinh(-pi * y0 / (2 * L))).^2 + (cosh(pi *
yl / (2 * L))).^2)...
    .*integral(@(z) 1./(1+(sin(pi*z/(2)).*sinh(pi*yl/
(2*L))+cos(pi*z/(2)).*sinh(-pi*y0/(2*L)))).^2), 0, 1);
  Pu3 = L * nco * sqrt((sinh(pi * y0 / (4 * L))).^2 + (cosh(pi *
yl / (2 * L))).^2)...
    .*integral(@(z) 1./(1+(sin(pi*z/(2)).*sinh(pi*yl/
(2*L))+cos(pi*z/(2)).*sinh(pi*y0/(4*L)))).^2), 0, 1);
  Pl3 = L * nco * sqrt((sinh(-pi * y0 / (4* L))).^2 + (cosh(pi *
yl / (2 * L))).^2)...
    .*integral(@(z) 1./(1+(sin(pi*z/(2)).*sinh(pi*yl/
(2*L))+cos(pi*z/(2)).*sinh(-pi*y0/(4*L)))).^2), 0, 1);

% Optical path of ray emerging from center of extended source
  P0 = L * nco * sqrt((cosh(pi * yl / (2 * L))).^2)...
    .* integral(@(z) 1./(1+(sin(pi*z/(2)).*sinh(pi*yl/
(2*L))).^2), 0, 1);

% Optical path difference between considered ray and central ray
  delta_pu1 = Pu1 - P0;
  delta_pl1 = Pl1 - P0;
  delta_pu3 = Pu3 - P0;
  delta_pl3 = Pl3 - P0;

% Amplitude distribution calculation
  p_thet = (sech(pi * yl / (2 * L))).^2;
  dthet_dyl = cosh(pi * yl / (2 * L)) * cosh(pi * y0 / (2 *
L))...
    /((sinh(pi * yl / (2 * L))).^2 + (cosh(pi * y0 / (2 *
L))).^2);
  dthet_dy0 = cosh(pi * yl / (2 * L)) / ((sinh(pi * yl / (2 *
L))).^2+1);
```

```
  t_n = 4 * nco * sech(pi * yl / (2 * L)) ./ (1 + nco * sech(pi *
yl / (2 * L))).^2;

% Amplitude contributions from ray trajectories
  A_yu1 = sqrt(p_thet .* dthet_dyl .* t_n) .* exp(-1j * k0 *
delta_pu1);
  A_yl1 = sqrt(p_thet .* dthet_dyl .* t_n) .* exp(-1j * k0 *
delta_pl1);
  A_yu3 = sqrt(p_thet .* dthet_dyl .* t_n) .* exp(-1j * k0 *
delta_pu3);
  A_yl3 = sqrt(p_thet .* dthet_dyl .* t_n) .* exp(-1j * k0 *
delta_pl3);

  A_y0 = sqrt(p_thet .* dthet_dy0 .* t_n);
  A_y = A_yu1 + A_yl1 + A_yu3 + A_yl3 +A_y0;

% Store the complex amplitude distribution
  A_mag(idx) = A_y;
end

% Far field calculation from aperture distribution A_mag(j)
lam= 3*10^-2; % Wavelength at 10 GHz in meter
lam = lam*100; % Wavelength in centimeter
theta_deg = linspace(-90, 90, 1000); % Angle in degrees, covering
full 360 degrees
theta_rad = deg2rad(theta_deg); % Convert degrees to radians
E_far = zeros(size(theta_rad));
ii = complex(0,1);

for i = 1:length(theta_rad)
  integral_result = zeros(size(yl_range));
  for j = 1:length(yl_range)
% integrand = A_mag(j) * exp((ii*2*pi*yl_range(j)/
lam)*sin(theta_rad(i)));
integrand = A_mag(j) * exp((ii*2*pi*yl_range(j)/
lam)*sin(theta_rad(i)));
integral_result(j) = integrand;
end

% Perform numerical integration
E_far(i) = trapz(yl_range, integral_result);
end

% Convert far field amplitude to dB
E_far = abs(E_far)/max(abs(E_far));
E_far = 20*log10(E_far);

plot(theta_deg, E_far, 'k:');
xlabel('\theta (degrees)');
ylabel('Far-field amplitude');
```

```
axis([-90 90 -45 0])
set(gca, 'xlim', [-90 90]);
set(gca, 'xtick', -90:30:90);
grid on
hold on
```

Problems for Students

10.1 Given a short-length phase-corrected horn antenna, the ray trajectory equation in a uniform refractive index medium is expressed as:

$$y(x) = y_0 + \tan(\theta_0)x$$

where y_0 and θ_0 are the initial position and angle of the ray, respectively.

(a) Modify the ray trajectory equation when the refractive index varies as $\eta(x) = 1 + kx^2$, where k is a constant.

(b) Determine the effect of this variation on the aperture field distribution.

10.2 The optical path length L for a ray traveling through a horn with a refractive $\eta(x) = 1 + ax^2$ is given by:

$$L = \int_{x1}^{x2} \eta(x)\sqrt{1+\left(\frac{dy}{dx}\right)^2}\, dx$$

(a) Derive the optical path length for a ray trajectory given by $y(x) = x2$ and $\alpha = 0.01$.

(b) Discuss how the variation in $n(x)n(x)n(x)$ impacts phase correction at the aperture.

10.3 The aperture amplitude distribution of a phase-corrected horn at the design frequency is given as:

$$A(x,y) = A\exp\left(-\frac{x^2}{2\sigma_x^2}\right)$$

(a) Derive the effective aperture area and the corresponding directivity for $\sigma = 0.05$m.

(b) Compare this aperture distribution to a uniform distribution of the same width.

10.4 A horn antenna operates in a hyperbolic secant refractive medium with $\eta(z) = \text{sech}(\alpha z)$.

(a) Derive the ray trajectory equation for rays emanating from an offset source at $z = z_0$.

(b) Calculate the aperture field distribution if $z_0 = 0.1$m and $\alpha = 0.5$.

10.5 The aperture field distribution at a modified frequency is proportional to the inverse of the frequency shift f_m/f_d, where f_m is the modified frequency and f_d is the design frequency.

(a) Calculate the aperture amplitude distribution at $f_m = 12$ GHz if the design frequency $f_d = 10$ GHz and the original distribution is Gaussian.
(b) Analyze the impact of the frequency shift on the far-field radiation pattern.

10.6 During the construction of a phase-corrected horn, the length L, aperture width W, and phase correction factor Φ are related by:

$$\Phi = \frac{2\pi}{\lambda}(L + W\tan(\theta))$$

where λ is the wavelength and θ is the flare angle.
(a) For a horn antenna designed at 5 GHz with $W = 0.3$m, $\theta = 15°$, calculate Φ.
(b) Discuss how variations in L affect the far-field pattern.

10.7 The far-field radiation pattern $F(\theta, \phi)$ of a horn antenna is approximated by:

$$F(\theta,\phi) = E_0\cos(\theta)\,\mathrm{sinc}\left(\frac{W}{\lambda}\sin(\theta)\right)$$

where W is the aperture width.
(a) Plot the radiation pattern for $W = 0.2$m and $\lambda = 0.05$m.
(b) Discuss the relationship between W, beamwidth, and sidelobe levels.

Problems for Instructors

10.1 Deriving the Ray Trajectory Equation: Using Section 10.1.1, derive the ray trajectory equation for a phase-corrected horn antenna with a given refractive index profile, $n(x) = 1 + ax^2$, where a is a constant. Describe how the trajectory changes with varying values of a, and plot the ray paths for at least two different values of a. Explain the significance of the trajectory in focusing the radiation at the horn's aperture.

10.2 Calculating the Optical Path for Ray Focusing: Given a horn antenna with an aperture length of 0.5m and a design frequency of 10 GHz, use the optical path equation (Section 10.1.2) to calculate the path length for rays entering at angles of 0°, 10°, and 20° relative to the horn axis. Explain how the path length variations impact phase correction at the aperture and how this enhances the focusing capability of the horn antenna.

10.3 Designing with Modified Refractive Index Variation: Using the ray trajectory equation with a modified refractive index variation from Section 10.1.4, design a phase-corrected horn with an exponential refractive index profile, $n(x) = e^{bx}$, where b is a constant. Determine the impact of this profile on aperture phase distribution and sidelobe suppression, and discuss how this design could be optimized for compactness without sacrificing performance.

10.4 Aperture Amplitude Distribution at Design Frequency: For a horn antenna operating at a design frequency of 12 GHz, calculate the aperture amplitude distribution using Section 10.1.6. Assume a Gaussian distribution profile at the aperture, and determine how this distribution influences the far-field radiation pattern. Plot the amplitude distribution and discuss its effect on the main lobe width and sidelobe levels.

10.5 Refocusing with a Hyperbolic Secant Refractive Medium: Analyze the refocusing properties of a horn antenna with a hyperbolic secant refractive medium, as discussed in Section 10.1.8. For a source offset by 0.1m from the central axis, calculate the new focal point location. Compare the efficiency of this medium in maintaining focus compared to a uniform medium, and discuss its potential advantages in achieving high directivity.

10.6 Far-Field Radiation Pattern Prediction: Using Section 10.3, predict the far-field radiation pattern of a short-length phase-corrected horn antenna designed for operation at 15 GHz. Calculate the angular width of the main lobe and the first sidelobe level. Compare the predicted pattern with that of a conventional horn antenna without phase correction, and discuss the benefits of phase correction on the antenna's radiation characteristics.

10.7 MATLAB Simulation of Aperture Amplitude for Extended Source: Using MATLAB, simulate the aperture amplitude distribution for a horn antenna fed by an extended source, as described in Section 10.1.10. Assume an extended source diameter of 0.1m and a frequency of 14 GHz. The simulation should:

(a) Display the calculated aperture amplitude distribution.

(b) Plot the resulting far-field radiation pattern.

(c) Analyze the impact of source extension on the amplitude distribution and explain any observed changes in the far-field radiation pattern.

Selected Bibliography

Ata, O. W., "Application of Ray Theory into Microwave Horn Antenna Design," PhD thesis, Cardiff University, UK, 1989.

Ata, O. W., T. M. Benson, and A. Marincic, "Application of Optical Ray Technique to the Design of Short Microwave Horn Antennas with Low Side Lobe Levels," *IEE Proc. H Microwav. Antennas Propag.*, Vol. 137, No. 2, April 1990, pp. 81–88.

Brown, J., "Non-Homogeneous Lenses—Part II," *Proceedings of the Conference on Centimetric Aerials for Marine Navigation Radar*, June 1950, pp. 74–79.

Brown, J., *Microwave Lenses*, London, UK: Methuen and CO, 1953.

Collin, R. E., *Antennas and Radiowave Propagation*, London: McGraw-Peregrinus, 1984.

Cornbleet, S., *Microwave and Optical Ray Geometry*, New York: John Wiley and Sons, 1984.

Marincic, A. S., "Beam Shaping of Sectoral and Pyramidal Horns with Stepped Amplitude Distribution," *Int. Conf. on Antennas and Propagation*, Part I, London, 1981, pp. 328–332.

Rizk, M., and G. Morris, "Broadband Dielectric Loaded Hybrid Mode Horns," *Proceedings of the Conference on Military Microwaves*, Brighton, UK, June 1986, pp. 267–272.

Rust, N. M., "The Phase Correction of Horn Radiators," *Journal of the Institution of Electrical Engineers-Part IIIA: Radiolocation*, Vol. 93, No. 1, 1946, pp. 50–51.

Sha, K., "Radiation Characteristics of Air-Lens E-Plane Fan-Shape Horn Antennas," *Transactions of the IECE*, Vol. 62-B, 1979, p. S57.

CHAPTER 11

Experimental Evaluation of the Antenna Far Field and Aperture Field

The validation of antenna designs requires thorough experimental evaluation, as theoretical models may not always precisely represent real-world performance. Rigorous testing and analysis allow engineers to verify the design's accuracy and make adjustments for optimal functionality. This chapter outlines a standard experimental procedure for evaluating any antenna with a far-field radiation pattern, typically measured inside an anechoic chamber.

As a case study, the HYSEC horn described in Chapter 10 serves as an example of this evaluation method. This antenna is particularly suited for experimental measurements, where its far-field radiation pattern is compared to predicted theoretical results. Additionally, this procedure facilitates the extraction of the aperture field from the measured far-field data. The original HYSEC horn features a smooth profile based on ray theory; however, for comparison and further exploration, a stepped-profile horn with similar dimensions and material has also been constructed.

A stepped-profile horn, shown in Figure 11.1, offers ease of manufacture and allows for the study of design modifications' effects on radiation patterns, particularly

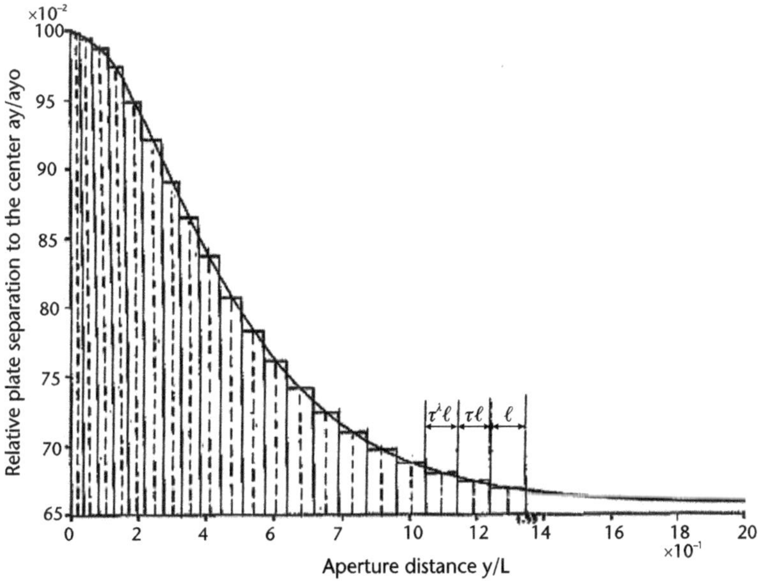

Figure 11.1 Plate separation a_y/a_{y0} versus aperture distance y/L showing the stepped profile on top of the smooth profile.

near the operational frequencies. The scaling factor chosen for the stepped-profile horn is $\tau = 0.95$, where the width of the first step at the edge is denoted as l. The width of each subsequent step closer to the center is scaled by τ, so the width of the second step is $\tau \times l$, the third is $\tau^2 \times l$, and so on. The size of each step is determined by the height from the middle of the step width meeting the horn's curve.

The core focus of this chapter is to experimentally verify the performance of the two HYSEC horns—one with a smooth profile and the other with a stepped profile. The experimental procedure employs a network analyzer to measure the far-field amplitude and phase distributions. This procedure takes place in an anechoic chamber, where the network analyzer is located in an adjacent room to ensure minimal signal interference and reflections from the surrounding environment.

Once the far-field measurements are captured, they are stored in computer files. These data are then processed using interpolation and inverse Fourier-transform techniques to predict the aperture amplitude and phase distributions. The results of this prediction process will be discussed and presented in subsequent sections of this chapter.

11.1 The Network Analyzer Facility

An anechoic chamber which constitutes part of the network analyzer facility is typically 4m long, 1.8m wide, and 2.4m high, which is covered with electromagnetic absorbing material along the walls and most of the floor, but is normally not covered along the ceiling. The effect is not serious but for broad vertical radiation patterns, reflections from the ceiling may alter the general pattern. With the absorber placed on the side walls, the effective width of the chamber that is the distance between the tips of the pyramid-shaped structures on the opposite side walls is only 1.2m.

Anechoic chambers employ specialized materials to absorb electromagnetic waves, preventing reflections and ensuring accurate testing and measurements. These materials, often a combination of carbon-loaded foam and pyramid-shaped structures, are designed to suppress electromagnetic energy across a broad frequency range, from low frequencies to the gigahertz range. The absorption properties of the material typically achieve significant attenuation, surpassing 40 dB or more, making them effective in absorbing radio frequencies, microwaves, and higher frequencies used in radar and communication systems. The pyramid shapes serve to increase surface area for absorption, diffuse incoming waves, and scatter them to prevent reflections, thereby minimizing interference and distortion within the chamber. Consequently, anechoic chamber materials offer effective suppression of electromagnetic waves across various frequencies, ensuring a controlled testing environment free from external interference for accurate measurements and testing in fields such as telecommunications, radar, and electromagnetic compatibility testing.

The measurement equipment, based in the neighboring room, which is of relevance to the far-field amplitude and phase measurements, typically comprises the following:

1. A complete microwave network measurement system composed of a network analyzer, reflection/transmission test set, programmable source, and computing controller. The network analyzer is based on a ratio meter that

uses both reference and test signal inputs. The reflection/transmission test provides the switching required for making transmission and reflection measurements with minimum changes required in the network setup. The system uses a single broadband source and test set that measures return loss and the transmitted signal in amplitude and phase over the 0.5–18-GHz range. The automatic frequency locking allows continuous multioctave sweeps and the conversion from RF to IF allows a sensitivity around 60 dB of dynamic range.
2. A desktop/laptop computer, with the appropriate software and peripherals, such as an external disc drive, a graphic display, and a printer.
3. An desktop/laptop interface that has analog inputs, an analog-to-digital (A/D) converter typically capable of 125,000 samples per second, digital lines that can be selected in pairs of inputs and outputs and a bus connection for use with the computer.
4. A rotary turntable coupled to a suitable stepper motor, thus enabling the former to be driven electrically.
5. A power unit, which drives the stepper motor of the turntable and interfaces the computer through its user-port terminal.

Figures 11.2 to 11.4 show a block diagram illustrating the connections of the equipment at the network analyzer facility, the test antenna, and the source antenna (a pyramidal horn) being appropriately positioned in the anechoic chamber.

The network analyzer was operated in transmission mode as the system was merely used as a source and a receiver throughout the measurements. The coaxial cables from the network analyzer to the setup in the chamber were provided with solid outers and had an overlength of about 12m giving a total attenuation 10 dB. Through the setup in the chamber, the transmitted signal was taken from the analog output ports of the transmission/reflection test set. The analog ouputs were connected to two of the four analog inputs of the interface, making use of its A/D capabilities. On the interface, there were four ranges of voltage sensitivities, which, in terms of the analog outputs of the network analyzer, were equivalent to ±0.1 dB, ±1 dB, ±10 dB, and ±100 dB. In terms of phase, the same ranges of sensitivities were equivalent to ±0.5°, ±5°, ±50°, and ±500°. Througout the measurments, the measured signal was set to allow these to be carried out in the two lower regions of high sensitivity. The sensitivities of the remaining higher regions were enhanced through the software by integrating the measured signal over a period of time. The integration was necessary when measuring the low side lobes of the radiation pattern of the test HYSEC horn antenna because the sidelobe levels were around the noise level.

The software associated with the measurement process in the network analyzer facility included routines for generating the required sequence of pulses, which were sent to the motor drive by the user port of the computer. The drive unit would then provide the proper phasing and the power for driving the stepper motor of the turntable. The software also included a correction for errors in the angular distance movements of the stepper motor of the turntable. The error was estimated to be less than 2° for a full 360° scan.

Other routines were included in the software, which provided common processing operations of the data, such as complex arithmetic operations, plotting, and printing routines. The main program, which contains all these routines, reads first

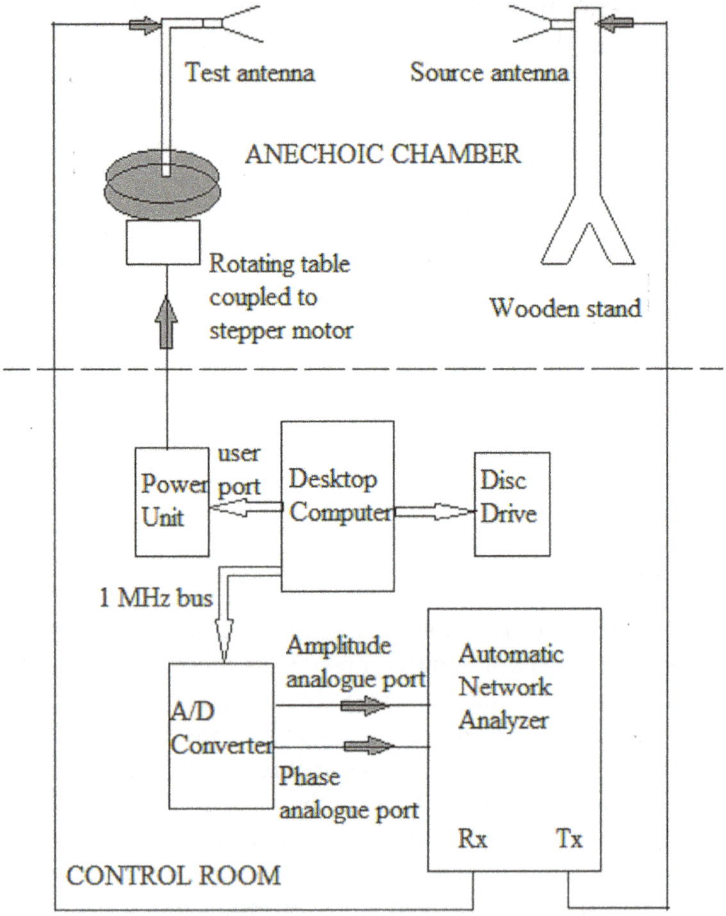

Figure 11.2 A block diagram showing the connection of the equipment at the network analyzer facility.

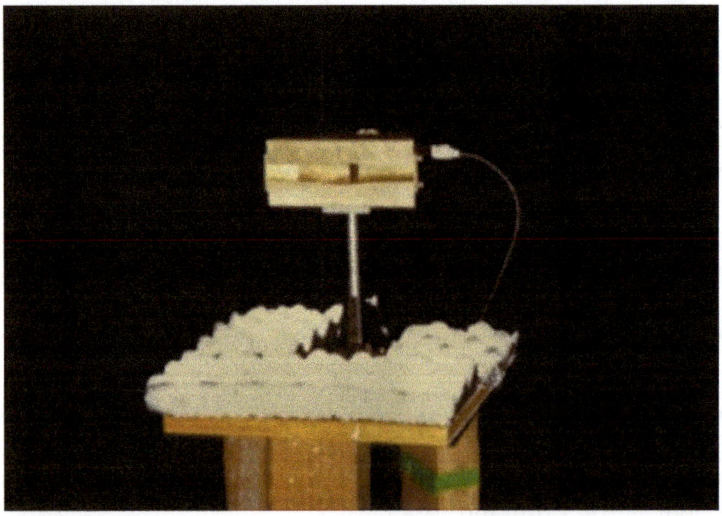

Figure 11.3 A photograph showing the HYSEC (test) horn, the off-setting micrometer, the wooden turntable, and the connecting coaxial cable to the feed inside the anechoic chamber. (Adapted from [1].)

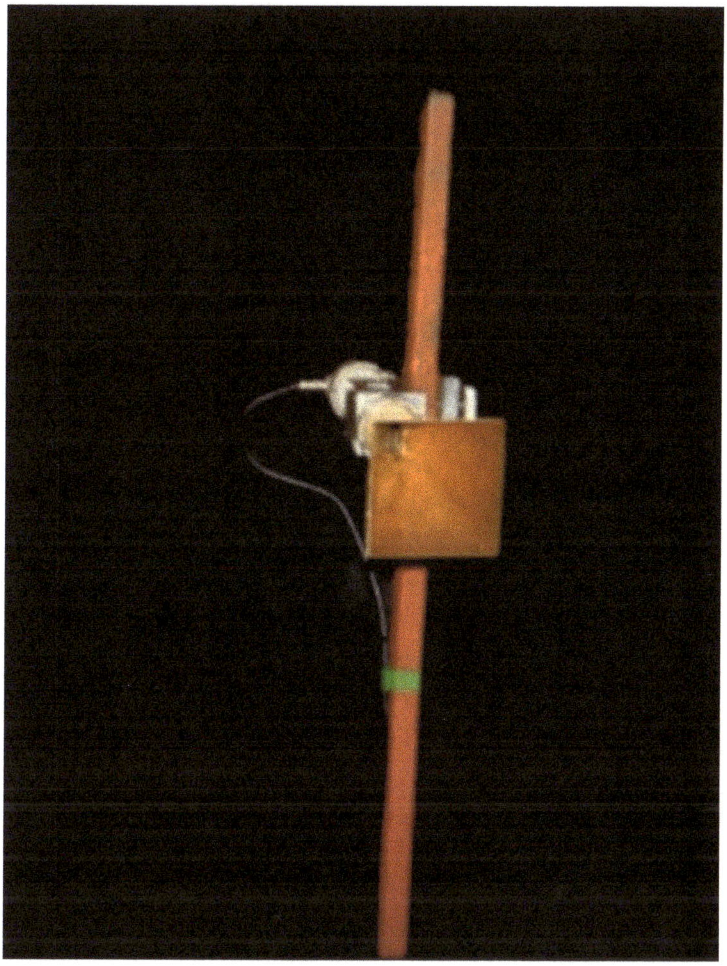

Figure 11.4 A photograph showing the pyramidal horn as a source antenna fixed to a wooden stand inside the anechoic chamber. (Adapted from [1].)

the required variables such as frequency, limits of the angular range at which the data is to be scanned, direction of scan (clockwise or counterclockwise) and the number of integration samples (for example, 257 samples) for improving the signal to noise level. The computer takes full command of the program, once the variables are input, and works out amplitude and phase in decibels and degrees and outputs the measured data onto a disc. Before stopping and returning control to the operator, a plot of the predicted far field is given on the graphic display.

11.2 Measurement of the E-Plane Radiation Pattern

In the measurement of the radiation patterns of the original and modified HYSEC horns, a pyramidal horn acting as the source antenna was fixed to a freestanding wooden post at one end of the anechoic chamber in a central position. At the opposite end of the chamber, the test antenna was mounted on top of a holder attached to a micrometer, with the center of the horn aperture lying over the determined center of rotation of a wooden stool, which is screwed, in turn, to the top of the rotating table.

Both horns were positioned at equal distance from the ground, approximately at middle height of the chamber. The distance between the apertures of the two horns was measured to be 240 cm, ensuring that the radiation distance exceeded the accepted $2D^2/\lambda$ criterion for the measurement of the far-field radiation pattern. Alignment of the vertical and horizontal planes of the rigidly supported source and test antennas was confirmed by the use of a spirit level.

Some of the measurements were taken with special absorber inserts, which were cut to the size and shape required to cover the back and short side plates of the original and modified HYSEC horns. The inserts were of the absorber type Eccosorb SF9.5 with chosen specifications to suit the required frequency range of operation. The purpose of using such absorber inserts was to reduce reflections inside the HYSEC horns and compare their effects, on the radiation pattern characteristics, to those with the inserts removed, at various frequencies of operation. Furthermore, the effect of shifting the aperture from the center of rotation, on the radiation patterns, was studied. That was undertaken by using the micrometer to offset the aperture by a known distance of less than a wavelength in the direction of propagation. All the required E-plane measurements were made through an angular range of 180° in 1° steps.

11.3 Experimental Verification of the Design

E-plane radiation pattern measurements were carried out in an anechoic chamber, utilizing an automatic network analyzer system for measuring far-field amplitude and phase distributions. Some of the measurements were taken with absorber inserts covering the back and short side plates of the horn. The inserts, as mentioned earlier, were of the absorber type Eccosorb SF9.5 whose absorption characteristics suit the frequency range studied. Figure 11.5 shows the measured radiation pattern of the smooth horn at the design frequency 10 GHz, with and without the absorbers placed at the back and the edge plates.

Placing or removing the absorbers would allow comparison with predicted patterns with or without edge reflections occurring due to rays emanating at relative higher angles. It is quite unusual, in the practical sense, for a horn antenna to be partially unbounded, and hence metallic flat plates are fixed at both edges to complete the boundary walls. These edge plates will give rise to ray reflections, which will affect the aperture distribution. The contribution of those edge plate reflections could be simply modeled in the computation of the amplitude distribution.

At the design frequency (Figure 11.5), with the absorber inserts removed, the sidelobe level is below −27 dB while the shoulders are below −18 dB. Broad agreement can be seen between the experimental behavior and that calculated using the ray theory approximation. The effect of edge-plate reflections can also be taken into account. Higher main-lobe shoulders and sidelobe levels are then observed. It can further be noticed that sidelobes appearing at ±60° with no absorber inserts in the design have been replaced by wide nulls, when the inserts are in place.

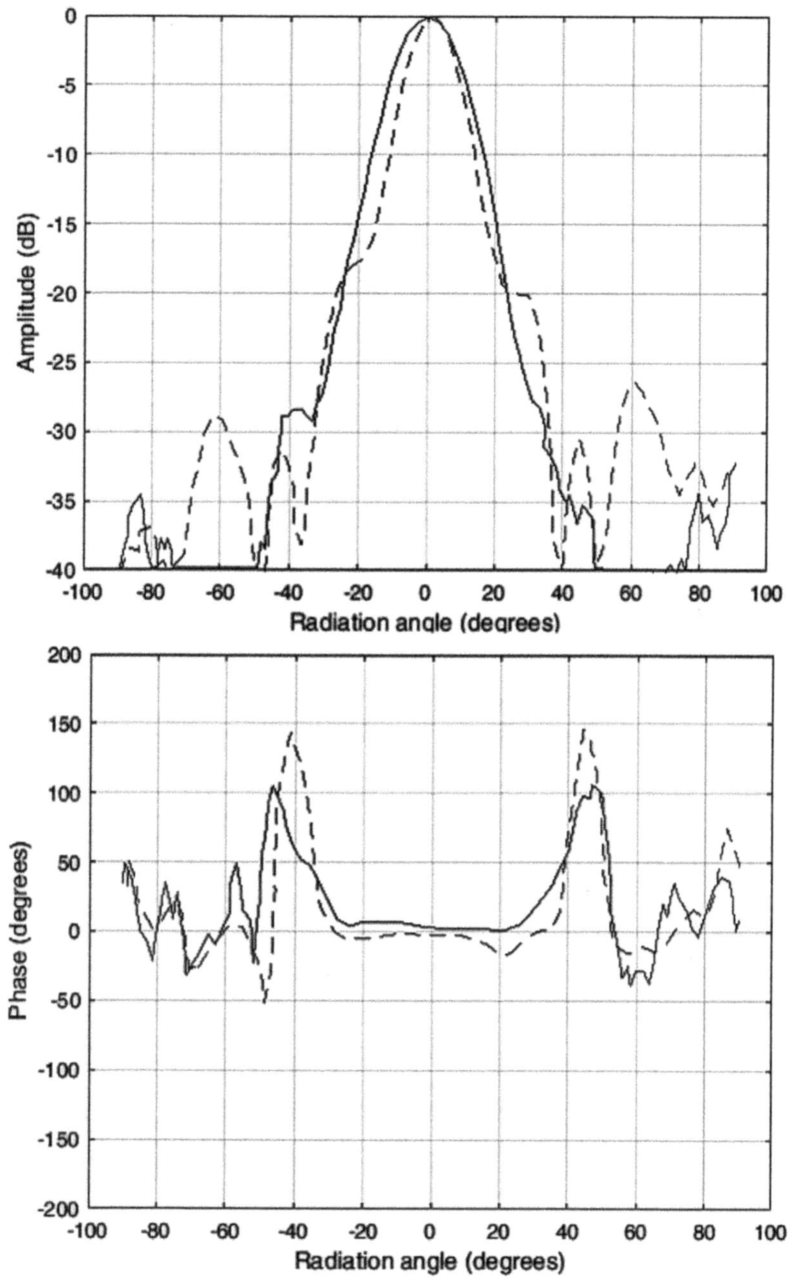

Figure 11.5 Radiation pattern of the smooth profile horn at 10 GHz: ——— with absorber inserts and – – – – without absorber inserts. (Adapted from [1].)

Figures 11.6 and 11.7 show the radiation patterns at other frequencies than those of the design, at 10.5 GHz and 9.5 GHz, respectively. It can be seen that the general effect of the absorber inserts is to decrease the HPBW and reduce the sidelobe level. Wide nulls were observed at 9.5 GHz, with absorbers, and raised shoulders higher than −20 dB were observed at 10.5 GHz, without absorbers. The sidelobe level at 10.5 GHz and with absorber inserts was approximately −25 dB.

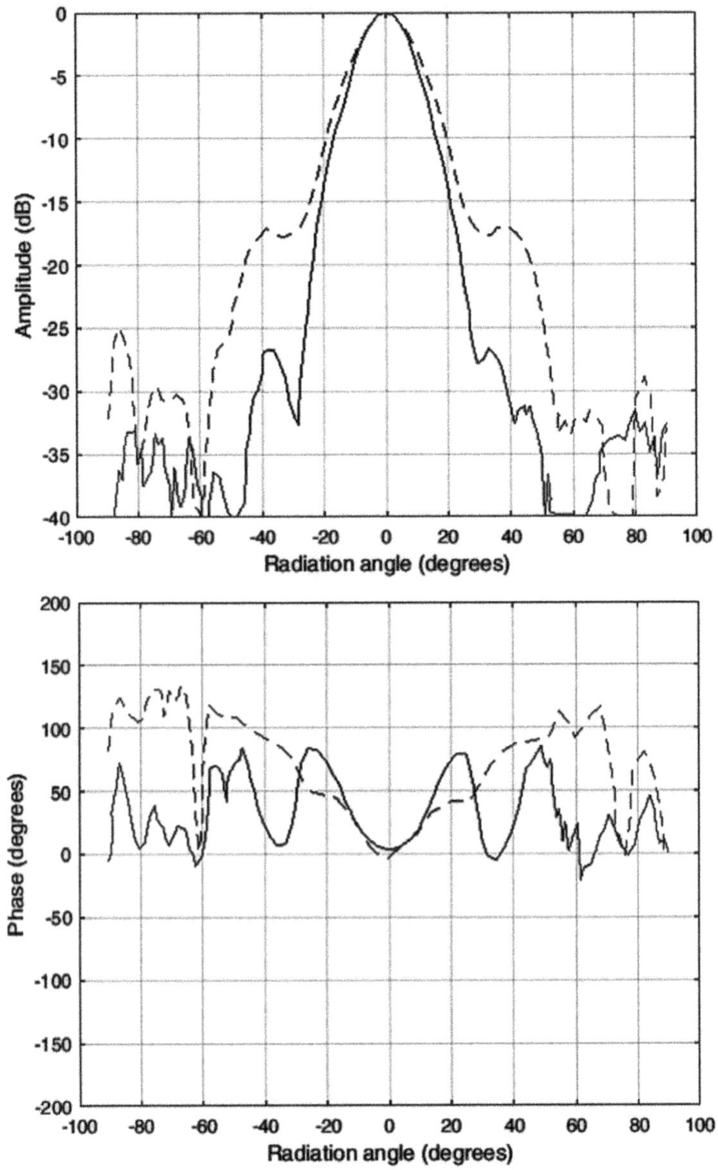

Figure 11.6 Radiation pattern of the smooth profile horn at 10.5 GHz: ——— with absorber inserts and – – – – without absorber inserts. (Adapted from [1].)

Stepping the profile of the horn plates introduces a physical tolerance, which seems to manifest itself as a small change in the radiation pattern at the design frequency as shown in Figure 11.8. Although the predicted shoulders of the main lobe in the step profile horn get smoothed out, the HPBW is close to that of the original horn. The effect of stepping the profile on the sidelobes is less significant. These were below –30 dB and their locations were reasonably well maintained. With the relative ease of manufacturing, the step profile horn is a good alternative to the smooth profile design. Figure 11.9 compares the radiation pattern of the stepped profile horn, without absorbers, at 10.5 GHz to that at 9.5 GHz.

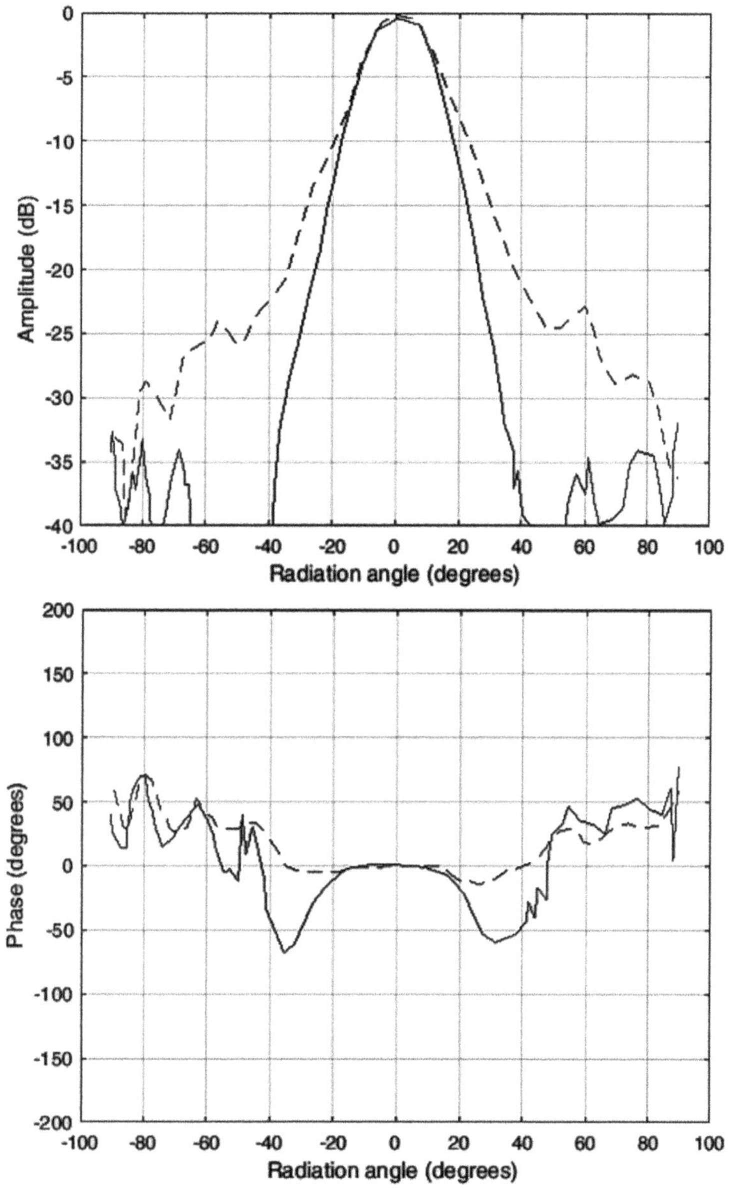

Figure 11.7 Radiation pattern of the smooth profile horn at 9.5 GHz: ——— with absorber inserts and — — — — without absorber inserts. (Adapted from [1].)

The H-plane radiation patterns from the original smooth profile horn at 10 GHz, 10.5 GHz, and 9.5 GHz are characterized by broad HPBWs of about 30°, 50°, and 60°, respectively. The change in peak power relative to that at 10 GHz is around –5 dB at 10.5 GHz and –8 dB at 9.5 GHz.

As far as aperture field evaluation, an inverse fast Fourier transform (FFT) routine was used to evaluate the aperture field from the experimental far-field data, which was measured for 10 increments of the radiation angle φ between ±90°. Accordingly the FFT routine is preceded by an interpolation routine to transform the φ domain into the sin φ domain required for the transform. In the anechoic chamber,

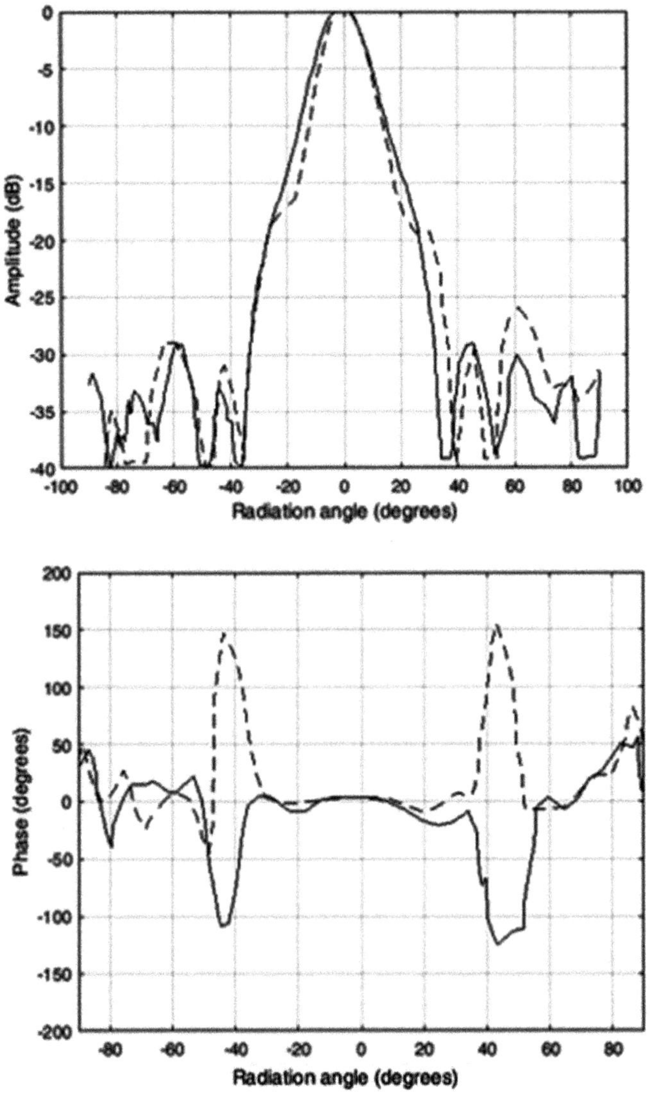

Figure 11.8 Radiation pattern of the test horn without absorber inserts at 10 GHz: ——— stepped profile horn, – – – – smooth-profile horn. (Adapted from [1].)

the distance between the source and test antenna was 240 cm. This corresponds to a 360° (δ/λ) phase retardation error due to the finite radiation distance, where $\delta = r' - r$ (r' is the radial distance from the edge of the horn in the y-direction to the far-field point P, and r = 240 cm) for the experimental horn. Now r' works out as 240.121 cm, λ = 3 cm (at the design frequency), and, hence, the phase retardation error equals 14.50. Figure 11.10 shows the aperture distribution of the smooth profile HYSEC horn with absorber inserts at 10, 9.5, and 10.5 GHz, evaluated from measured far-field data. The E-plane aperture extends between ±7.62 cm. The linear phase effect, observed at 10 GHz is probably due to either a lateral shift in the aperture from the center of rotation or asymmetry in the manufacture or separation

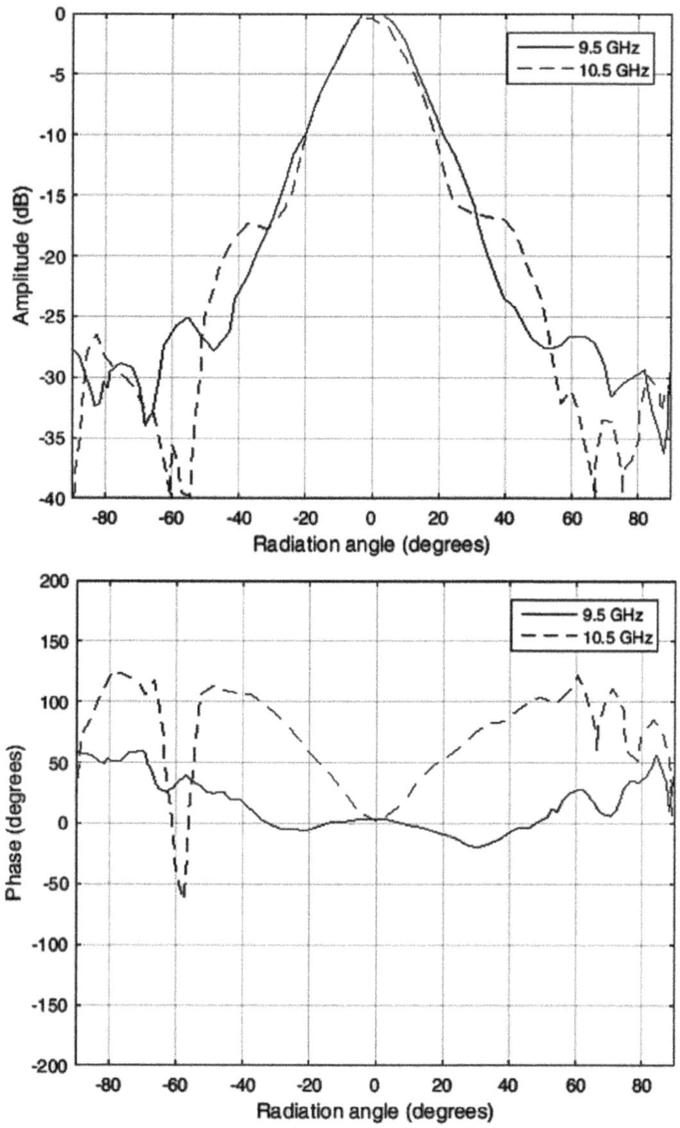

Figure 11.9 Radiation pattern of the stepped profile horn without absorber inserts at 9.5 and 10.5 GHz. (Adapted from [1].)

of the horn plates. Apart from this effect, the phase distribution across the aperture is within the phase retardation error.

The amplitude distribution tapers down to approximately −20 dB at the edges, in good agreement with the theoretical value. At 9.5 GHz, the aperture amplitude distribution looks smoother than at the design frequency and tapers steeply to below −30 dB at the aperture edge. The aperture phase distribution remains within 20°, which is mostly associated with the phase retardation error. Absorber inserts placed on the edge and back plates of the horn increased the tapering and smoothed the amplitude distribution at the design frequency but decreased the gain, from around

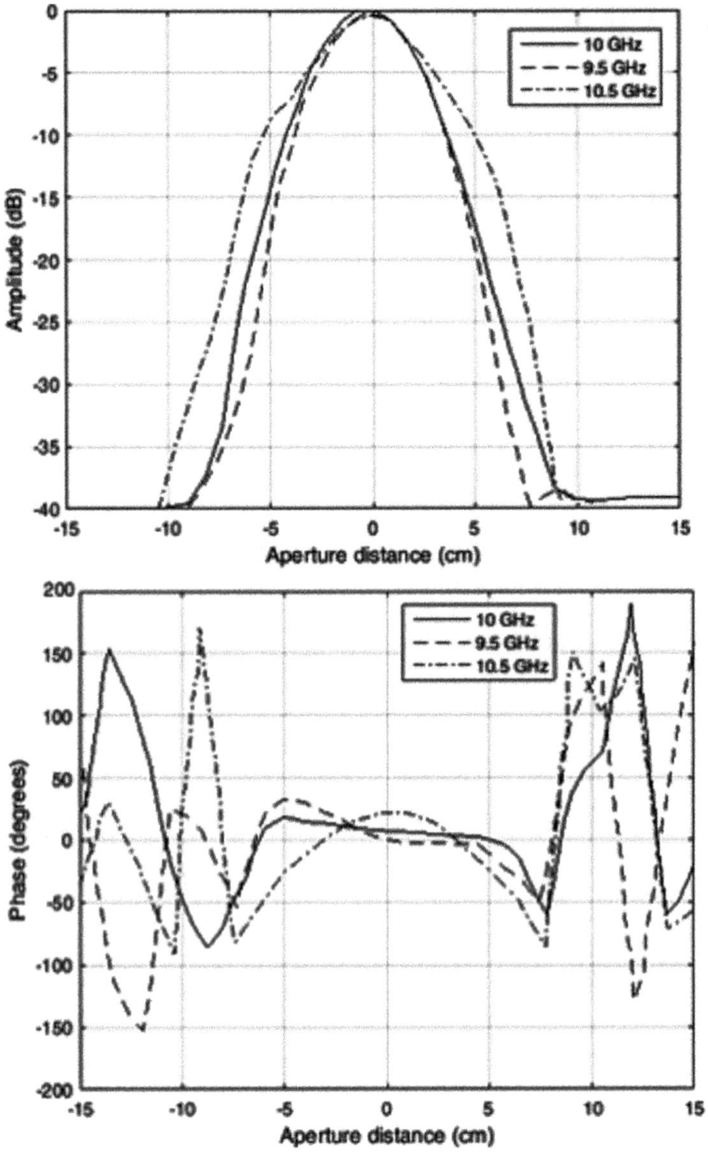

Figure 11.10 Aperture distribution of the smooth profile HYSEC horn with absorber inserts at 10, 9.5, and 10.5 GHz. (Adapted from [1].)

15 to 14. With a primary feed application in mind, the loss in gain is not of major concern. Further, Figure 11.11 shows the amplitude distribution of a stepped profile horn at the design frequency with the aperture plane over the center of rotation and another with a small shift of 8 mm.

11.4 Optical Ray Technique Versus Finite-Difference Method

The work presented in this chapter has experimentally verified the short-focus HYSEC horn, explored in Chapter 10. The broad agreement between the theoretical predictions, based on a ray analysis treatment, and the experimental aperture

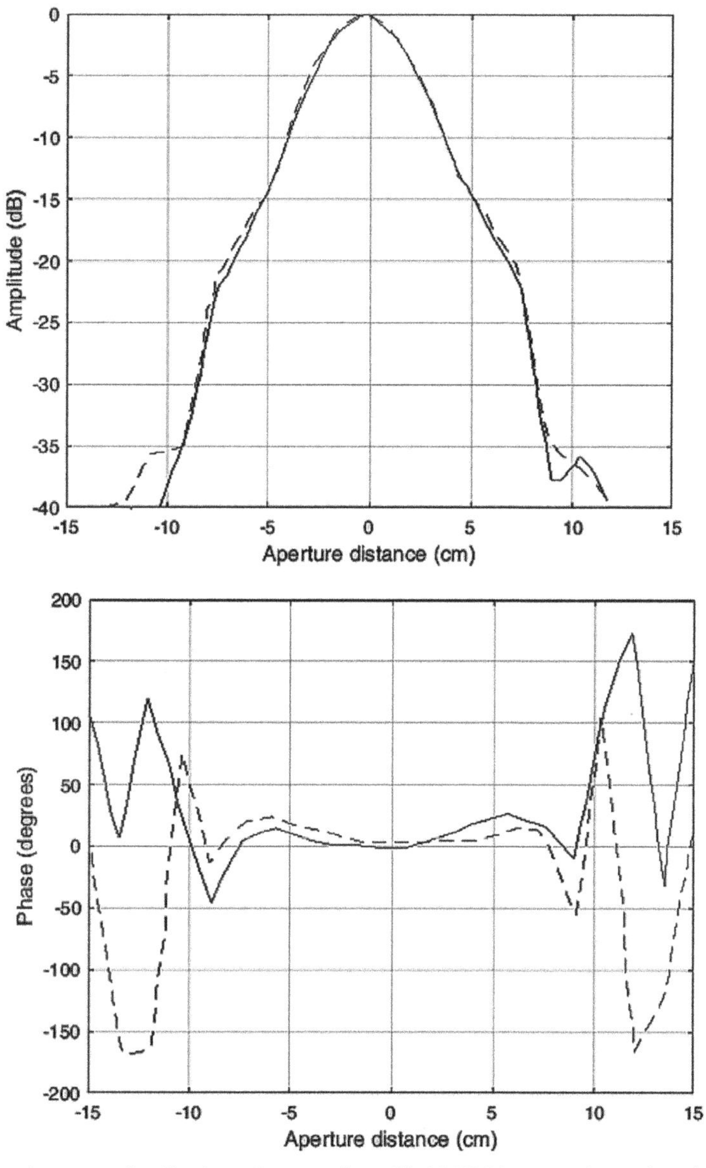

Figure 11.11 Aperture distribution of stepped-profile HYSEC horn without absorber inserts at 10 GHz: ——— aperture plane over the center of rotation, — — — — aperture plane shifted by 8 mm from the center of rotation away from the source antenna. (Adapted from [1].)

and far-field distributions suggests that the application of optical ray theory into microwave horn antenna design is realistic. Point-by-point comparisons have been made of the taper ratio, the aperture-phase compensation, the radiation HPBW, the sidelobe level, and the general radiation-pattern form. A ray theory approach can be much simpler than an electromagnetic one, based on Maxwell's equations, in the modification of far-field radiation patterns for achieving the desired low sidelobe levels. Although the radiation-pattern modification of a horn is largely dominated by its aperture geometry, as well as its illumination, an analytical solution of an electromagnetic nature is limited to the simple standard rectangular, circular, and elliptical apertures. Once a simple modification to the aperture geometry is introduced, such as in the channel-insert E-sectoral horn studied in Chapter 9, the computational finite-difference technique becomes indispensable as the level of mathematics in any possible analytical solution is highly complex.

The reduction of the electromagnetic field expression describing the double-region aperture of the E-sectoral horn into a line integral that involves the varying field along the rim eases the task of computation and makes the finite-difference technique very attractive. However, the lowest sidelobe level in the far-field radiation pattern cannot be predicted beforehand. A series of computations, which involves a range of different channel sizes, must be performed in the theoretical predictions. The run time needed for each computation depends on the size of the mesh and the number of iterations required. Furthermore, it has been shown experimentally that the channel profile (smooth or stepped) between the throat and the aperture of the horn affects the radiation sidelobe level. Consequently, the combined channel size and profile will have to be submitted to a detailed experimental study for achieving the lowest possible sidelobe level in the radiation pattern.

The advantages of the optical ray technique over the finite-difference method are in that the computation of the aperture field, from which the far-field pattern can be predicted, is relatively simpler and requires much fewer units of computer run time. In addition, it provides the tool to produce a phase-corrected aperture with the necessary modified geometry at a shorter axial horn length than the known standard horns even when these are loaded with phase-correcting lenses or metallic fins. The simplicity in applying the ray-tracing technique to the HYSEC horn lies in the fact that the effective refractive index function needed in designing the relevant plates to the required profile is 1D. Furthermore, a sensible approximation of the polar power distribution at the feed eliminates the high complexity of a possible exact solution.

Despite the success of ray theory in the HYSEC horn design, a notable drawback is its narrow bandwidth. Because the horn's internal structure is on the order of a wavelength, frequency shifts alter the effective refractive index, introducing aperture phase errors. For instance, at 10.5 GHz, these errors become pronounced, as predicted earlier. At 9.5 GHz, the performance improves slightly, but the effective aperture physically shrinks. This shrinkage aligns with predictions in Chapter 10, where rays do not propagate effectively beyond certain lateral boundaries. These boundaries, at lower frequencies, tend to pull towards the central axis, eventually nearing the lower limit of the X-band (WG16 waveguide cutoff at 6.56 GHz).

Another challenge lies in ray reflections from the horn's edge and back plates, which become increasingly complex and frequency-dependent. These reflections can elevate shoulder and sidelobe levels in the radiation pattern. Experimentally, absorber inserts proved effective in mitigating these reflections, reducing sidelobes significantly even with limited absorber reflectivity (e.g., −12 dB at 10.5 and 11 GHz). However, it is unclear whether edge or back reflections dominate at different frequencies. Careful cutting and placement of absorber inserts, especially along back plates, are essential to preserve E-plane symmetry in the radiation pattern.

At some frequencies near the design point, the smooth profile HYSEC horn exhibited anomalous radiation behavior, such as broader main lobes with high-level shoulders. This behavior is attributed to multireflections within the horn aperture's end regions. Absorber inserts help to suppress these resonant effects, as shown in the 9.8-GHz radiation pattern.

The transition from the inhomogeneous refractive horn region to free space can also introduce unwanted aperture plane reflections, although these were not specifically addressed in the experiments. Future work may include adding thin metal strips near the aperture edges to control impedance matching and reflection minimization.

Stepped profile plates, an alternative to the smooth design, introduce small physical tolerances that minimally affect radiation patterns. Their simpler manufacturing process makes them a practical option, with negligible impact on sidelobe levels at the design frequency.

Lastly, the measurements taken in the anechoic chamber are subject to experimental uncertainties. Despite careful setup, factors such as uncovered ceilings, the wooden stool holding the antenna, the coaxial cable's length and tension, and its movement during antenna rotation could introduce reflection artifacts. These setup-induced reflections potentially distort the measured radiation patterns, underscoring the need for highly controlled testing environments.

Two major design approaches for microwave horn antennas are discussed: ray tracing and the finite-difference method. The finite-difference approach, although computationally intensive depending on mesh size, is highly flexible and essential for modeling complex geometries, particularly for high-frequency sectoral horns. In contrast, the optical ray technique is simpler and faster for certain configurations, especially when dealing with 1D refractive index variations.

Experimental results based on Table 11.1 show strong performance from horns designed using ray techniques. Two horn profiles, each with dimensions 15.24 cm × 2.286 cm and axial length 5.72 cm, were tested. The stepped profile horn achieved sidelobe levels better than −23 dB, and the smooth profile horn exceeded −27 dB at the design frequency. At sub-10-GHz frequencies, the stepped horn with absorber inserts achieved sidelobe levels of −32 dB, demonstrating effective suppression of internal reflections and distortion. Even at frequencies above 10 GHz, performance remained solid, confirming the value of both the ray-based design method and absorber technology in improving horn antenna performance.

The results demonstrate the effectiveness of applying ray theory into the design of E-plane horns, as well as the effectiveness of channel absorbers in combatting unwanted ray reflection inside the horn.

Table 11.1 Measured HPBW and Sidelobe/Shoulder Level (SLL) of the HYSEC Horn

	Stepped Profile Horn (No Absorber Inserts)		*Stepped Profile Horn (with Absorber Inserts)*		*Smooth Profile Horn (No Absorber Inserts)*		*Smooth Profile Horn (with Absorber Inserts)*	
Frequency (GHz)	*HPBW (degree)*	*SLL (degree)*	*HPBW (degree)*	*SLL (degree)*	*HPBW (degree)*	*SLL (degree)*	*HPBW (degree)*	*SLL (degree)*
9.0	35	−14	35	−32	33	−13	37.5	−23
9.5	21.5	25	20	32	21.5	23	20	33
10.0	16	−29	19.5	−23	13.5	−27 shoulder level (−18)	19	−28
10.5	18.5	−27 shoulder level (−17)	17	−24	19	−25 shoulder level (−17)	17	−27
11.0	25.5	−13	19.5	−21	26	−13	22	−28

Table 11.2 compares the HPBW and sidelobe level of two horns of aperture size $A \times B$ = 20.96 cm × 2.286 cm and a semi-flare angle of 18° (axial length = 32.25 cm). Channel inserts measured a/A = 0.15 and b/B = 0.05. The feed was a WG16 waveguide.

Introducing channel inserts, by applying numerical techniques to an E-sectoral horn, has helped the realization of a design that produced effectively suppressed sidelobes, better than −32 dB, at the design frequency, better than −23 dB at 7 GHz and better than −26 dB at 12.4 GHz.

Moreover, the cross-polarization for the horn with the channel inserts is still maintained at all frequencies, 7.0, 10.0, and 12.4 GHz, to a level better than −23 dB, compared to a level better than −29 dB for a similar horn but with no channel inserts. Overall, numerical techniques utilized in designing optimized channel inserts are effective but can relatively be less easy than applying a ray technique in the design of horns with suppressed sidelobe level.

Problems for Students

11.1 A vector network analyzer is used to measure the reflection coefficient S_{11} of an HYSEC horn antenna.
   (a) If the measured S_{11} is −15 dB at 10 GHz, calculate the voltage standing wave ratio.
   (b) Discuss whether this value of S_{11} is acceptable for an antenna operating in this frequency range.

11.2 The calibration of a network analyzer involves using standard calibration kits with known open, short, and load impedances.
   (a) Explain the purpose of each calibration standard in ensuring accurate measurements of the HYSEC horn's far-field pattern.
   (b) Describe a possible error introduced if the calibration plane is not properly defined and its impact on radiation pattern measurements.

11.3 The E-plane radiation pattern of an HYSEC horn antenna is measured at 10 GHz using a rotational stage.
   (a) For the measured pattern, $E(\theta) = E_0\cos(\theta)$, calculate the beamwidth between the first nulls.

Table 11.2 Measured HPBW, Sidelobe/Shoulder Level (SLL) and Cross-Polarization of E-Plane Sectoral Horn

	E-Plane Sectoral Horn (No Channel Inserts)			*E-Plane Sectoral Horn (with Channel Inserts)*		
Frequency (GHz)	*HPBW (degree)*	*SLL (dB)*	*Cross-Polarization (dB)*	*HPBW (degree)*	*SLL (dB)*	*Cross-Polarization (dB)*
7.0	10.9	−4.15	−29.8	15.75	−23.85	−23
10.0	24.7, ripple on top of main beam	−10.75 shoulder level (−5.08)	−34.9	9.28	−32.6	−24.5
12.4	33 dip on top of main beam	−4.5 shoulder level (+0.78)	−32.7	7.71	−26.1	−29

(b) Determine the directivity if the main lobe power accounts for 90% of the total radiated power.

11.4 The far-field distance R is determined using the Fraunhofer criterion $R \geq 2D^2/\lambda$, where D is the aperture diameter and λ is the wavelength.
   (a) Calculate the minimum far-field distance for an HYSEC horn with $D =$ 0.5m at 15 GHz.
   (b) Verify if measurements conducted at $R = 10$m meet the far-field criterion.

11.5 During the experimental verification of an HYSEC horn antenna's design, the aperture field distribution is measured and compared to the theoretical Gaussian distribution.
   (a) If the measured aperture field amplitude deviates by 10% from the theoretical value, discuss the potential causes.
   (b) Suggest improvements in the fabrication process to minimize this deviation.

11.6 The cross-polarization levels of the HYSEC horn are measured to ensure compliance with design specifications.
   (a) If the measured cross-polarized radiation is −25 dB relative to the copolarized radiation, evaluate the antenna's suitability for high-purity polarization applications.
   (b) Calculate the isolation between copolarized and cross-polarized components.

11.7 The gain of an HYSEC horn antenna is measured using the substitution method, with a reference horn of known gain $G_{\text{ref}} = 15$ dB.
   (a) If the received power difference between the test and reference antennas is $\Delta P = 3$ dB, calculate the gain of the HYSEC horn.
   (b) Discuss the advantages and limitations of the substitution method for gain measurement.

Problems for Instructors

11.1 Understanding the Network Analyzer Facility: Based on Section 11.1, explain the primary components of a network analyzer facility used in the evaluation of horn antennas. Describe the steps for calibrating the network analyzer and discuss the importance of each calibration step for achieving accurate measurement of the horn antenna's far-field characteristics.

11.2 E-Plane Radiation Pattern Measurement: In Section 11.2, the E-plane radiation pattern is measured using a network analyzer. Set up an experiment to measure the E-plane pattern for an HYSEC horn antenna at a frequency of 10 GHz. Describe the setup, including the positioning of the antenna, the equipment used, and the data collection procedure. Plot the expected radiation pattern, and explain how you would verify whether the measured results match theoretical expectations.

11.3 Analysis of Measurement Errors in an E-Plane Radiation Pattern: Identify at least three sources of error that may arise during the experimental measurement of the E-plane radiation pattern. Discuss how each error could affect the accuracy of the measurements and propose methods for mitigating these errors. Calculate the potential impact of a $\pm 1°$ misalignment in the measurement setup on the E-plane pattern for a horn antenna with a 20° beamwidth.

11.4 Experimental Verification of Design Theory: Using the procedures in Section 11.3, describe an experimental approach to verify the theoretical design of a HYSEC horn antenna. Detail the steps involved in comparing the measured results with the theoretical design specifications, such as gain, beamwidth, and sidelobe level. Explain how you would interpret discrepancies between the theoretical and experimental results.

11.5 Aperture Field Measurement and Analysis: Outline a procedure to measure the aperture field distribution of the HYSEC horn antenna at 10 GHz. Describe how the aperture field measurement can be used to predict the far-field radiation pattern. Use the measured data to calculate and plot the expected main lobe and sidelobe levels in the far-field pattern, comparing them to the theoretical pattern.

11.6 Far-Field Pattern Prediction from Aperture Field Data: Suppose you have measured the aperture field distribution and recorded the amplitude and phase at various points across the aperture. Develop a method to predict the far-field radiation pattern from this aperture data using Fourier transform techniques. Discuss the assumptions made in this approach and the expected limitations.

11.7 MATLAB Simulation for Far-Field Prediction Based on Aperture Field: Using MATLAB, simulate the far-field radiation pattern of the HYSEC horn antenna based on a given set of aperture field measurements at 12 GHz. Assume an aperture size of 0.3m and a Gaussian amplitude distribution. The simulation should:
   (a) Generate a plot of the simulated far-field radiation pattern.
   (b) Compare the simulated pattern to an idealized pattern for a horn with the same aperture size.
   (c) Analyze the effects of amplitude taper on sidelobe levels in the far field.

Reference

[1] Ata, O. W., "Application of Ray Theory into Microwave Horn Antenna Design," PhD thesis, Cardiff University, UK, 1989.

Selected Bibliography

Chapra, S., and R. Canale, *Numerical Methods for Engineers*, New York: McGraw-Hill, 1985.

Clarke, R. H., and J. Brown, *Diffraction Theory and Antennas*, Ellis Harwood Ltd, 1980.

Cross, A. W., *Experimental Microwaves*, Marconi Instruments Ltd., 1977, p. 90.

ECCOSORB SF Thin Resonant Absorbers, Technical Bulletin 8-2-18, Canton, MA: Emerson & Cumming Inc.

Hewlett Packard Ltd., *HP/8408S Automatic Network Analyser System.*

Johnson, R. C., and H. Jasik, (eds.), *Antenna Engineering Handbook*, Second Edition, New York: McGraw-Hill, 1984.

CHAPTER 12

Novel Phase-Corrected Compact Antenna

This chapter introduces a novel short-length, phase-corrected compact antenna, designed to overcome limitations associated with traditional horn antennas and other conventional radar antenna types. Unlike standard horn antennas that typically function as passive feed elements or waveguide transitions, the proposed structure serves as a self-contained antenna system, engineered with compactness, efficiency, and precision in mind. Its potential application in marine radar systems stems from its ability to deliver a narrow beamwidth and low sidelobe levels while minimizing issues like feed blockage commonly found in cheese antennas.

The design leverages the truncation of a near-Gaussian aperture distribution to provide strong control over sidelobe levels in the radiation pattern. Experimental results from the HYSEC horn configuration achieved sidelobe suppression below −27 dB for a normalized aperture height of $A_y/L = 2.67$. However, to meet the higher directivity demands of radar applications, an increase in effective aperture size is necessary. This requirement traditionally entails a proportional increase in horn length, given the interdependence of the y and z dimensions in the electric field equation (10.15). To address this challenge, a new effective refractive-index function has been introduced, one that maintains the desirable aperture distribution while significantly shortening the horn length.

By doing so, the proposed antenna design advances beyond the realm of typical horn structures and enters the domain of high-performance compact antennas. Compared to slotted-waveguide arrays, widely used in radar systems, the novel design offers improved robustness against second-order effects such as slot-length inaccuracies, internal reflections, and higher-order mode coupling. These imperfections can lead to substantial degradation in sidelobe levels, sometimes by as much as 7 dB from a designed target of −35 dB, particularly if tight fabrication tolerances are not met.

Although the design phase considered the use of advanced dielectric materials, including loaded foams and artificial dielectrics, financial and time constraints prevented immediate fabrication and testing. As such, a direct performance comparison with commercial slotted arrays remains inconclusive. Nevertheless, the proposed concept demonstrates strong theoretical promise and experimental viability. Further experimental evaluation, along with efforts toward batch manufacturing and prototyping, is strongly recommended to validate its performance and explore its full potential in real-world radar applications.

12.1 Shortened Length Horn Design

Consider a stratified medium in which the effective refractive index is of the type:

$$\eta^2(y,\theta) = \frac{\eta_{co}^2}{b^2\cosh^2\left(\frac{\pi y}{2L}\right)} + d\cdot h_\theta^2 \tag{12.1}$$

where $b = L'/L$ (L' being the new horn length), d is a constant that depends on b, and $h_\theta = \eta_{co}\cos\theta$ for a particular ray.

The trajectory of a ray emerging from the feeding point at a given angle θ is calculated from:

$$z(y) = \int_0^y \frac{h_\theta dy}{\left(\eta^2(y,\theta) - h_\theta^2\right)^{1/2}} \tag{12.2}$$

As shown in Chapter 10, such integral form can be solved by substituting for $\eta^2(y, \theta)$ from (12.1) into the above integral (12.2). After some calculation, the following is obtained:

$$\sinh\left(\frac{\pi y}{2L}\right) = \tau\cdot\sinh\left(\frac{\pi z}{2bL}\right) \tag{12.3}$$

where

$$\tau = \left(\frac{1 - b^2(1-d)\cos^2\theta}{b^2(1-d)\cos^2\theta}\right)^{1/2} \tag{12.4}$$

If $b^2(1 - d)$ is made equal to 1, then

$$\tau = \tan\theta \tag{12.5}$$

$$d = 1 - \frac{1}{b^2} \tag{12.6}$$

Hence, (12.3) transforms into

$$\sinh\left(\frac{\pi y}{2L}\right) = \tan\theta\cdot\sinh\left(\frac{\pi z}{2bL}\right) \tag{12.7}$$

It is obvious that (12.7) is similar in form to (10.15) except that the new focal length has become $2bL$. The original design corresponds to $b = 1$ and $d = 0$. In the new design, b can be less than 1 and d then becomes a negative constant.

The design equation for implementing the refractive index distribution $\eta(y, \theta)$ is based on the calculation of an effective refractive index from the propagation constant for the TE_{01} mode of a dilectric-filled parallel-plate waveguide. Hence, the phase coefficient for the TE_{01} mode is

$$\beta^2 = \omega^2 \mu_0 \varepsilon_0 \varepsilon_r - \left(\frac{\pi}{a_{y,z}}\right)^2 = \left(\frac{\omega\eta(y,\theta)}{c}\right)^2 \tag{12.8}$$

where ε_r is the relative permittivity of the waveguide medium, $a_{y,z}$ (i.e., $a(y, z)$) is the relevant plate separation, and other symbols have their usual meaning.

The variation of a_y will provide the required change in η.

Substituting for $\eta^2(y, \theta)$ from (12.1) into (12.8) yields:

$$\varepsilon_r - \left(\frac{\pi c}{a_{y,z}\omega}\right)^2 = \eta_{co}^2 \left(\frac{\operatorname{sech}^2\left(\frac{\pi y}{2L}\right)}{b^2} + d \cdot \cos^2\theta\right) \tag{12.9}$$

Employing the trignometric relation $1 + \tan^2\theta = 1/\cos^2\theta$, and substituting for $\tan\theta$ from (12.7) gives

$$\cos^2\theta = \frac{\sin^2\left(\frac{\pi z}{2bL}\right)}{\sin^2\left(\frac{\pi z}{2bL}\right) + \sinh^2\left(\frac{\pi y}{2L}\right)} \tag{12.10}$$

except for $(z, y = (0, 0))$ where the ratio becomes undefined.

MATLAB Program 21 calculates and plots the squared refractive index function and the plate separation variation, versus z/L, described by (12.9) and (12.10), as shown in Figures 12.1 and 12.2, for the case $b = 0.5$ and $\eta^2_{co} = 0.5697$ ($\varepsilon_r = 1$, $a_0 = 2.286$ cm, and $f = 10$ GHz). The same figures show that $\eta^2(y, z) < 4\eta^2_{co}$ and so $a(y, z)$ can be found to give the required $\eta^2(y, z)$ provided $\varepsilon_r > 4\eta^2_{co}$.

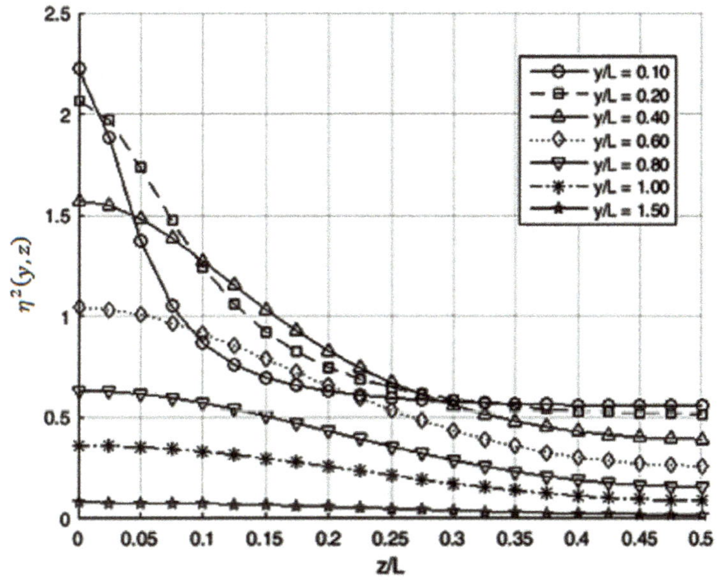

Figure 12.1 Variation of $\eta^2(y, z)$ with z/L at various levels of y/L.

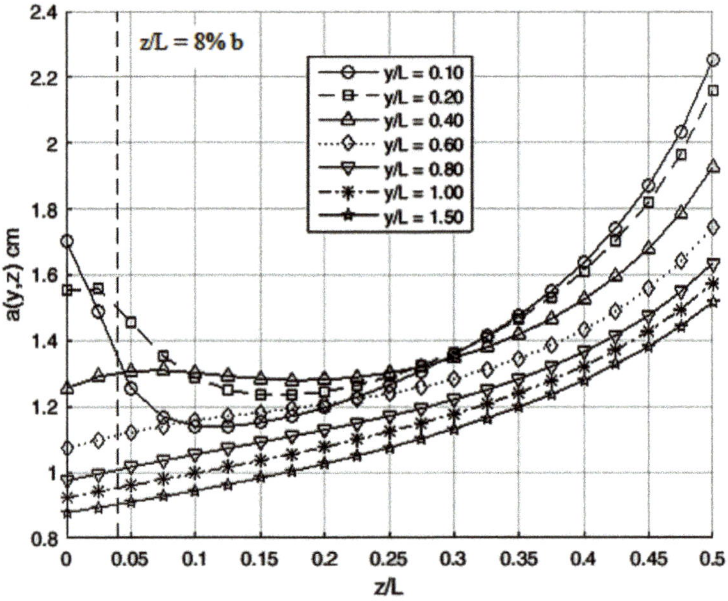

Figure 12.2 Variation of $a(y, z)$ in centimeters with z/L at various levels of y/L.

MATLAB Program 21

```
% Matlab program to calculate and plot the 1- square of the
refractive index % vs. z/L and 2- the plate separation variation
vs. z/L (at various values of % y/L values) of the novel
shortened phase-corrected aperture horn.

% Define the constants
c = 3*10^10; % speed of light in cm
w0 = 2*pi*10^10; % design frequency in radians
b = 0.5;
d = 1 - 1/b^2;
% The refractive index (squared) at the center of the horn y/L=0
etaz_sq = 0.5694;

zl_range = 0:0.025:0.5;
yl_values = [0.1, 0.2, 0.4, 0.6, 0.8, 1.0, 1.5];

% Define the combined line styles and symbols
line_styles = {'-', '--', '-', ':', '-', '-.', '-'};
symbols = {'o', 's', '^', 'd', 'v', '*', 'p'};
figure(1)
hold on
xlabel('z/L');
ylabel('\eta_{sq}');
legend_labels = cell(length(yl_values), 1);
```

```
figure(2)
hold on
xlabel('z/L');
ylabel('a(y,z) cm');
legend_labels = cell(length(yl_values), 1);

% Loop over each yl value
for i = 1:length(yl_values)
  yl = yl_values(i);

  % Calculate Th (i.e., cos^2(theta)) and eta_sq for each zl in
the range
  eta_sq = zeros(size(zl_range));
  a_yz = zeros(size(zl_range)); % Initialize a_yz array

  for j = 1:length(zl_range)
    zl = zl_range(j);
    Th = (sin(pi * zl / (2 * b)).^2) ./ (sin(pi * zl / (2 *
b)).^2 + sinh(pi * yl / 2).^2);
    eta_sq(j) = etaz_sq * ((sech(pi * yl / 2)).^2 / b^2 + d *
Th);
    a_yz(j) = sqrt(pi^2 * c^2 / (w0^2 * (3 - 4 * zl
- eta_sq(j))));
  end

  % Plot the results with different line styles and symbols
  figure(1)
  plot(zl_range, eta_sq, 'k', 'LineStyle', line_styles{i},
'Marker', symbols{i});
  legend_labels{i} = sprintf('y/L = %.2f', yl);

  figure(2)
  plot(zl_range, a_yz, 'k', 'LineStyle', line_styles{i},
'Marker', symbols{i});
  legend_labels{i} = sprintf('y/L = %.2f', yl);
end

figure(1)
legend(legend_labels, 'Location', 'best');
grid on;
hold off

figure(2)
legend(legend_labels, 'Location', 'best');
grid on;

ylim([0.8 2.4]); % Set y-axis limit
vertical_line_position = 0.08 * b;
% Define the vertical line position as 8% * b * L (with L
normalized to 1)
```

```
line([vertical_line_position vertical_line_position], ylim,
'Color', 'k', 'LineStyle', '--', 'LineWidth', 1); % Add thin
dashed black line
legend(legend_labels, 'Location', 'best');
grid on;
hold off
```

Directivity in the H-plane of the antenna is desirable and can be achieved using a graded horn load dielectric whose permittivity decreases towards the aperture.

Figure 12.2 shows the variation of the plate separation $a(y, z)$ with z/L at various levels of y/L for the $\eta^2(y, z)$ of Figure 12.1 with linear grading of the form:

$$\varepsilon_r = 3 - \frac{4z}{L} \tag{12.11}$$

where ε_r at the new aperture ($z/L = 0.5$) becomes equal to 1. The linear grading of ε_r offers a smooth variation of $a(y, z)$, a degree of flare in the H-plane and relative ease of manufacture. Calculated ray trajectories show that the initial 8% of the horn refractive region ($z < z_p$) can be replaced with a nonrefractive region with $a = a(0, z_p)$ and $\varepsilon_r = \varepsilon_r(0, z_p)$ with little effect on the aperture distribution. Figure 12.3 shows a comparison between the predicted ray trajectories in the initial 8% of the horn, and the radial rays in a similar nonrefractive region. The deviation error is better than 4% for a maximum allowed angle of emergence of $\theta = 75°$.

The same graded dielectric material is suggested for the waveguide feed as it will enable tapering of the rectangular waveguide dimensions from those of WG16 ($\varepsilon_r = 1$) to those required at $z = 0$ while maintaining the same TE_{10} propagation constant.

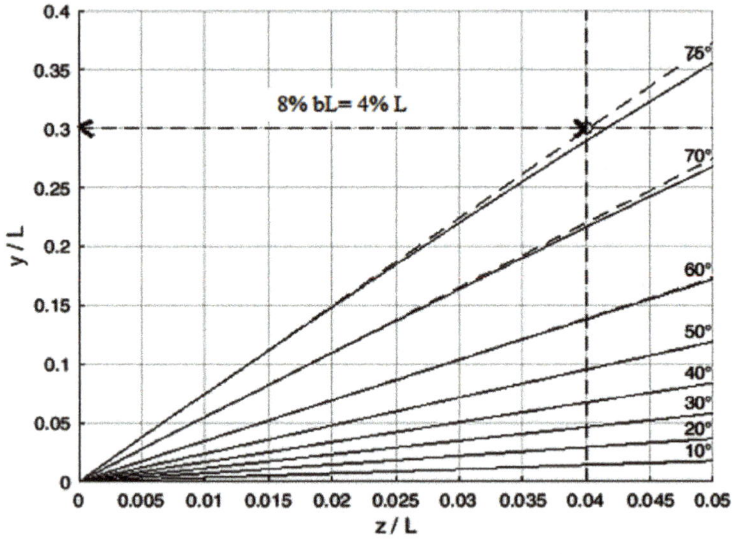

Figure 12.3 Ray trajectories (———) in the initial 8% of the new horn length in the actual HYSEC refractive region, compared to radial rays (– – – –) in a similar nonrefractive region.

MATLAB Program 22

```
% Matlab program to compute and plot the ray trajectories in the
initial 8% of the new horn length in the actual HYSEC refractive
region, compared to a non-refractive region.

% Define the parameters
b = 0.5;
thet_deg = [10, 20, 30, 40, 50, 60, 70, 75];
zl = linspace(0, 0.05, 100); % Vary zl from 0 to 0.05

% Convert theta to radians
thet_rad = deg2rad(thet_deg);

% Initialize figure
figure;
hold on;

% Loop over each value of theta for the first set of curves
for i = 1:length(thet_rad)
  thet = thet_rad(i);

  % Calculate yl
  yl = (2/pi) * asinh(tan(thet) .* sinh(pi * zl / (2 * b)));

  % Plot yl versus zl with solid black lines
  plot(zl, yl, 'k');

  % Add degree labels at the end of each solid curve
  text(zl(end), yl(end), sprintf('%d°', thet_deg(i)),
'VerticalAlignment', 'bottom', 'HorizontalAlignment', 'right');
end

% Loop over each value of theta for the second set of curves
for i = 1:length(thet_rad)
  thet = thet_rad(i);
  % Calculate the new expression
  y2 = zl / b * tan(thet);

  % Plot zl/b*tan(thet) versus zl with dashed black lines
  plot(zl, y2, 'k--');
end

% Add a vertical line at zl = 0.04
line([0.04 0.04], ylim, 'Color', 'k', 'LineStyle', '--');

% Add a horizontal line at yl = 0.3
line(xlim, [0.3 0.3], 'Color', 'k', 'LineStyle', '--');
```

```
% Add a point where the vertical and horizontal lines meet
plot(0.04, 0.3, 'ko');

% Finalize the plot
hold off;
xlabel('z / L');
ylabel('y / L');
grid on;
```

12.2 Proposed Construction of a Compact Antenna

The experimental verification of the HYSEC horn has earlier suggested that the aperture ratio $A_y/L = 2.67$ is indeed a good basis for design. If $b = 0.5$, then $A_y/L' = 5.34$. Choosing a practical size of an antenna aperture such as $A_y = 33\lambda$ ($\lambda = 3$ cm at 10 GHz) will consequently result in a physical axial length of $L' = 6.18\lambda$.

In the same way as a Luneberg lens is constructed from concentric spherical shells of dielectric material to give the required refractive index variation, planar dielectric layers are proposed in the construction of the compact antenna. These are arranged in equal thicknesses and with a permittivity changing linearly with the order of the layer in the whole structure. The layers can be fused on top of each other and then machined according to the profile suggested in Figure 12.2. It might be easier to cut each layer according to the required profile at a constant z/L, which relates its order or position in the bulk. Accordingly, the delicate fusion process of the layer may follow after cutting them to the required profile. Once the profiled sides of the fused layers are well established, they could be covered with in-wall metalized fiber glass or coated with a suitable metallic spray. The other sides may be covered with microwave absorber material. Absorber inserts in the HYSEC horn have been found to improve the voltage standing wave ratio (VSWR) and effectively reduce the radiation sidelobe level, particularly at the higher frequencies above the design. It is worth mentioning that, in one example of a Luneberg lens, 10 concentric spherical shells were arranged one within the other to give the required variation in permittivity from 2.0 to 1.1 increments of 0.1. The lens operated in the X-band of the microwave frequency range. In our case, 20 planar dielectric layers may be sufficient to give the required permittivity variation between 3 and 1 along the axis. The same stepped-index dielectric material has already been suggested for the waveguide feed. Figure 12.4 shows a schematic design of the proposed antenna and its feed. The compactness of the antenna makes it a good candidate for marine radar applications. The predicted E-plane radiation pattern of the antenna (Figure 12.5) shows a 3-dB beamwidth of approximately 2° and sidelobe levels below −35 dB.

The radiation pattern can be computed by running MATLAB Program 19 after some modifications. The waveguide feed needed is WG16 and hence its narrow dimension at the design frequency (10 GHz) is $2y0 = 0.34\lambda$. The aperture, $A_y = 2.67 \times 5.72$ cm, is scaled up by 6.482 folds to 99 cm and, similarly, the length of the horn to $L = 37.07$ cm. The dielectric load would only affect the compactness of the horn by making the axial length shorter by 50%, without affecting the aperture distribution pattern.

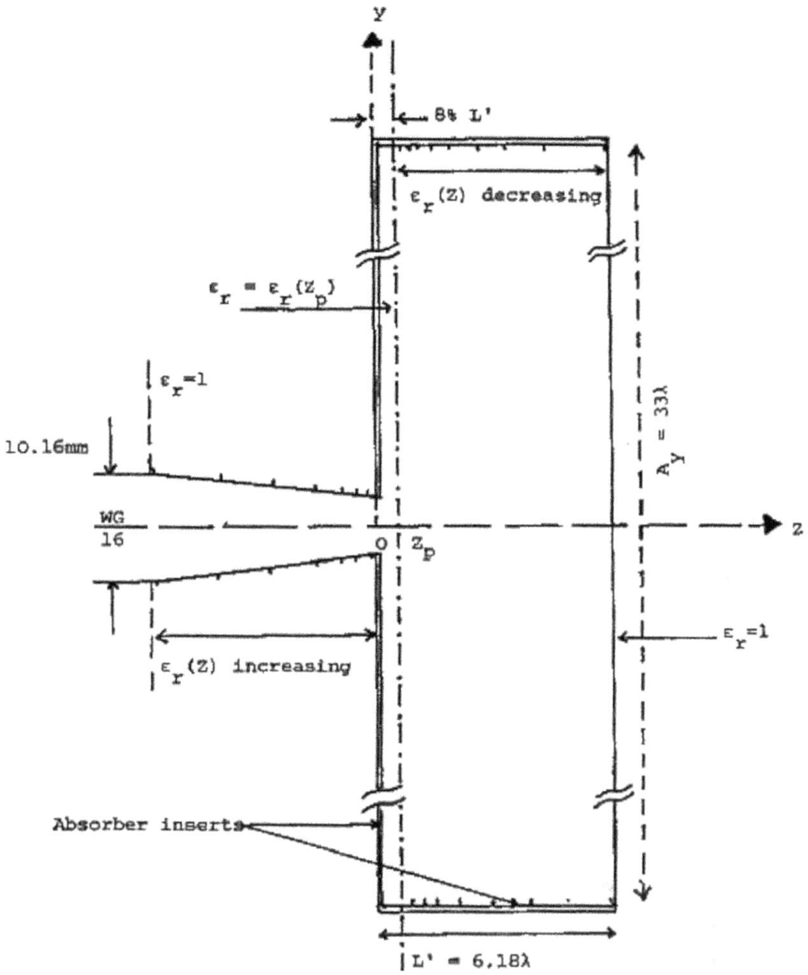

Figure 12.4 Schematic design of the proposed dielectric-loaded shortened axial-length antenna and its feed. (Adapted from [1].)

Problems for Students

12.1 A phase-corrected horn antenna is designed with a reduced length of 50% compared to a conventional horn.

(a) Discuss how the phase correction mechanism compensates for the shorter horn length.

(b) Calculate the phase error at the aperture for a conventional horn and the phase-corrected horn if the horn length is reduced from 2λ to λ.

12.2 Compare the gain of a novel phase-corrected horn antenna with a conventional horn antenna of the same aperture size at 10 GHz.

(a) Given that the aperture efficiency of the phase-corrected horn is 85% and the conventional horn's efficiency is 70%, calculate the gain for both. Assume that aperture diameter $D = 0.3$m.

(b) Discuss how the compact design affects sidelobe levels and beamwidth.

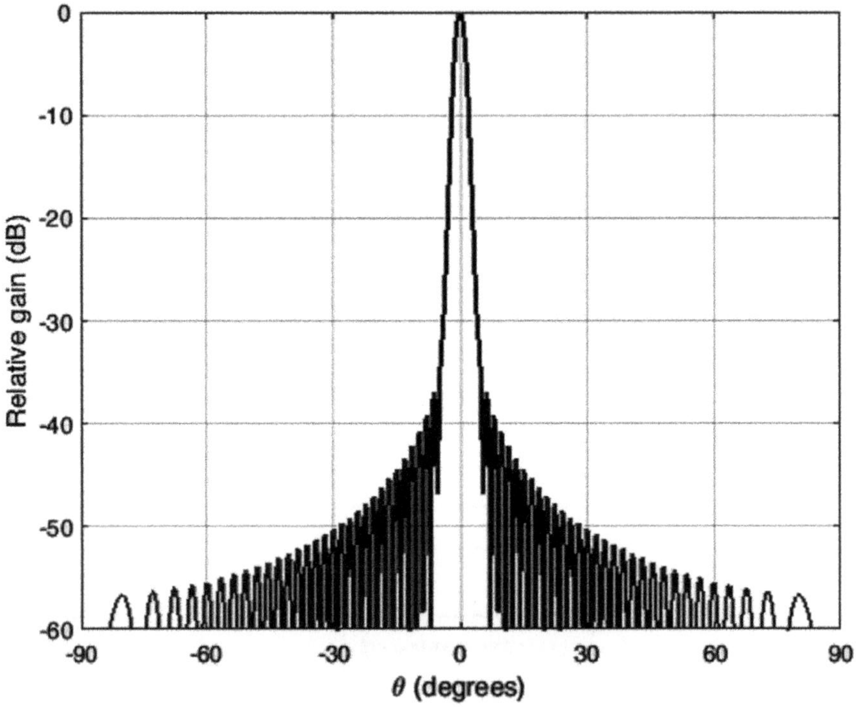

Figure 12.5 Predicted radiation pattern of the proposed antenna for marine applications (HPBW = 1.98°).

12.3 The proposed compact horn antenna uses a hyperbolic refractive index distribution for phase correction.
   (a) Derive the equation for the phase distribution across the aperture.
   (b) Evaluate the improvement in far-field radiation pattern symmetry due to this phase correction.

12.4 During the fabrication of the phase-corrected horn, a misalignment occurs in the aperture-shaping process.
   (a) Quantify the impact of a 5% aperture misalignment on the directivity and sidelobe levels at 15 GHz.
   (b) Propose design and fabrication techniques to mitigate these errors.

12.5 The compact horn antenna is designed to operate over a frequency range of 8 GHz to 12 GHz.
   (a) Calculate the beamwidth variation across the frequency range if the aperture size remains constant.
   (b) Discuss the implications of phase correction on maintaining a consistent radiation pattern over this bandwidth.

12.6 A novel phase-corrected horn is simulated using a finite-difference time-domain (FDTD) method.
   (a) Outline the steps involved in setting up the simulation for the horn antenna.
   (b) Discuss how simulation results (e.g., gain, beamwidth, and sidelobe levels) validate the theoretical design equations.

Problems for Instructors

12.1 Analysis of Shortened-Length Horn Design: In Section 12.1, the shortened-length horn design was discussed. Explain why reducing the length of a horn antenna can be beneficial for compact applications. Then derive the key parameters (such as aperture size and flare angle) for a shortened horn antenna design operating at 10 GHz, and compare the expected radiation efficiency and gain with those of a traditional horn antenna.

12.2 Phase Correction Techniques for Horn Antennas: Describe the primary phase correction techniques applied in compact horn antenna designs. Discuss how phase correction improves the performance of a shortened horn antenna and analyze the impact on the main lobe and sidelobe levels. Assume a design frequency of 8 GHz, and propose a specific phase correction technique to optimize the far-field pattern.

12.3 Design Optimization of a Compact Horn Antenna: Using the principles discussed in Section 12.2, propose a construction design for a compact phase-corrected horn antenna with an aperture size of 0.2m. Specify the materials and geometric modifications needed to achieve phase correction and calculate the predicted gain, beamwidth, and sidelobe suppression at 12 GHz. Explain your design choices and how they contribute to the antenna's compactness and performance.

12.4 Evaluation of Phase Correction on Radiation Pattern: Simulate the radiation pattern of a compact phase-corrected horn antenna using MATLAB or another simulation tool, assuming a design frequency of 15 GHz and an aperture taper with Gaussian amplitude distribution. Analyze the effects of phase correction on beam sharpness, sidelobe levels, and overall gain. Compare the results with a standard horn antenna of the same aperture size.

12.5 Trade-Offs in Compact Antenna Design: List and evaluate the primary trade-offs involved in designing a compact, phase-corrected horn antenna. Consider factors such as impedance matching, bandwidth, radiation efficiency, and mechanical stability. Provide a hypothetical design scenario where you prioritize compactness over gain and explain the impact on overall performance metrics.

12.6 Construction Challenges in Compact Antenna Design: Identify three common construction challenges when designing and implementing compact, phase-corrected horn antennas. For each challenge, propose a solution or adjustment to the design that can help to mitigate its impact. Examples may include challenges related to phase stability, material selection, and aperture efficiency in small form factors.

12.7 Comparative Analysis of Compact Antenna Designs: Perform a comparative analysis between two compact, phase-corrected horn antennas, one using a stepped-section phase correction and the other using a gradual curvature phase correction method. Discuss the advantages and disadvantages of each design in terms of ease of manufacturing, radiation pattern consistency, and phase stability. Which design would you recommend for high-frequency applications above 20 GHz, and why?

Reference

[1] Ata, O. W., "Application of Ray Theory into Microwave Horn Antenna Design," PhD thesis, Cardiff University, UK, 1989.

Selected Bibliography

Ata, O. W., T. M. Benson, and A. Marincic, "Application of Optical Ray Technique to the Design of Short Microwave Horn Antennas with Low Side Lobe Levels," *IEE Proc. H Microwav. Antennas Propag.*, Vol. 137, No. 2, April 1990, pp. 81–88.

Byers, H. G., and M. Katchky, "Slotted Waveguide Array for Marine Radar," *Electronics*, 1958, pp. 94–96.

Johnson, R. C., and H. Jasik, (eds.), *Antenna Engineering Handbook*, Second Edition, New York: McGraw-Hill, 1984.

Kiely, D. G., A. E. Collins, and G. S. Evans, "Cheese Aerials for Marine Navigational Radar," *Proceedings of the IEE-Part III: Radio and Communication Engineering*, Vol. 98, No. 51, 1951, pp. 37–45.

Skolnik, M. I., *Introduction to Radar Systems*, New York: McGraw-Hill, 1962.

Ward, H. T., W. O. Puro, and D. M. Bowie, "Artificial Dielectrics Utilizing Cylindrical and Spherical Voids," *Proceedings of the IRE*, Vol. 44, No. 2, 1956, pp. 171–174.

CHAPTER 13

Radio Wave Propagation Through the Atmosphere

The study of antenna techniques is intrinsically linked to the understanding of radio wave propagation through the atmosphere, as these two fields jointly define the efficacy of wireless communication systems. Antennas serve as the primary interface between the transmitter and the medium through which the signal travels: the atmosphere. Radio wave propagation is influenced by various atmospheric conditions and phenomena, including reflection, refraction, diffraction, absorption, and scattering. Atmospheric layers such as the troposphere and ionosphere, as well as weather conditions and geographical factors, dictate these effects. A profound comprehension of these propagation characteristics enables engineers to design antennas that can maximize signal strength, minimize interference, and ensure reliable communication over varying distances. By tailoring antenna parameters such as gain, radiation pattern, polarization, and frequency, it is possible to mitigate adverse atmospheric impacts and enhance the overall performance of communication systems.

Consequently, the synergy between antenna technology and atmospheric propagation studies is essential for advancing both theoretical research and practical applications in telecommunications, satellite communications, radar systems, and remote sensing. This chapter aims to highlight the critical relationship between the radio wave signal and atmospheric propagation, and associated mechanisms that the radio wave signal experiences through the tropospheric and ionospheric layers.

13.1 The Earth's Atmosphere

The Earth's atmosphere is a dynamic and multilayered system, each layer playing a vital role in maintaining life, regulating climate, and enabling technologies such as radio communication. The main atmospheric layers—troposphere, stratosphere, mesosphere, and ionosphere—exhibit distinct characteristics and influence radio wave propagation in different ways.

The troposphere, the lowest atmospheric layer, extends up to approximately 15 km (9 miles) and contains around 75% of the atmosphere's total mass. It is the region where weather phenomena occur, including clouds, precipitation, and storms. Temperature decreases with altitude in this layer. In terms of radio communication, the troposphere primarily affects signals through tropospheric scattering and refraction, allowing VHF and UHF radio waves to propagate slightly beyond the line of sight. However, variable weather conditions such as rain and fog can degrade signal quality, leading to fading and interference.

Above it lies the stratosphere, stretching from about 15 to 50 km (9 to 31 miles) in altitude. This layer features a temperature inversion, with temperatures increasing with height due to the presence of the ozone layer, which absorbs ultraviolet (UV) radiation. While the stratosphere does not directly impact radio wave propagation, it plays an important indirect role by filtering harmful UV radiation, which affects ionization levels in the upper atmosphere, thereby influencing radio wave behavior in higher layers.

The mesosphere, situated between 50 and 85 km (31 to 53 miles) above Earth, is characterized by decreasing temperatures with altitude, reaching the coldest atmospheric temperatures. It is the region where meteors commonly disintegrate upon entry. Although it has limited direct interaction with radio waves, the mesosphere supports the formation of sporadic E layers within the ionosphere, occasionally enhancing VHF signal propagation.

The ionosphere, ranging from about 85 to over 1,000 km (53 to 620 miles), is rich in free electrons and ions generated by solar radiation. It is subdivided into D, E, and F layers, each with varying electron densities. The ionosphere is critical for long-distance radio communication, particularly in the high-frequency (shortwave) band. It reflects and refracts radio waves back toward Earth, enabling signal transmission far beyond the horizon. However, its effectiveness varies with solar activity, geographic location, and time of day.

In conclusion, while all atmospheric layers contribute to radio wave dynamics, the troposphere and ionosphere are most influential. The former supports short-range propagation through scattering, and the latter enables long-range communication via reflection and refraction.

13.2 Radio Wave Propagation: Frequency Bands and Their Suitability

In radio wave propagation, different frequency bands are strategically used to harness the distinct properties of the Earth's atmospheric layers, particularly the troposphere and ionosphere. Tropospheric propagation primarily utilizes VHF and UHF bands, ranging from 30 MHz to 3 GHz. These frequencies are well suited for line-of-sight communication and benefit from scattering and refraction within the troposphere, which extends up to about 15 km. As VHF and UHF signals are largely unaffected by ionospheric conditions, they are ideal for short-range to medium-range applications such as FM radio, television broadcasting, and mobile communication. The troposphere's relative stability ensures reliable and predictable signal transmission.

In contrast, ionospheric propagation relies on the high-frequency band, spanning 3 MHz to 30 MHz. The ionosphere, located between 85 km and 1,000 km, contains ionized layers (mainly E and F) that reflect and refract high-frequency waves back to Earth, enabling long-distance, over-the-horizon communication through multiple hops. These layers are influenced by solar radiation, varying in density and height with the time of day and solar activity. As a result, the high-frequency band supports global communication, commonly used in amateur radio, maritime communication, and international broadcasting.

Table 13.1 summarizes the regions of propagation with their allocated frequency spectrum, propagation mechanism, and systems for each band.

13.3 Refraction of Radio Waves in the Troposphere

The refraction of radio waves in the troposphere, the layer of the Earth's atmosphere immediately above the surface, is a significant phenomenon governed by Snell's law of refraction. This law states that the angle of incidence of a wave as it enters a new medium is related to the angle of refraction through the ratio of the wave velocities in the two media. Figure 13.1 demonstrates two refractive media separated by an interface where $n_1 > n_2$.

Snell's law is expressed as:

$$n_1 \cos\theta_1 = n_2 \cos\theta_2 = h_\theta \tag{13.1}$$

where n_1 and n_2 are the refractive indices of the two media, θ_1 and θ_2 are called the grazing angles that fall between the rays and the interface, and h_θ is a constant.

Table 13.1 Regions of Propagation

Frequency	*Band*	*Propagation Mechanism*	*Systems*
300 Hz–3 kHz	ELF		No uses
3 kHz–30 MHz	VLF	Earth-ionosphere waveguide	Worldwide, military, and navigation
30 kHz–300 kHz	LF	Surface wave	Long-wave band, stable transmission, distances up to 1,500 km
300 kHz–3 MHz	MF	Surface wave (short ranges), sky wave (long range)	Medium-wave broadcasts, sky wave over longer distances, subject to fading
3 MHz–30 MHz	High frequency	Sky wave	Short-wave communication band, 3–6-MHz continental, 6–30-MHz intercontinental inside skip distance, short distance ground waves skip distance, short-distance ground waves possible, mobile land communication, skip-shore communication
30 MHz–300 MHz	VHF	Space wave, scatter wave	Line-of-sight VHF, links over short distances, >50-MHz ionospheric scatter, 2,000 km over long distances
300 MHz–3 GHz	UHF	Space wave, scatter wave	Line of sight, short distances, >500 MHz upwards, tropospheric scatter over long distances, 600m
3 GHz–30 GHz	SHF (microwave)	Space wave	Space vehicle communication (e.g., satellite links microwave radio relays)
30 GHz–300 GHz	EHF (millimeter-wave)		Line-of-sight millimeter links, satellite links, and possible local distribution systems

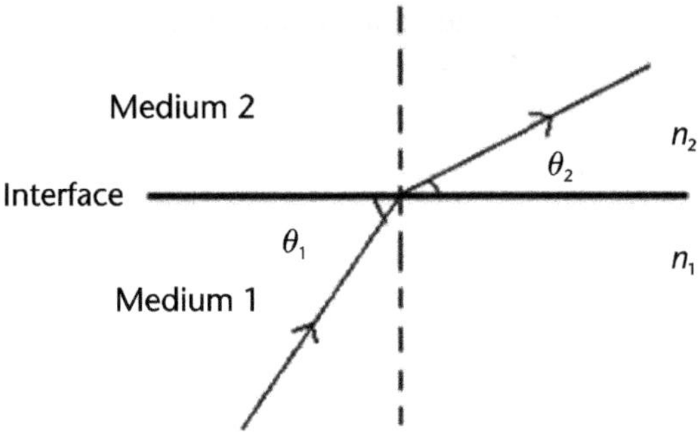

Figure 13.1 Two refractive media separated by an interface where $n_1 > n_2$.

When applied to radio waves propagating through the troposphere, Snell's law elucidates how changes in atmospheric conditions, such as temperature, pressure, and humidity, affect the wave's path. As radio waves travel through the troposphere, they encounter varying refractive indices due to these changing conditions, causing the waves to bend or refract. This bending effect is crucial for understanding long-distance radio communications and signal behavior. Figure 13.2 demonstrates a stratified medium with decreasing refractive index that shows a bended ray.

The troposphere can be considered a stratified medium where the refractive index changes with altitude. This stratification is primarily due to the gradient in atmospheric density, which decreases with height. The refractive index of the Earth's atmosphere for radio waves is influenced by several factors, including the electron density and the molecular composition of the air. In the troposphere, the refractive index generally decreases with altitude, causing radio waves to bend towards the Earth's surface. This bending can extend the radio horizon, allowing signals to be received beyond the line of sight. Understanding the stratified nature of the troposphere and its impact on radio wave propagation is essential for optimizing communication systems, particularly for aviation, maritime, and remote-sensing applications.

The refractive index in the troposphere is a crucial parameter for radio wave propagation. It is typically close to but slightly more than 1. The refractive index changes from 1.000324 at the surface to 1.0000423 at a height of 15 km. The variations in the refractive index are relatively small but significant enough to affect the trajectory of radio waves. These variations are influenced by atmospheric conditions such as temperature, humidity, and pressure, which can cause the refractive index to change diurnally and seasonally. For instance, during the daytime, the heating of the Earth's surface can lead to a decrease in the refractive index at lower altitudes,

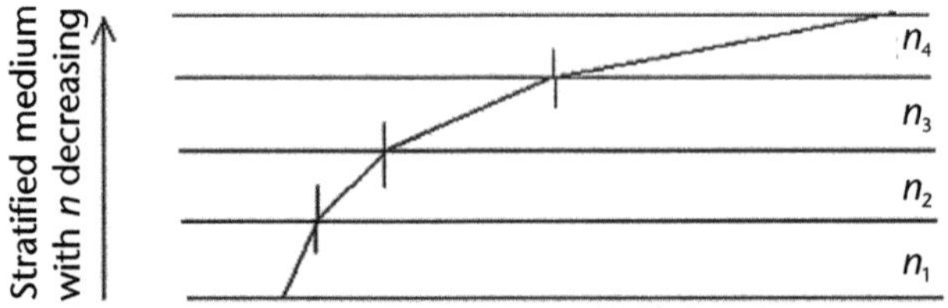

Figure 13.2 A stratified medium with decreasing refractive index that shows a bended ray.

enhancing the bending of radio waves. Conversely, at night, cooling can increase the refractive index near the surface, potentially reducing the bending effect. Accurate modeling of these refractive index variations is essential for predicting and mitigating potential issues in radio communication and navigation systems.

As we ascend from the Earth's surface, we find that:

1. Pressure decreases and therefore the density of gas (O_2 and N_2) decreases.
2. Temperature decreases.
3. Water vapor content and vapor pressure change.

Table 13.2 provides typical values for the above quantities:

It is convenient to have an empirical model to calculate the value of n from the values of p, e, and T. The model is called the Standard Atmosphere, which describes the mean behavior of the Earth's atmosphere.

$$n = 1 + \left(\frac{79p}{T} - \frac{11e}{T} + \frac{3.8 \times 10^5 e}{T^2} \right) \times 10^{-6} \tag{13.2}$$

where n is the refractive index (dimensionless), p is the atmospheric pressure (in mb = N/m^2), e is water vapor pressure (in mb), and T is absolute temperature in Kelvin (T°C + 273).

13.4 Radio Wave Curved Path and Surface Range Expression

Consider a wavefront, AC, launched at an angle Ψ with the Earth's surface, through the troposphere. After a time t, the new wavefront is BE, where the two rays, AB and CE, arrive simultaneously, as a result of the decreasing refractive index of the stratified medium with the height dh.

From the geometry of the diagram (Figure 13.3), it is possible to calculate the radius of curvature of path, ρ, of a radio wave.

$$AB = s = vt \tag{13.3}$$

where v is the velocity of the radio wave at the Earth's surface and t is the time duration.

$$n = \frac{c}{v} \tag{13.4}$$

Table 13.2 Typical Atmospheric Figures (p, e, T, and n) for Various Altitudes

Altitude h (feet)	*Atmospheric Pressure p (mb)*	*Water Vapor Pressure e (mb)*	*Temperature T (C°)*	*Refractive Index n*
0	1,013	10.2	15	1.000324
2,000	942	7.9	11	1.000299
10,000	697	2.5	–4.8	1.000218
50,000	116	0.0	–55	1.0000423

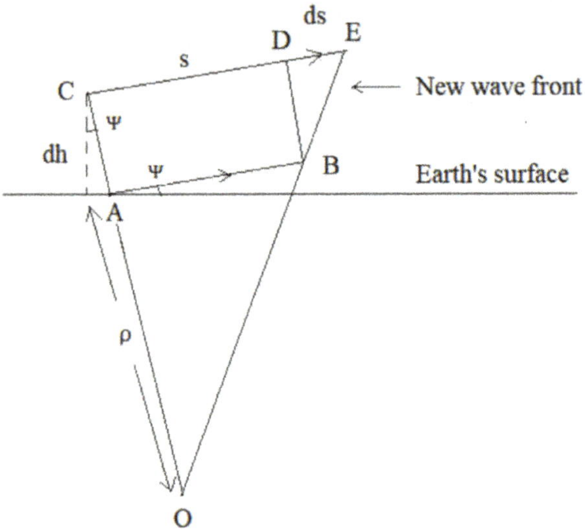

Figure 13.3 A diagram showing the curvature of path of a radio wave.

where c is the free space velocity of the radio wave.

Substituting for v from (13.4) into (13.3) and differentiating s with respect to n yields:

$$ds = ct \cdot d\left(\frac{1}{n}\right) = -ct \cdot \frac{dn}{n^2} \tag{13.5}$$

The two triangles OAB and BDE in Figure 13.3 are similar and therefore

$$\frac{s}{\rho} = \frac{ds}{DB} = \frac{ds}{\dfrac{dh}{\cos\Psi}} \tag{13.6}$$

$$\rho = \frac{s \cdot dh}{ds\cos\Psi} = \frac{\dfrac{ct}{n} \cdot dh}{\dfrac{-ct}{n^2} \cdot dn\cos\Psi} = \frac{-n}{\dfrac{dn}{dh}\cos\Psi} \tag{13.7}$$

Consider $\Psi \approx 0$, so that $\cos\Psi \approx 1$ and since $n \approx 1$, (13.7) becomes:

$$\rho \approx \frac{-1}{\dfrac{dn}{dh}} \tag{13.8}$$

It is possible to substitute for dn/dh from Table 13.2 for heights between 0 and 2,000 feet so that $d\eta/dh = -25 \times 10^{-6}/2{,}000$ feet^{-1}.

Hence ρ works out to be equal to 8×10^7 feet, which is approximately 4 times the radius of the Earth.

Consider Figure 13.4 where a transmitter is situated at C, at the top of a tower of height h above the surface of the Earth. The curved path of the radio wave touches the Earth's surface at point S. The range d is the length of the arc SC.

Using the cosine rule in the triangle OBC gives:

$$\begin{aligned}\rho^2 &= (a+h)^2 + (\rho - a)^2 - 2(a+h)(\rho - a)\cos(\pi - \theta) \\ &= a^2 + 2ah + h^2 + \rho^2 - 2\rho a + a^2 + 2(a\rho - a^2 + h\rho - ha)\cos\theta\end{aligned} \quad (13.9)$$

Canceling ρ^2 on both sides and dropping the small quantity h^2 gives:

$$0 = 2a^2 + 2ah - 2a\rho + 2(a\rho - a^2 + h\rho - ha)\cos\theta \quad (13.10)$$

Also, because θ is small, $\cos\theta \approx 1 - \theta^2/2$ and $\theta = d/a$ and therefore

$$d^2 = \frac{2a^2 h\rho}{(-a^2 - ha + h\rho + a\rho)} = \frac{2ah}{\left(-\frac{a}{\rho} - \frac{h}{\rho} + \frac{h}{a} + 1\right)} \quad (13.11)$$

Because $h << \rho$ and $h << a$, then

$$d^2 \approx \frac{2ah}{\left(1 - \frac{a}{\rho}\right)} \quad (13.12)$$

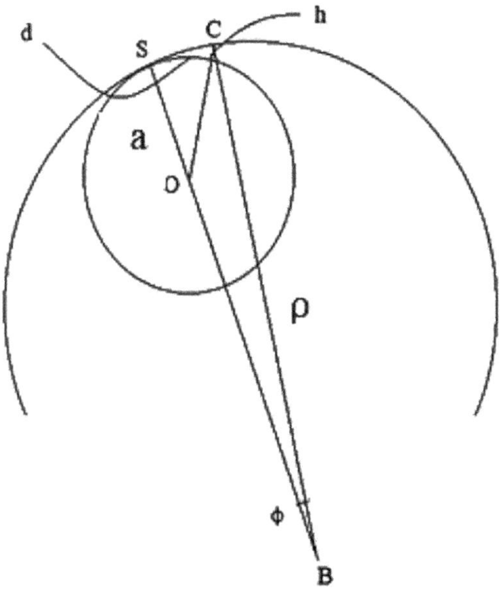

Figure 13.4 A diagram showing a radio wave launched from the top of a tower situated at point C and touches the Earth's surface at point S.

We know that $a/\rho = 1/4$; hence,

$$d^2 \approx \frac{4}{3} \times 2ah \tag{13.13}$$

13.5 4/3rd Earth Radius Diagram

The outcome expressed by (13.13), which related the range d to only the radius of the Earth a and the height of the tower h, motivates the following question. If we could now imagine the Earth, without an atmosphere, so that a radio wave travels in a straight line, what would be the new radius of our hypothetical planet for a similar range d, as in Figure 13.4, which would results from a radio wave traveling along a straight line, from the top of the tower, height h, until it touches the surface of the Earth?

From Figure 13.5, we need to find a relationship between d, a', and h.

For a small θ,

$$\cos\theta \approx 1 - \frac{\theta^2}{2} = \frac{a'}{a' + h} \tag{13.14}$$

After manipulating (13.14),

$$\theta^2 = \frac{2h}{a' + h} \tag{13.15}$$

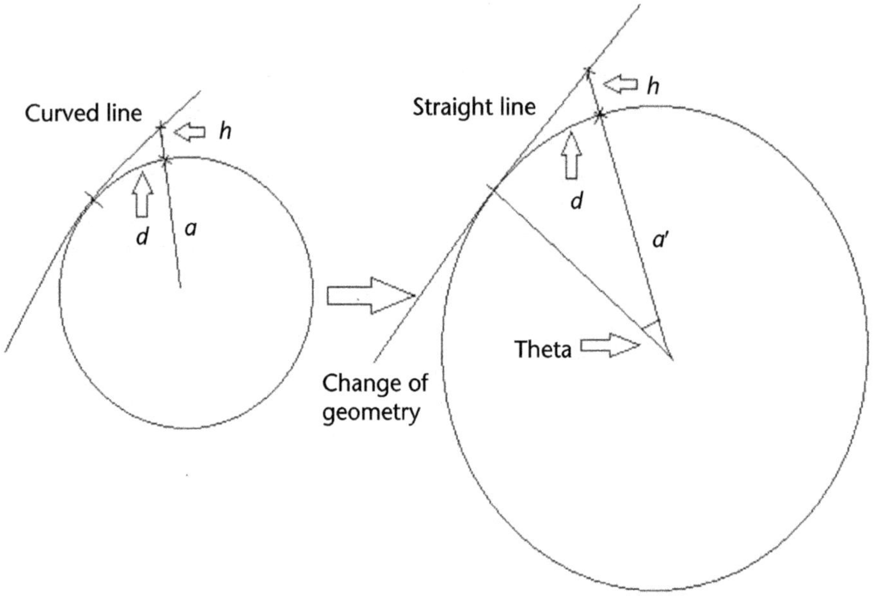

Figure 13.5 Change of geometry from planet Earth with an actual radius a to a hypothetical planet with radius a'.

Since $h << a'$ and $\theta = d/a'$, (13.15) becomes

$$\frac{d^2}{a'^2} = \frac{2h}{a'} \tag{13.16}$$

and hence

$$d^2 \approx 2ha' \tag{13.17}$$

Comparing (13.17) with (13.13),

$$a' = \frac{4}{3}a \tag{13.18}$$

MATLAB Program 23 simulates and plots the radio wave propagation behavior through the troposphere for different launch angles, as shown in Figure 13.6.

MATLAB Program 23

```
% This code simulates the propagation of radiowaves in
the troposphere
% for different launch angles. The goal is to trace the curved
ray paths
% as they travel through the atmosphere, considering the
varying refractive
% index with height. The code also computes the maximum height
and horizontal
% distance at which the ray reaches its maximum height for each
launch angle.
% It uses an exponential model for the refractive index and
calculates the
% ray's trajectory based on the curvature of the path due to
the varying
% refractive index in the troposphere.
%
% The exponential refractive index model used in this code:
% n(h) = 1 + A * exp(-B * h) is formulated to approximate the
% experimental data of the refractive index presented in Table
13-2 of
% the reference text. The values of A and B are determined from
this model
% to match the observed refractive index profile with height in
the troposphere.
% Constants
Re = 6371e3;          % Radius of Earth in meters (constant)
h_max = 15240;        % Max height to simulate (in meters, 15.24
km i.e. 50,000 foot))
dh = 10;              % Altitude step size in meters (spatial
resolution for ray tracing)
```

```
% Exponential refractive index model: n(h) = 1 + A * exp(-B * h)
A = 0.00032665;             % Constant A for refractive
index model
B = 0.00018413;             % Constant B for refractive
index model
n_exp = @(h) 1 + A * exp(-B * h); % Function for refractive index
as a function of height
dn_dh_exp = @(h) -A * B * exp(-B * h); % Derivative of refractive
index with respect to height, used to calculate radius
of curvature

% Launch angles to be tested (in degrees)
psi_deg = [0.5, 1, 1.25];        % Angles for 0.5°, 1°, and 1.25°
launch angles
colors = ['r', 'g', 'b'];    % Colors for plotting the different
curves (red, green, blue)

% Initialize the figure for plotting
figure;

% Loop over the different launch angles to simulate the
ray propagation
for i = 1:length(psi_deg)
  % Current launch angle (in degrees)
  psi0_deg = psi_deg(i);        % Set the current launch angle
  psi0 = deg2rad(psi0_deg);      % Convert the launch angle
to radians

  % Initialize ray path parameters
  x = 0;                        % Initial horizontal position (starts
at 0)
  y = 0;                        % Initial vertical position (height
starts at 0)
  theta = pi/2 - psi0;          % Zenith angle (relation to
launch angle)
  step = 10;                    % Step size for ray path
(in meters)
  xray = []; yray = []; angles = [];  % Arrays to store ray path
coordinates and angles

  % Start ray tracing while the ray is below the max height
  while y < h_max && y >= 0
    % Calculate refractive index and its vertical derivative at
current height
    n_curr = n_exp(y);          % Get current refractive index at
height y
    dn_dy = dn_dh_exp(y);         % Get the derivative of
refractive index at height y

    % Radius of curvature: rho = n / (dn/dy)
```

```
    rho = -n_curr / dn_dy;        % Calculate radius of curvature
for ray path

    % Calculate change in angle (dtheta) based on arc length step
and curvature
    dtheta = step / rho;      % Angle change per step based
on curvature

    % Update ray's position (x, y) and angle (theta)
    x = x + step * sin(theta);   % Update horizontal position
based on angle
    y = y + step * cos(theta);   % Update vertical position
(height)

    % Update the angle
    theta = theta + dtheta;      % Update the angle after step

    % Store updated values for plotting
    xray(end+1) = x;            % Store horizontal position
    yray(end+1) = y;            % Store vertical position (height)
    angles(end+1) = rad2deg(pi/2 - theta); % Store grazing angle
(in degrees)
  end

  % Plotting the ray path and grazing angle on separate subplots
  subplot(2, 1, 1); % First subplot for ray path
  plot(xray / 1000, yray / 1000, colors(i), 'LineWidth', 1.5); %
Plot ray path in km (convert meters to km)
  hold on;

  subplot(2, 1, 2); % Second subplot for grazing angle vs.
horizontal distance
  plot(xray / 1000, angles, colors(i), 'LineWidth', 1.5); % Plot
grazing angle (in degrees)
  hold on;
% Print out results for each launch angle
  max_height = max(yray) / 1000; % Convert max height
to kilometers
  horizontal_distance = xray(end) / 1000; % Convert horizontal
distance to kilometers
  fprintf('Launch angle: %.2f°, Max height: %.2f km, Horizontal
distance covered: %.2f km\n', ...
    psi0_deg, max_height, horizontal_distance);
end

% Formatting the ray path plot
subplot(2, 1, 1);
xlabel('Horizontal Distance (km)'); % Label x-axis (horizontal
distance in km)
ylabel('Height (km)');        % Label y-axis (height in km)
```

```
title('Curved Ray Path in the Troposphere for Different
Launch Angles');
grid on;                     % Enable grid for the plot
legend('0.5°', '1°', '1.25°'); % Legend for different
launch angles
hold off;

% Formatting the grazing angle plot
subplot(2, 1, 2);
xlabel('Horizontal Distance (km)'); % Label x-axis (horizontal
distance in km)
ylabel('Graze Angle (degrees)');    % Label y-axis (grazing angle
in degrees)
title('Variation of Graze Angle with Horizontal Distance');
grid on;                     % Enable grid for the plot
legend('0.5°', '1°', '1.25°'); % Legend for different
launch angles
hold off;
Launch angle: 0.50°, Max height: 0.67 km, Horizontal distance
covered 315.19 km
Launch angle: 1.00°, Max height: 3.41 km, Horizontal distance
covered 874.73 km
Launch angle: 1.25°, Max height: 7.08 km, Horizontal distance
covered 1669.10 km
```

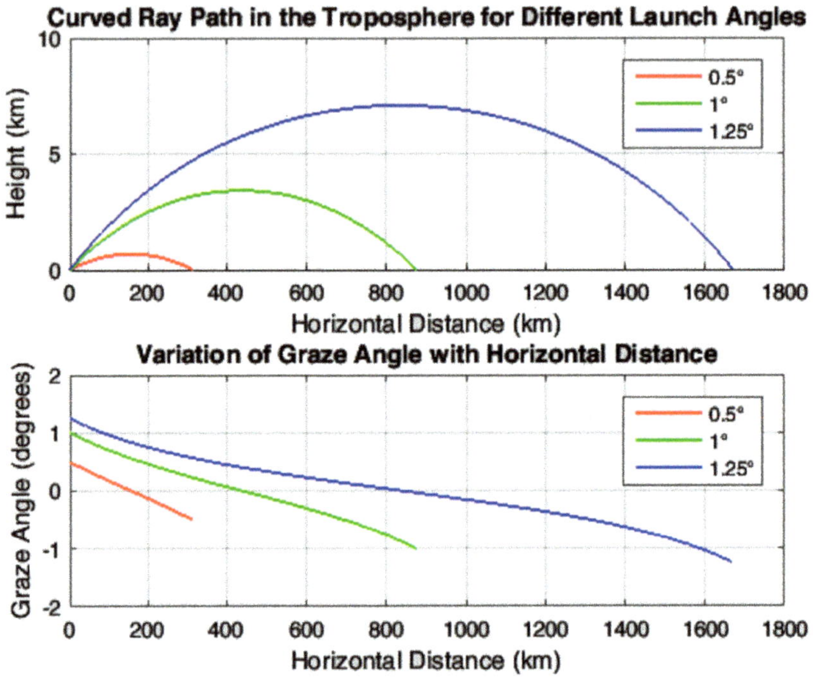

Figure 13.6 Radio wave propagation behavior through the troposphere.

13.6 Ionospheric Propagation and Dielectric Behavior

A plane wave is defined as a propagating wave with a surface of a constant phase and a plane that is perpendicular to the direction of propagation.

In a plane wave of an electromagnetic radiation, the electric or magnetic field can be expressed as:

$$E = E_0 \cdot e^{j(\omega t - kz)} \tag{13.19}$$

where $\omega = 2\pi f$ is the angular frequency, f is the frequency, $k = 2\pi/\lambda$ is the wave vector, and λ is the wavelength.

Now $v = f\lambda = c/n$, the velocity of the electromagnetic wave, n is the refractive index, and c is the free space wave velocity.

Alternatively,

$$E = E_0 \cdot e^{jw\left(t - \frac{k}{\omega} z\right)} = E_0 \cdot e^{j\omega\left(t - \frac{n}{c} z\right)} \tag{13.20}$$

where

$$\frac{k}{\omega} = \frac{1}{v} = \frac{n}{c}$$

The question that may arise is: What is the value of the refractive index n?

Since $c = 1/\sqrt{\mu_0 \varepsilon_0}$ and $v = 1/\sqrt{\mu_0 \mu \varepsilon_0 \varepsilon}$, then $n = c/v = \sqrt{\mu \varepsilon}$.

The magnetic permeability μ is approximately 1 for gases and hence = $\sqrt{\varepsilon}$, where ε is the dielectric constant or the relative permittivity.

Equation (13.20) is transformed to

$$E = E_0 \cdot e^{j\omega\left(t - \frac{k}{w} z\right)} = E_0 \cdot e^{j\omega\left(t - \frac{\sqrt{\varepsilon}}{c} z\right)} \tag{13.21}$$

The ionosphere consists of ionized gas at a low pressure and we shall calculate, in the next section, the effective dielectric constant for an ionized gas.

At this point, it is imperative to know that, in the Earth's atmosphere, there are layers of ionized gas, and so there are free electrons, which, because of their low mass, respond easily to the electric field of radio waves. The value N, which is defined as the density of free electrons per cubic meter, has maxima, as shown in Figure 13.7, where E_1, F_1, and F_2 represent layers of the ionosphere.

Consider a single electron of mass m and charge $-e$, in an electric field that varies with time as e^{jwt}. The following equation of motion applies, ignoring collisions:

$$m\frac{dv}{dt} = -eE \tag{13.22}$$

Since $\boldsymbol{E}$ is proportional to e^{jwt}, $\boldsymbol{v}$ is also expected to be proportional to e^{jwt}, and therefore

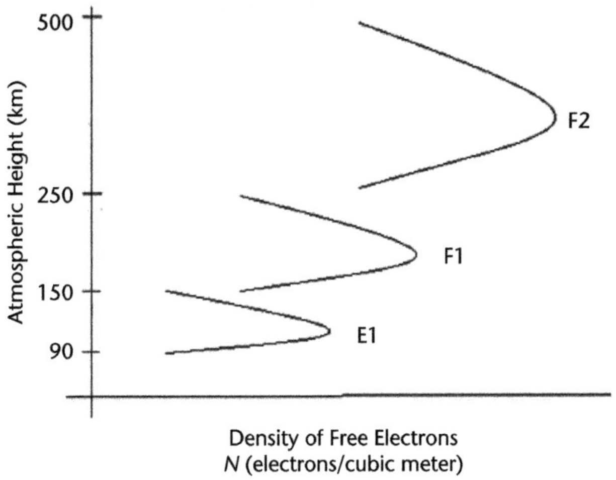

Figure 13.7 Structure of the ionosphere.

$$\frac{dv}{dt} = j\omega v \tag{13.23}$$

Substituting (13.23) into (13.22) yields:

$$v = \frac{-eE}{j\omega m} \tag{13.24}$$

For N electrons/unit volume of gas, the current density J is given by:

$$J = -eNv = \frac{Ne^2E}{j\omega m} \tag{13.25}$$

Substituting for J and for $\mathbf{D}$ in the following Maxwell's equation:

$$\text{curl } \mathbf{H} = J + \frac{\partial \mathbf{D}}{\partial t} = \frac{Ne^2\mathbf{E}}{j\omega m} + j\omega\varepsilon\varepsilon_0\mathbf{E} \tag{13.26}$$

We can define an effective dielectric constant ε' of the ionized gas by combining the two terms on the right side of the equation, as a single term, $\varepsilon'\varepsilon_0\mathbf{E}$. Hence,

$$j\omega\varepsilon'\varepsilon_0 = \frac{Ne^2}{j\omega m} + j\omega\varepsilon\varepsilon_0 \tag{13.27}$$

$$\varepsilon' = \varepsilon - \frac{Ne^2}{\varepsilon_0 m\omega^2} \tag{13.28}$$

Hence,

$$\text{curl } \boldsymbol{H} = j\omega\varepsilon'\varepsilon_0 \boldsymbol{B} \tag{13.29}$$

Define a plasma frequency w_p by means of the following equation

$$\omega_p^2 = \frac{Ne^2}{\varepsilon_0 m} \tag{13.30}$$

Substituting (13.30) into (13.28) and considering $\varepsilon \approx 1$, which is the dielectric constant of an ionized background gas, as in the ionosphere, gives:

$$\varepsilon' = 1 - \frac{\omega_p^2}{\omega^2} \tag{13.31}$$

The electric field of a plane wave propagating in the ionosphere can now be represented as:

$$E = E_0 \cdot e^{j\omega\left(t - \frac{\sqrt{\varepsilon'}}{c} z\right)} \tag{13.32}$$

This is similar to (13.21), except for the effective dielectric constant.

MATLAB Program 24 simulates and plots the effective dielectric constant versus the electron density at 10-MHz frequency, as shown in Figure 13.8. It further plots the effective dielectric constant versus frequency at N = 10^{12} electrons/m.

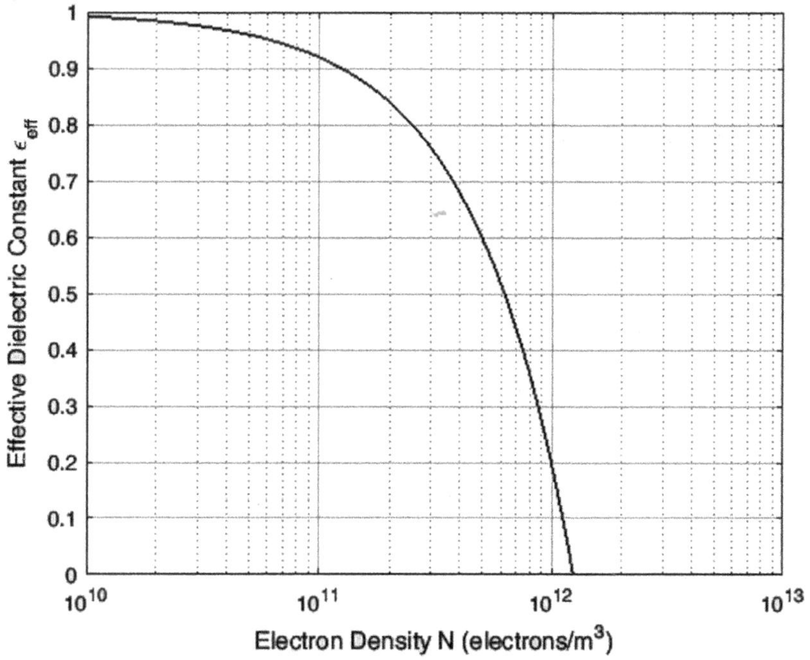

Figure 13.8 Effective dielectric constant versus electron density at f = 10 MHz.

MATLAB Program 24

```
% This MATLAB program calculates and plots the **effective
dielectric constant (ε_eff)**
% of an ionized gas (such as the ionosphere) based on the plasma
frequency formulation.
% The effective dielectric constant is derived from the relation:
%    ε_eff = 1 - (ω_p^2 / ω^2)
% where ω_p is the plasma angular frequency determined by the
electron density.
%
% The program generates two plots:
% 1. ε_eff versus **electron density** (N) for a fixed
wave frequency.
% 2. ε_eff versus **wave frequency** for a fixed
electron density.
%
% Physically, ε_eff determines how electromagnetic waves
propagate through
% ionized media. When ε_eff drops to zero or becomes negative,
the wave is
% reflected or does not propagate. To ensure physical realism,
the code
% prevents ε_eff from dropping below zero.

clc;
clear;

% Physical constants
e = 1.602e-19;      % Electron charge (C)
m = 9.109e-31;      % Electron mass (kg)
epsilon_0 = 8.854e-12; % Vacuum permittivity (F/m)

%% 1. ε_eff vs Electron Density
f_fixed = 10e6;                % Fixed frequency = 10 MHz
omega_fixed = 2 * pi * f_fixed;  % Angular frequency

N = logspace(10, 13, 500);      % Electron density range: 1e10 to
1e13 electrons/m^3
omega_p_sq = (N * e^2) / (epsilon_0 * m);  % Plasma
frequency squared
epsilon_eff_N = 1 - omega_p_sq ./ (omega_fixed^2);

% Ensure ε_eff >= 0 (physically meaningful)
epsilon_eff_N(epsilon_eff_N < 0) = NaN;

% Plot ε_eff vs N
figure;
semilogx(N, epsilon_eff_N, 'k', 'LineWidth', 1.5);
xlabel('Electron Density N (electrons/m^3)');
```

```
ylabel('Effective Dielectric Constant \epsilon_{eff}');
title(['\epsilon_{eff} vs Electron Density at f = ' num2str(f_
fixed/1e6) ' MHz']);
grid on;

%% 2. ε_eff vs Frequency
N_fixed = 1e12;                    % Fixed electron density
f = logspace(6, 10, 500);            % Frequency from 1 MHz to
10 GHz
omega = 2 * pi * f;
omega_p_sq_fixed = (N_fixed * e^2) / (epsilon_0 * m);
epsilon_eff_f = 1 - omega_p_sq_fixed ./ (omega.^2);

% Ensure ε_eff >= 0
epsilon_eff_f(epsilon_eff_f < 0) = NaN;

% Plot ε_eff vs frequency
figure;
semilogx(f/1e6, epsilon_eff_f, 'k', 'LineWidth', 1.5);
xlabel('Frequency (MHz)');
ylabel('Effective Dielectric Constant \epsilon_{eff}');
title(['\epsilon_{eff} vs Frequency at N = ' num2str(N_fixed,
'%.2e') ' electrons/m^3']);
grid on;
```

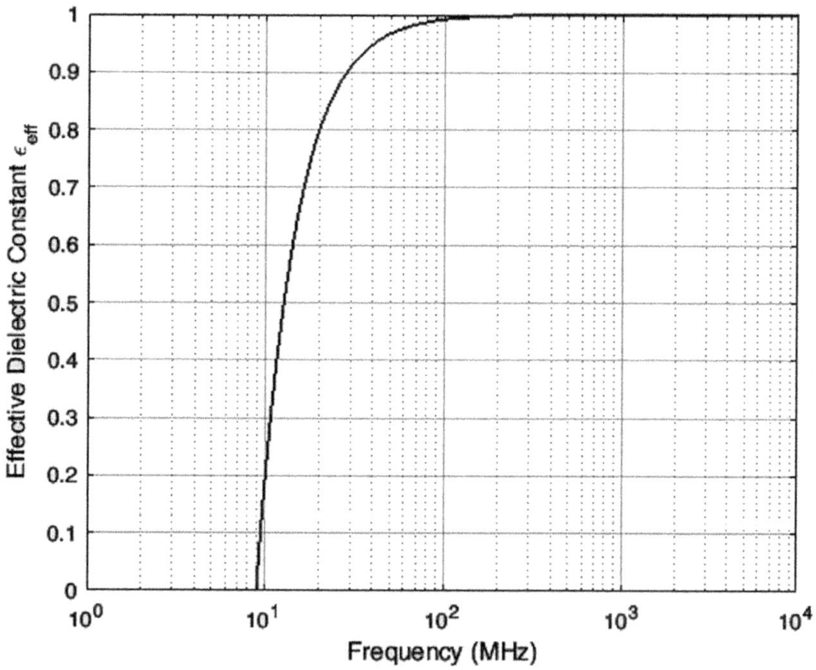

Figure 13.9 Effective dielectric constant versus frequency at $N = 10^{12}$ electrons/m^3.

13.7 Ionospheric Wave Incidence and Skip Distance Calculations

In the normal incidence case, consider an electromagnetic wave, launched normally from a point at the Earth's surface, as shown in Figure 13.10, towards the ionosphere, which is in the z-direction. The electric field of the wave can be expressed by (13.30) to (13.32).

At the Earth's surface, as shown in Figure 13.10, $N = 0$, $w_p = 0$ and $\varepsilon' = 1$. As the wave begins to enter the ionosphere, N increases from 0, w_p^2 increases, and ε' decreases from the value 1. The effect of this is to increase the wavelength, as follows:

$$\lambda = \frac{v}{f} = \frac{c}{nf} = \frac{c}{\left(\sqrt{\varepsilon'}\right)f} \tag{13.33}$$

Therefore, λ increases because of the decrease in the effective dielectric constant. With further increase of N with increasing height, there would be a point where $\varepsilon' \to 0$ and then $\varepsilon' < 0$. The effect of this is that $\lambda \to \infty$; hence, the wave would experience an exponential decay and an attenuation of the electric field. The wave starts to reflect downwards towards Earth. As $\varepsilon' \to 0$, $\omega_p^2 = Ne^2/\varepsilon_0 m = \omega^2$.

It is interesting to know that the height of the ionosphere can be estimated from the time taken by a signal to return to Earth after reflection from the ionosphere.

In the oblique incidence case, consider a wave launched with an angle towards the ionosphere. Figure 13.11 shows a transmitted wave and a reflected wave and, in the electromagnetic theory, expressions are derived for the following relation.

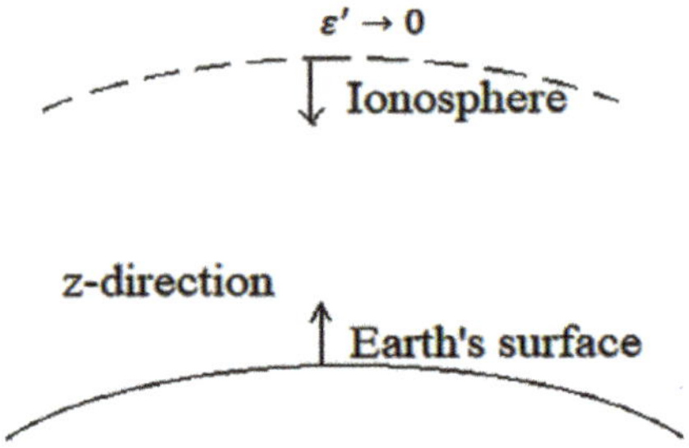

Figure 13.10 Normal incidence of an electromagnetic wave launched towards the ionosphere.

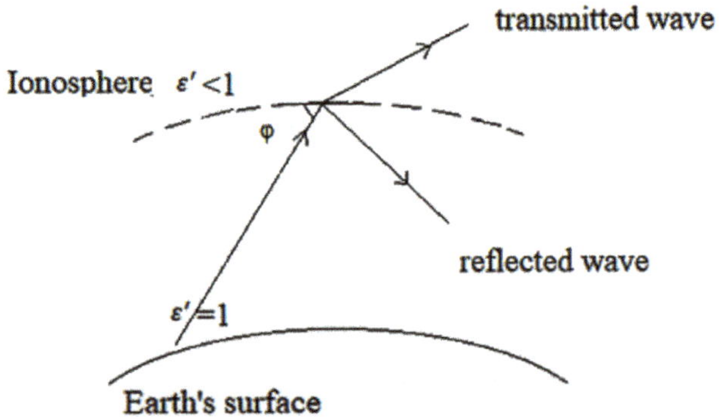

Figure 13.11 Oblique incidence of an electromagnetic wave launched towards the ionosphere.

The reflection coefficient

$$R_E = \frac{\text{electric field amplitude in reflected beam}}{\text{electric field amplitude in incident beam}} \tag{13.34}$$

For the cases of the electric field vector, which is perpendicular and in the plane of incidence, relatively

$$R_{E_\perp} = \frac{\dfrac{\sin\varphi}{\sqrt{n^2 - \cos^2\varphi}} - 1}{\dfrac{\sin\varphi}{\sqrt{n^2 - \cos^2\varphi}} + 1} \tag{13.35}$$

$$R_{E_\parallel} = \frac{\dfrac{n^2\sin\varphi}{\sqrt{n^2 - \cos^2\varphi}} - 1}{\dfrac{n^2\sin\varphi}{\sqrt{n^2 - \cos^2\varphi}} + 1} \tag{13.36}$$

For the special case of normal incidence, $\varphi = 90°$; hence,

$$R_{E_\perp} = \frac{1-n}{1+n} \approx 0 \tag{13.37}$$

and

$$R_{E_\parallel} = \frac{n-1}{n+1} \approx 0 \tag{13.38}$$

At oblique incidence, there would be a value of φ when

$$n^2 - \cos^2\varphi_c = 0 \tag{13.39}$$

At the critical angle φ_c, both numerators and denominators in (13.34) and (13.35) tend to infinity and hence $R_{E_\perp} \approx 1$ and $R_{E_\parallel} \approx 1$.

It follows by substituting for $n^2 = \varepsilon'$ from (13.31) that

$$1 - \frac{\omega_p^2}{\omega^2} = \cos^2\varphi_c = 1 - \sin^2\varphi_c \tag{13.40}$$

hence

$$\sin\varphi_c = \frac{\omega_p}{\omega} = \cos i_c \tag{13.41}$$

where i_c is the complement angle of φ_c.

The skip distance, shown in Figure 13.12, is the length of the zone of silence. θ is a launch angle of the wave from a radio transmitter and θ_c is the value launch angle corresponding to the case where $\varphi = \varphi_c$.

The value of the skip distance can be found by considering the triangle *ITO* in Figure 13.12.

$$\gamma + \frac{\pi}{2} + \theta_c + i_c = \pi \tag{13.42}$$

Substituting for $i_c = \pi/2 - \varphi_c$ into (13.42),

$$\gamma = \varphi_c - \theta_c \tag{13.43}$$

$$d = 2a\gamma = 2a(\varphi_c - \theta_c) \tag{13.44}$$

θ_c can be worked out by applying the sine rule of triangle *ITO*.

$$\frac{\sin i_c}{a} = \frac{\sin\left(\frac{\pi}{2} + \theta_c\right)}{a+h} = \frac{\cos\theta_c}{a+h} \tag{13.45}$$

$$\cos\theta_c = \frac{a+h}{a}\sin i_c = \frac{a+h}{a}\cos\varphi_c = \frac{a+h}{a}\left(1 - \frac{\omega_p^2}{\omega^2}\right)^{1/2} \tag{13.46}$$

Substituting for φ_c from (13.41) and for θ_c from (13.46) into (13.44) yields

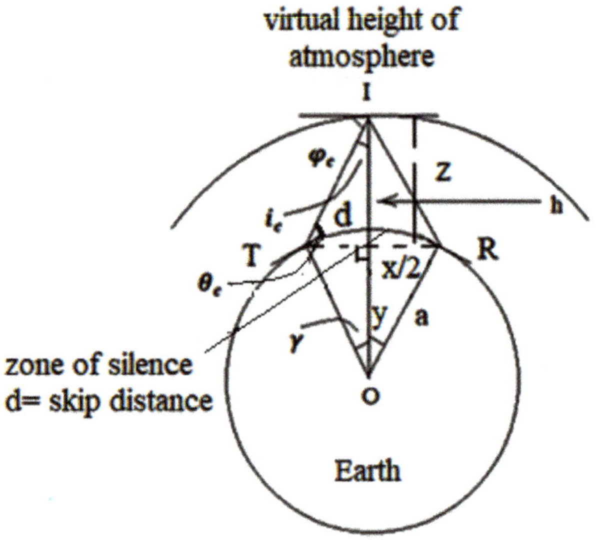

Figure 13.12 Zone of silence and skip distance *d* as a result of wave reflection from the ionosphere.

$$d = 2a\left[\sin^{-1}\left(\frac{\omega_p}{\omega}\right) - \cos^{-1}\left\{\frac{a+h}{a}\left(1-\frac{\omega_p^2}{\omega^2}\right)^{1/2}\right\}\right] \tag{13.47}$$

The maximum skip distance can be reached by bouncing a wave once off the ionosphere and is called the maximum range d_m. This can be calculated by reference to Figure 13.13.

$$\cos\gamma_m = \frac{a}{a+h} \tag{13.48}$$

hence

$$d_m = 2a\gamma_m = 2a\cdot\cos^{-1}\left(\frac{a}{a+h}\right) \tag{13.49}$$

The following MATLAB program proposes a set of plasma frequencies for a range of atmospheric heights between 90 km and 500 km, covering the sublayers E_1, F_1, and F_2. It further presumes different broadcasting frequencies in the high-frequency range and computes the anticipated electron densities, launching angles of the radio wave and the skip distance.

MATLAB Program 25 computes the electron density, launching angle and skip distance at various heights. The results are displayed in Table 13.3.

MATLAB Program 25

```
% Matlab program to compute  the electron density (N) and the
skip distance (d) for radio waves
% broadcasted at different frequencies for various ionospheric
layers, given different heights.
```

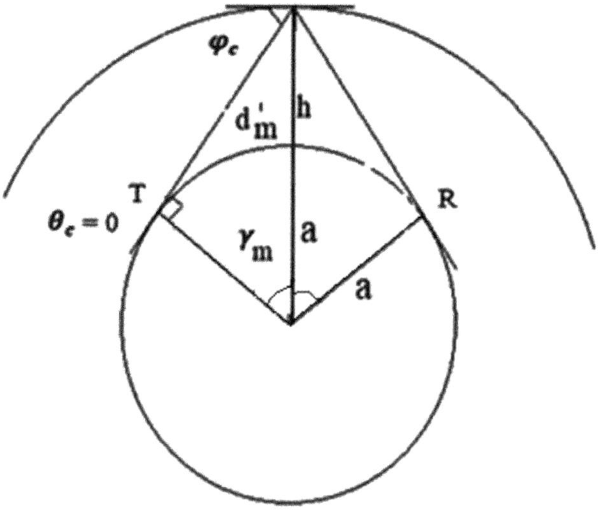

Figure 13.13 Maximum range d_m as a result of wave reflection from the ionosphere.

Table 13.3 Computed Electron Density, Launching Angle, and Skip Distance at Various Heights

Atmospheric Height (km)	*Typical Plasma Frequency (Hz)*	*Broadcast Frequency (Hz)*	*Electron Density (e/m³)*	*Launch Theta (°)*	*Skip Distance (km)*
90	5e+5	3e+6	3.1016e+09	0.6	1,997.7
150	3e+6	1e+7	1.1166e+11	12.4	1,108.6
250	5e+6	1e+7	3.1016e+11	25.8	924.9
300	1e+7	1.5e+7	1.2406e+12	38.6	692.1
400	1.2e+7	1.5e+7	1.7865e+12	50.3	611.2
500	1.4e+7	1.5e+7	2.4316e+12	67.2	386.9

```
% It also calculates and displays the theta in degrees in a
nicely formatted table.

% Define constants
e = 1.602e-19; % Coulombs
epsilon = 8.854e-12; % F/m
m = 9.109e-31; % kg
a = 6371; % Earth's radius in km

% Heights (in km) and corresponding plasma frequencies (in MHz)
heights = [90, 150, 250, 300, 400, 500];
plasma_frequencies_mhz = [0.5, 3, 5, 10, 12, 14]; % in MHz

% Convert plasma frequencies to Hz
plasma_frequencies_hz = plasma_frequencies_mhz * 1e6; % Hz

% Broadcast frequencies in MHz based on height
f_mhz = [3, 10, 10, 15, 15, 15]; % Updated frequency
f_hz = f_mhz * 1e6; % Convert to Hz

% Preallocate arrays for N, d, and theta
N_values = zeros(size(heights));
d_values = zeros(size(heights));
theta_degrees = zeros(size(heights));

% Compute N and d for each height
for i = 1:length(heights)
  h = heights(i); % height in km
  fp = plasma_frequencies_hz(i); % plasma frequency in Hz
  omega_p = 2 * pi * fp;

  % Determine the broadcast frequency for current height
  f_current_hz = f_hz(i);
  omega = 2 * pi * f_current_hz;

  % Calculate electron density N
  N_values(i) = (omega_p)^2 * epsilon * m / e^2;
```

```
    % Calculate distance d
    ratio_wp_w = omega_p / omega;
    if abs(ratio_wp_w) <= 1
      phi = asin(ratio_wp_w);
    else
      phi = NaN; % Invalid calculation
    end

    theta_argument = (a + h) / a * sqrt(1 - min(max(ratio_wp_w^2,
  0), 1));
    if abs(theta_argument) <= 1
      theta = acos(theta_argument);
      theta_degrees(i) = rad2deg(theta);
    else
      theta = NaN; % Invalid calculation
      theta_degrees(i) = NaN;
    end

    if isnan(phi) || isnan(theta)
      d_values(i) = NaN; % Invalid calculation
    else
      d_values(i) = 2 * a * (phi - theta);
    end
  end

  % Create and display results table
  results_table = table(heights', plasma_frequencies_hz', f_hz',
  N_values', theta_degrees', d_values', ...
    'VariableNames', {'Height_km', 'Plasma_Frequency_Hz',
  'Broadcast_Frequency_Hz', 'Electron_Density_m3', 'Theta_Degrees',
  'Skip_Distance_km'});
  disp(results_table)
```

For a single-hop skip distance versus a launching angle, in radio broadcasting, it is important to predict in advance the launching angle of the radiated radio wave, which would result in a targeted skip distance along the Earth's surface *d*.

In reference to Figure 13.12:

$$\begin{aligned} x^2 &= 2\left(2a^2 - 2a^2\cos^2(\gamma)\right) = 2\left(2a^2 - a^2(1+\cos 2\gamma)\right) \\ &= 2a^2 - 2a^2\cos 2\gamma \end{aligned} \tag{13.50}$$

$$x = \sqrt{2}a\sqrt{(1 - \cos 2\gamma)} \tag{13.51}$$

$$z = h + a - y = h + a\left(1 - \cos(\gamma)\right) \tag{13.52}$$

but

$$\tan\varphi_c = \frac{2z}{x} \tag{13.53}$$

Substituting (13.51) and (13.52) into (13.53) gives:

$$\varphi_c = \text{atan}\left[\frac{\sqrt{2}\left\{h + a\left(1 - \cos(\gamma)\right)\right\}}{a\sqrt{(1 - \cos 2\gamma)}}\right] \tag{13.54}$$

Substituting (13.54) into (13.43), the launching angle yields:

$$\theta_c = \text{atan}\left[\frac{\sqrt{2}\left\{h + a\left(1 - \cos\left(\frac{d}{2a}\right)\right)\right\}}{a\sqrt{(1 - \cos 2\gamma)}}\right] - \frac{d}{2a} \tag{13.55}$$

where $\gamma = d/2a$. Furthermore, $\cos^2\gamma = \cos^2\gamma - \sin^2\gamma$:

$$\theta_c = \text{atan}\left[\frac{h + a\left(1 - \cos\left(\frac{d}{2a}\right)\right)}{a \cdot \sin\left(\frac{d}{2a}\right)}\right] - \frac{d}{2a} \tag{13.56}$$

Thus, it is possible to plot the radiation launching (elevation/radiation) angle versus the skip distance for a range of atmospheric heights of the ionosphere, bearing in mind that the E_1 sublayer falls between approximately, 90 to 150 km, F_1 between 150 to 250 km, and F_2 between 250 to 500 km. This should provide a predictive guideline to radio skywave broadcasters. However, the actual distances depend on the state of the electronic density profiles of the ionospheric sublayers, the atmospheric conditions, the seasons, the geographic locations, and the times of the day. The highest frequency that can be reflected by the ionosphere increases with the electron density, which is influenced by solar activity and the atmospheric heights. These heights can be measured in real time using a sounding technique similar to radar, called an ionosonde. An ionosonde transmits radio pulses vertically into the ionosphere. These pulses are reflected back to the ground from different layers within the ionosphere. By measuring the time delay between transmission and reception, the ionosonde can determine the virtual height of the reflecting layers. The frequency of the transmitted pulses is gradually varied over a wide range, typically from about 1 to 30 MHz. Data from an ionosonde is used to generate an ionogram, which is a graph that shows the virtual height of the ionospheric layers as a function of frequency. The y-axis represents the virtual height (in kilometers), and the x-axis represents the frequency (in megahertz). MATLAB Program 26 simulates and plots the launching (radiation) angle (in degrees) versus skip distance (in kilometers) for a range of atmospheric heights, as shown in Figure 13.14.

MATLAB Program 26

```
% Matlab program to plot launching Angle (alpha-deg)  vs. Skip
distance d (km) for different atmospheric heights
% Parameters
```

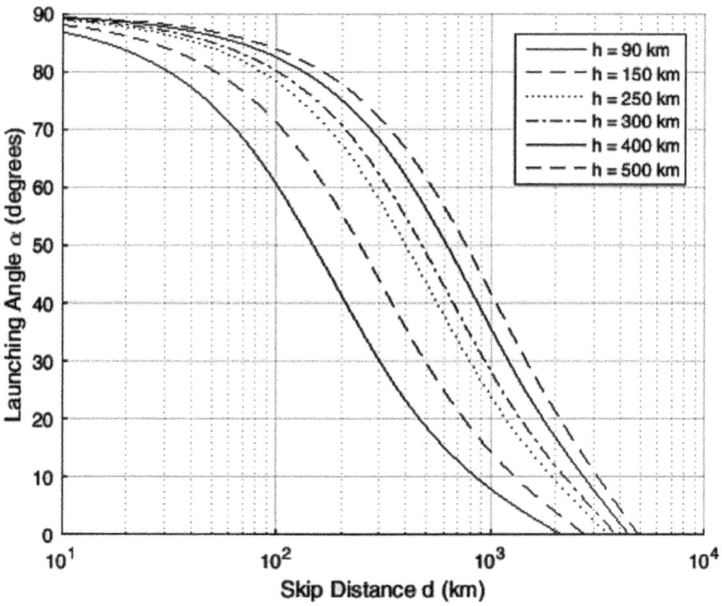

Figure 13.14 Launching (radiation) angle (degrees) versus skip distance (kilometers) for a range of atmospheric heights.

```
R = 6471; % Radius of the Earth in km
h = [90, 150, 220, 300, 400, 500]; % Atmospheric heights in km
d = logspace(1, 4, 1000); % Skip distance in km (10 to 10000 km,
logarithmically spaced)

% Initialize array to store alpha values
alpha_deg = zeros(length(h), length(d));

% Calculate alpha for each height and distance
for i = 1:length(h)
  H = h(i);
  alpha_rad = atan((H + R * (1 - cos(d / (2 * R)))) ./ (R * sin(d
/ (2 * R)))) - d / (2 * R);
  alpha_deg(i, :) = rad2deg(alpha_rad); % Convert to degrees
end

% Plot the results
figure;
hold on;
line_styles = {'-', '--', ':', '-.', '-', '--'}; % Different
black line styles
legends = {}; % Initialize an empty cell array for legend entries
for i = 1:length(h)
  % Only plot values of alpha between 0 and 90 degrees
  valid_indices = alpha_deg(i, :) >= 0 & alpha_deg(i, :) <= 90;
  if any(valid_indices) % Check if there are valid indices
to plot
```

```
    plot(d(valid_indices), alpha_deg(i, valid_indices), 'k',
'LineStyle', line_styles{i}, 'DisplayName', ['h = ' num2str(h(i))
' km']);
    legends{end+1} = ['h = ' num2str(h(i)) ' km']; % Add to
legend entries
  end
end
hold off;
set(gca, 'XScale', 'log'); % Set x-axis to logarithmic scale
xlabel('Skip Distance d (km)');
ylabel('Launching Angle \alpha (degrees)');
if ~isempty(legends) % Only display legend if there are
valid entries
  legend(legends);
end
grid on;
```

13.8 Effect of the Earth's Magnetic Field on Radio Wave Propagation in the Ionosphere

In a zero magnetic field, the motion of an electron under the influence of an electric force can be expresses by the following equation

$$m\frac{dv}{dt} = -eE \tag{13.57}$$

where the time-varying electric field of the radio wave $\boldsymbol{E} = E_0 e^{j\omega t}$.

Under the influence of a magnetic flux density $\boldsymbol{B} = \mu\boldsymbol{H}$, where $\boldsymbol{H}$ is the magnetic field, the above equation of motion transforms into

$$m\frac{dv}{dt} = -eE - e\mathbf{v} \times \mathbf{B} \tag{13.58}$$

The force that combines the electric and magnetic fields is called the Lorentz force and the magnetic field $\boldsymbol{H}$ is static, just like the Earth's magnetic field.

Because $\boldsymbol{E}$ is time-dependent, $\boldsymbol{v}$ in (13.58) is expected to be time-dependent too. Therefore,

$$\frac{dv}{dt} = j\omega v \tag{13.59}$$

Equation (13.58) becomes

$$m(j\omega v) = -e(E + \mathbf{v} \times \mathbf{B}) \tag{13.60}$$

Considering $\boldsymbol{B}$ at a point in the ionosphere in the z-direction

$$\upsilon \times B = \begin{vmatrix} i & j & k \\ \upsilon_x & \upsilon_y & \upsilon_z \\ 0 & 0 & B_z \end{vmatrix} = i\upsilon_y B_z - j\upsilon_x B_z \tag{13.61}$$

Substituting back into (13.59) and writing out the equation of motion in its component form:

$$j\omega\upsilon_x = -\frac{e}{m}\left(E_x + \upsilon_y B_z\right) \tag{13.62a}$$

$$j\omega\upsilon_y = -\frac{e}{m}\left(E_y - \upsilon_x B_z\right) \tag{13.62b}$$

$$j\omega\upsilon_z = -\frac{e}{m}E_z \tag{13.62c}$$

In matrix form, the three equations, (13.62(a–c)), are rewritten as:

$$\begin{pmatrix} j\omega & \frac{eB_z}{m} & 0 \\ -\frac{eB_z}{m} & j\omega & 0 \\ 0 & 0 & j\omega \end{pmatrix}\begin{pmatrix} \upsilon_x \\ \upsilon_y \\ \upsilon_z \end{pmatrix} = -\frac{e}{m}\begin{pmatrix} E_x \\ E_y \\ E_z \end{pmatrix} \tag{13.63}$$

To derive an expression for υ, we need to invert the 3 × 3 matrix.

$$\upsilon = \frac{e}{jm\omega\left(\omega_c^2 - \omega^2\right)}\begin{pmatrix} \omega^2 & -j\omega\omega_c & 0 \\ j\omega\omega_c & \omega^2 & 0 \\ 0 & 0 & \omega^2 - \omega_c^2 \end{pmatrix}\begin{pmatrix} E_x \\ E_y \\ E_z \end{pmatrix} \tag{13.64}$$

where $\omega_c = -eB_z/m$ is called the cyclotron frequency.

The velocity of an electron is hence expressed in combined electric and magnetic fields. We can also write this as:

$$\upsilon = \mu \cdot E \tag{13.65}$$

where μ is termed a mobility matrix.

The current density flowing if there are N electrons per unit volume are:

$$J = -Ne\upsilon = -Ne\mu \cdot E = \sigma \cdot E \tag{13.66}$$

where $\sigma = -Ne\mu$ is termed a conductivity tensor.

We substitute $\boldsymbol{J}$ into one of Maxwell's equations and then derive a wave equation.

$$\sigma = \frac{Ne^2}{jm\omega\left(\omega^2 - \omega_c^2\right)}\begin{pmatrix} \omega^2 & -j\omega\omega_c & 0 \\ j\omega\omega_c & \omega^2 & 0 \\ 0 & 0 & \omega^2 - \omega_c^2 \end{pmatrix} \tag{13.67}$$

Applying Maxwell's equation (13.26) gives:

$$\begin{aligned} \text{curl } \boldsymbol{H} &= \sigma \cdot \boldsymbol{E} + j\omega\varepsilon\varepsilon_0 \cdot I \cdot \boldsymbol{E} \\ &= j\omega\left(\frac{1}{j\omega\varepsilon_0}\sigma + \varepsilon I\right) \cdot \varepsilon_0 \boldsymbol{E} = j\omega\varepsilon'' \cdot \varepsilon_0 \boldsymbol{E} \end{aligned} \tag{13.68}$$

where

$$\boldsymbol{D} = \varepsilon\varepsilon_0 \boldsymbol{E} \text{ and } I = \begin{pmatrix} 1 & 0 & 0 \\ 0 & 1 & 0 \\ 0 & 0 & 1 \end{pmatrix}$$

ε is known as the dielectric constant of background gas of the ionosphere. ε'' is now expressed as:

$$\varepsilon'' = \begin{pmatrix} \left(1 + \dfrac{\omega_p^2}{\omega_c^2 - \omega^2}\right) & -\dfrac{j\omega_p^2\omega_c/\omega}{\omega_c^2 - \omega^2} & 0 \\ \dfrac{j\omega_p^2\omega_c/\omega}{\omega_c^2 - \omega^2} & \left(1 + \dfrac{\omega_p^2}{\omega_c^2 - \omega^2}\right) & 0 \\ 0 & 0 & \left(1 - \dfrac{\omega_p^2}{\omega^2}\right) \end{pmatrix} \tag{13.69}$$

where ω_p and ω_c were earlier defined as the plasma and cyclotron frequencies, respectively.

Consider the following Maxwell's equation

$$\text{curl } \boldsymbol{E} = -\frac{\partial \boldsymbol{B}}{\partial t} = -\mu\mu_0 \frac{\partial \boldsymbol{H}}{\partial t} = -j\omega\mu\mu_0 \boldsymbol{H} \tag{13.70}$$

Along with (13.26), both describe two unknowns $\boldsymbol{E}$ and $\boldsymbol{H}$.

Taking the curl of (13.70),

$$\text{curl curl } \boldsymbol{E} = -j\omega\mu\mu_0 \text{curl } \boldsymbol{H} \tag{13.71}$$

Substituting for *curl* $\boldsymbol{H}$ from (13.68) into (13.71) gives:

$$\text{curl curl } \boldsymbol{E} = -j\omega\mu\mu_0\left(j\omega\varepsilon'' \cdot \varepsilon_0 \boldsymbol{E}\right) = \mu\mu_0\varepsilon_0\omega^2\varepsilon'' \cdot \boldsymbol{E} \tag{13.72}$$

Since curl ≡ grad div – ∇^2, then

$$\text{curl curl } \boldsymbol{E} = \text{grad div } \boldsymbol{E} - \nabla^2\boldsymbol{E} = -\nabla^2\boldsymbol{E} \tag{13.73}$$

where grad div E = grad$(\rho/\varepsilon\varepsilon_0) = 0$, since the volume charge density ρ is a constant.

Substituting (13.73) into (13.72) gives:

$$\nabla^2\boldsymbol{E} + \mu\mu_0\varepsilon_0\omega^2\varepsilon'' \cdot \boldsymbol{E} = 0 \tag{13.74}$$

which is known as the wave equation.

In order to solve the wave equation, consider a traveling wave along the z-direction with a wavefront in the xy-plane that contains $\boldsymbol{E} = (E_x, E_y, 0)$. The spatial variation of $\boldsymbol{E}$ is with the z-direction and (13.66) transforms to:

$$\frac{\partial^2}{\partial z^2}\begin{pmatrix} E_x \\ E_y \\ 0 \end{pmatrix} + \mu\mu_0\varepsilon_0\omega^2\varepsilon'' \cdot \begin{pmatrix} E_x \\ E_y \\ 0 \end{pmatrix} = 0 \tag{13.75}$$

The above equation yields the following two equations:

$$\frac{\partial^2 E_x}{\partial z^2} + \mu\mu_0\varepsilon_0\omega^2\left(\varepsilon''_{11}E_x - j\varepsilon''_{12}E_y\right) = 0 \tag{13.76}$$

$$\frac{\partial^2 E_y}{\partial z^2} + \mu\mu_0\varepsilon_0\omega^2\left(j\varepsilon''_{12}E_x + \varepsilon''_{11}E_y\right) = 0 \tag{13.77}$$

where

$$\varepsilon''_{11} = 1 + \frac{\omega_p^2}{\omega_c^2 - \omega^2} \quad \text{and} \quad \varepsilon''_{12} = \frac{\omega_p^2\omega_c/\omega}{\omega_c^2 - \omega^2}$$

Here E_y appears in E_x equation and E_x appears in E_y equation. These are called coupled differential equations.

To solve the coupled differential equations, these need be transformed into two independent equations. This is enabled by considering the required combinations $E_x + jE_y$ and $E_x - jE_y$. Equation (13.77) is multiplied by j and then is added to (13.76) to get a new equation. To produce a second new equation, (13.77), after multiplying it with j, is subtracted from (13.76).

The following two independent equations result:

$$\frac{\partial^2}{\partial z^2}\left(E_x + jE_y\right) + \mu\mu_0\varepsilon_0\omega^2\left(\varepsilon''_{11} - \varepsilon''_{12}\right)\left(E_x + jE_y\right) = 0 \tag{13.78}$$

$$\frac{\partial^2}{\partial z^2}(E_x - jE_y) + \mu\mu_0\varepsilon_0\omega^2(\varepsilon''_{11} + \varepsilon''_{12})(E_x - jE_y) = 0 \tag{13.79}$$

where $E^+ = E_x + jE_y$ and $E^- = E_x - jE_y$.

What are E^+ and E^-? They are rotating electric filed vectors representing circularly polarized modes of propagation, rotating in opposite directions (one clockwise and another counterclockwise). These are the normal modes for propagation in an ionized plasma (or ionized gas) with a steady magnetic field present.

Solutions of the independent equations are:

$$E^+ = E_0 e^{j(\omega t - k^+ z)} \tag{13.80}$$

$$E^- = E_0 e^{j(\omega t - k^- z)} \tag{13.81}$$

where

$$k^+ = \omega\left[\mu\mu_0\varepsilon_0(\varepsilon''_{11} - \varepsilon''_{12})\right]^{1/2}$$
$$k^- = \omega\left[\mu\mu_0\varepsilon_0(\varepsilon''_{11} + \varepsilon''_{12})\right]^{1/2}$$

The k^+ and k^- are called wave vectors related to wavelength λ by $k = 2\pi/\lambda$.

This means that the wavelengths of the clockwise and counterclockwise modes are different. It follows that the velocities of the clockwise and counterclockwise modes are different too.

Since $\mu = 1$ and the velocity of wave in free space $c = 1/\sqrt{\mu_0\varepsilon_0}$, and substituting further for ε''_{11} and ε''_{12} in k^+ and k^- yields:

$$k^+ = \frac{\omega}{c}\left[1 - \frac{\omega_p^2}{\omega(\omega + \omega_c)}\right]^{1/2} \tag{13.82}$$

$$k^- = \frac{\omega}{c}\left[1 - \frac{\omega_p^2}{\omega(\omega - \omega_c)}\right]^{1/2} \tag{13.83}$$

If $B_z = 0$, then $\omega_c = 0$ and both (13.75) and (13.76) reduce to $k = (\omega/c)[1 - (\omega_p^2/\omega^2]^{1/2}$ as derived earlier for an ionized gas in zero magnetic field. This means that when $B_z = 0$, both modes have the same wavelength and velocity.

For the circularly polarized modes where $B_z \neq 0$, the velocities of the two modes are different and can be derived from $v = \omega/k^{\pm}$. It follows that:

$$v^{\pm} = \frac{\omega}{k^{\pm}} = c\left[1 - \frac{\omega_p^2}{\omega(\omega \pm \omega_c)}\right]^{-1/2} \tag{13.84}$$

13.8.1 Fading of Radio Waves

The wave transmitted as a plane, a polarized wave with an E vector vertical, say, travels as two oppositely rotating circularly polarized modes that have different velocities in the ionosphere because of the Earth's magnetic field (B_z). This means that the phase of polarization rotates as the wave progresses and reaches the receiving antenna with the E vector at some angle, θ, from the vertical. If the receiving aerial (antenna) is vertical, there is a fading of the signal. To correct for this, rotate the receiving areal so as to be parallel with the E vector; automatically or otherwise.

13.9 Derivation of the Faraday Rotation Angle

Faraday rotation occurs when a linearly polarized electromagnetic wave passes through a medium with a magnetic field. The plane of polarization rotates due to the difference in the refractive indices for right-handed and left-handed circularly polarized components of the wave. Figure 13.15 demonstrates the transformation of the radio wave polarized mode from linearly polarized to circularly polarized with different velocities in the ionosphere.

In a plasma, the cyclotron resonance causes different refractive indices for the two polarization states. For a weak magnetic field compared to the plasma frequency ($\omega \gg \omega_c$), the refractive indices for the right-handed (n^+) and left-handed (n^-) circularly polarized waves are approximately:

$$n_{\pm}^2 = 1 - \frac{\omega_p^2}{\omega(\omega \pm \omega_c)} \tag{13.85}$$

For $\omega \gg \omega_c$, we can approximate:

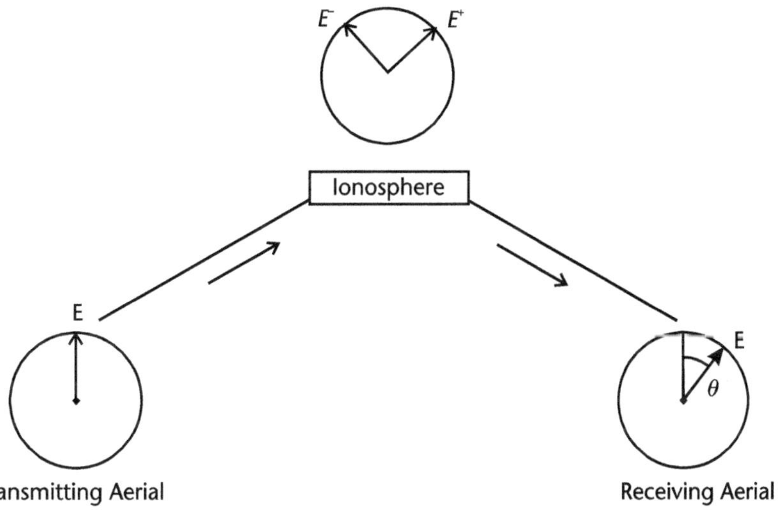

Figure 13.15 Transformation of the radio wave polarized mode from linearly polarized to circularly polarized with different velocities in the ionosphere.

$$n_+ \approx \sqrt{1 - \frac{\omega_p^2}{\omega^2}\left(1 - \frac{\omega_c}{\omega}\right)} \tag{13.86}$$

$$n_- \approx \sqrt{1 - \frac{\omega_p^2}{\omega^2}\left(1 + \frac{\omega_c}{\omega}\right)} \tag{13.87}$$

The difference in the refractive indices Δn is then:

$$\Delta n \approx n_+ - n_- \tag{13.88}$$

The difference in the refractive indices causes a phase shift $\Delta\phi$ between the two circular polarizations over a distance L. The phase shift can be approximated by:

$$\Delta\phi = \frac{2\pi\Delta n L}{\lambda} = \frac{2\pi\Delta n L f}{c} \tag{13.89}$$

For small ω_c/ω, the difference in refractive indices is:

$$\Delta n \approx \frac{\omega_p^2\omega_c}{\omega^3} = \frac{\omega_c}{\omega^3}\left(\frac{e^2N}{\varepsilon_0 m}\right) \tag{13.90}$$

Furthermore, because $\omega_c = eB/m$,

$$\Delta n \approx \frac{\omega_c\omega_p^2}{\omega^3} = \frac{\frac{eB}{m}}{\omega^3}\left(\frac{e^2N}{\varepsilon_0 m}\right) = \frac{e^3BN}{\omega^3m^2\varepsilon_0} \tag{13.91}$$

Substituting Δn, the phase shift (13.89) gives:

$$\Delta\phi = \frac{2\pi Lf}{c}\frac{e^3BN}{\omega^3m^2\varepsilon_0} \tag{13.92}$$

The rotation angle ϕ, in radians, of the plane of polarization is one-half the phase difference between the right circularly polarized (RCP) and left circularly polarized (LCP) components. Hence:

$$\phi = \frac{1}{2}\cdot\frac{2\pi Lfe^3BN}{\left(8f^3\pi^3\right)m^2\varepsilon_0 c} = \frac{e^3BNL}{8\pi^2f^2m^2\varepsilon_0 c} \tag{13.93}$$

This expression shows the dependence of the Faraday rotation angle on the magnetic field strength B, electron density N, path length L, frequency of wave f, and fundamental constants.

MATLAB Program 27 calculates and plots the variation of Faraday rotation angle with radio wave frequency, as shown in Figure 13.16.

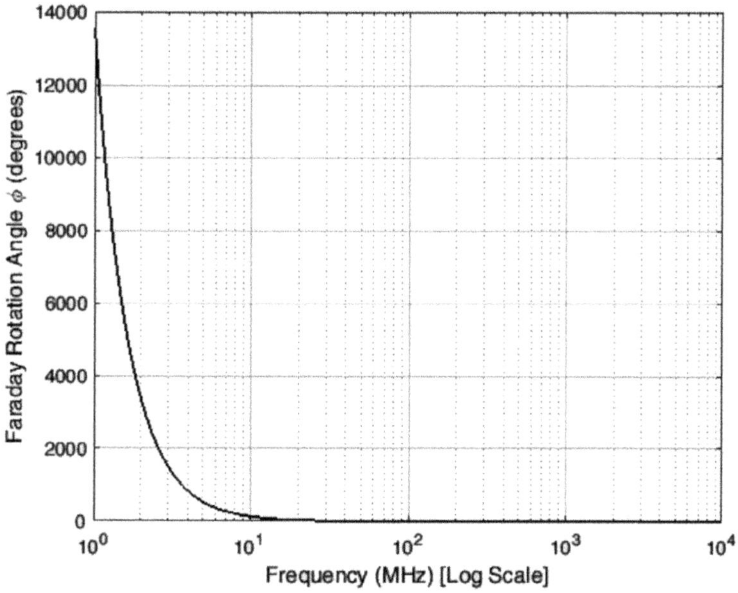

Figure 13.16 Variation of a Faraday rotation angle with a radio wave frequency ($B = 10\ \mu T$, $L =$ 1,000m).

MATLAB Program 27

```
% This MATLAB script calculates and plots the Faraday
rotation angle
% experienced by a linearly polarized electromagnetic wave as
it propagates
% through a magnetized plasma medium, such as the
Earth's ionosphere.
%
% Faraday rotation is the rotation of the polarization plane due
to the
% difference in phase velocities (or refractive indices) of the
right-handed
% and left-handed circularly polarized components in the presence
of a
% magnetic field.
%
% The rotation angle is computed using the analytical expression:
%    ϕ = (e^3 * B * N * L) / (8 * π^2 * f^2 * m^2 * ε0 * c)
% where:
%  e   = electron charge,
%  m   = electron mass,
%  ε0  = vacuum permittivity,
%  c   = speed of light,
%  B   = magnetic field strength,
%  N   = electron density,
```

```
%  L   = path length through the plasma,
%  f   = wave frequency.
%
% The frequency axis is plotted on a logarithmic scale from 1 MHz
to 10 GHz.
% The output plot helps visualize how the Faraday rotation
angle decreases
% with increasing frequency.
%
% This simulation is useful for antenna designers and
communication engineers
% to determine how much the receiving antenna should be rotated
to maintain
% alignment with the incoming wave's electric field, thus
minimizing power
% loss due to polarization mismatch.
%---------------------------------------------------------
clc;
clear;

% Physical Constants
e = 1.602e-19;      % Electron charge (C)
m = 9.109e-31;      % Electron mass (kg)
epsilon_0 = 8.854e-12; % Vacuum permittivity (F/m)
c = 3e8;            % Speed of light (m/s)
pi_val = pi;

% Plasma Parameters
N = 1e12;          % Electron density (electrons/m^3)
L = 1000;          % Path length (m)
B = 1e-5;          % Magnetic field strength (Tesla), e.g., 10 μT

% Frequency Range (1 MHz to 10 GHz)
f = logspace(6, 10, 500);  % 1e6 Hz to 1e10 Hz

% Faraday rotation angle (in radians)
phi = (e^3 * B * N * L) ./ (8 * pi_val^2 * f.^2 * m^2 * epsilon_0
* c);

% Convert to degrees
phi_deg = rad2deg(phi);

% Plot
figure;
semilogx(f/1e6, phi_deg, 'k', 'LineWidth', 1.5);  % Frequency
in MHz
xlabel('Frequency (MHz) [Log Scale]');
ylabel('Faraday Rotation Angle \phi (degrees)');
%title(['Faraday Rotation vs Frequency (B = ', num2str(B*1e6), '
μT, L = ', num2str(L), ' m)']);
grid on;
```

So if the Faraday rotation was 1,100°, for example, it would be required to rotate the receiving feed of the parabolic reflector by exactly the same angle.

So correction angle works out as 1,100° – 6 × 180° = 20°

This is because linear polarization repeats every 180° and therefore rotating a linearly polarized antenna by 180° is indistinguishable in received power.

Problems for Students

13.1 (a) Derive an expression of the skip distance on Earth, upon launching a radio wave at an angle towards the ionosphere, in terms of radio wave and plasma frequencies (w and w_p, respectively), the Earth's radius (a), and the height of the ionosphere (h)? Draw your diagram clearly.

(b) Calculate the skip distance (in kilometers) and the launching angle (in degrees), if a = 6,371 km, h = 300 km, f = 8 MHz, $N = 2.4 \times 10^{11}$ electrons/m^3, $\varepsilon_0 = 8.85 \times 10^{-12}$ F/m, $e = 1.6 \times 10^{-19}$ C, and $m = 9.1 \times 10^{-31}$ kg. What is the launching angle?

(c) If the frequency of operation was kept at 8 MHz, but the launching angle was changed to 0° (i.e., horizontal launching), determine the maximum skip distance.

13.2 (a) Derive an expression for the effective dielectric constant, ε¢, in the ionosphere and show that it is related to the launched and plasma frequencies.

(b) What would be the value of the free electron density in the ionosphere layer, at a point, when a vertically launched 25-MHz radio wave from the Earth's surface, starts reflecting back to Earth?

(c) What would be the maximum skip distance if the radio wave was launched horizontally and reflected from an atmospheric height of 150 km? Note that $e = 1.6 \times 10^{-19}$ C, $m = 9.1 \times 10^{-31}$ kg, $\varepsilon_0 = 8.85 \times 10^{-12}$ F/m, and Earth radius = 6,371 km.

13.3 (a) Derive an expression for the critical frequency f_c in the ionosphere, in terms of electron density N, elementary charge e, and permittivity of free space ε_0.

(b) Calculate the critical frequency for an ionospheric layer where $N = 10^{12}$ electrons/m^3.

(c) If a radio wave with a frequency of 10 MHz is transmitted vertically upwards towards this ionospheric layer, determine whether it will be reflected back to Earth or not.

13.4 (a) Explain the concept of ionospheric absorption of radio waves and its dependence on frequency and solar activity.

(b) Calculate the absorption loss in decibels for a 20-MHz radio wave traveling through the D-layer of the ionosphere during daytime conditions, given specific absorption rate values.

13.5 (a) Define the Faraday rotation in the context of radio wave propagation through the ionosphere.

(b) Derive an expression for the rotation angle ϕ in degrees, considering the magnetic field strength B and the electron density N (electrons/m^3).

(c) Calculate the tilt angle of the receiving antenna for a 30-MHz radio wave passing through the ionosphere with B = 50 mT and $N = 10^{11}$ electrons/m^3, assuming a path length of 1,500 km.

13.6 A radio wave with a frequency f is launched vertically towards the atmosphere. Given the parameters of magnetic field strength, $B = 20\ \mu T$, the electron density $N = 1.2^{13}$ electron/m^3, electron charge, $e = 1.6 \times 10^{-19}$ C, electron mass, $m = 9.11 \times 10^{-31}$ kg, and frequency of the radio wave, $f = 30$ MHz:
   (a) Determine if the radio wave will be reflected or transmitted.
   (b) If $B = 0$, what will be the status of the radio wave?

13.7 (a) Derive an expression for the refractive index gradient required to produce a curved radio wave path that matches the Earth's curvature.
   (b) For a troposphere with a refractive index decreasing linearly with height at a rate of 4×10^{-8} per meter, calculate the corresponding radius of curvature of the radio wave path and compare it to the Earth's effective radius (4/3 Earth's radius).

13.8 (a) Extend the derivation of the skip distance formula to include the effects of an irregular ionospheric layer, where the electron density varies quadratically with height.
   (b) Calculate the new skip distance and launching angle when the electron density profile follows

$$N(h) = N_0(1 - h/h_m)^2$$

   where $N_0 = 2.4 \times 10^{11}$ electrons/m^3, $h_m = 300$ km, and $f = 10$ MHz. Assume Earth's radius $a = 6{,}371$ km.

13.9 (a) Derive an expression for the total path length of a multihop radio wave propagation, considering a single-hop skip distance d_s and the reflection height h in the ionosphere.
   (b) For a radio wave operating at 12 MHz, calculate the total path length and number of hops required to communicate over a distance of 3,000 km, assuming an ionospheric height of 250 km and a skip distance of 800 km per hop.
   (c) Discuss the factors affecting signal strength and delay in multihop communication.

Problems for Instructors

13.1 Frequency Band Selection for Atmospheric Propagation: Discuss the characteristics of different frequency bands used for radio wave propagation in the Earth's atmosphere. Evaluate their suitability based on parameters such as atmospheric absorption, ionospheric interaction, and refractive effects in the troposphere. Which frequency bands are most appropriate for satellite communication, and why?

13.2 Tropospheric Refraction Effects on Radio Signals: Calculate the refracted path of a radio wave as it travels through a tropospheric layer with a refractive index that varies linearly with altitude. Given an initial refractive index of 1.0003 at sea level and a reduction rate of 0.00002 per kilometer, determine how the wave's path curvature changes with increasing height and how this affects its range on Earth.

13.3 Curvature of Radio Wave Paths: Using the concept of an effective Earth radius, derive the modified radio wave path curvature for a 4/3 Earth radius model. Explain the conditions under which this approximation is valid and calculate the effect of this model on a signal transmitted horizontally at sea level.

13.4 Propagation in the Ionosphere and Dielectric Constant of Ionized Gas: Assume a simplified model of the ionosphere where electron density varies exponentially with height. Derive an expression for the effective dielectric constant of this ionized layer, considering parameters such as electron density and collision frequency. How does this dielectric constant impact the behavior of radio waves in high-frequency and VHF bands?

13.5 Normal and Oblique Incidence in the Ionosphere: For a radio wave incident at normal and oblique angles on an ionospheric layer, derive the conditions for reflection and transmission based on electron density and frequency. How does oblique incidence affect the skip distance compared to normal incidence? Calculate the skip distance for a 10-MHz wave launched at a 30° angle into the ionosphere with a critical frequency of 7 MHz.

13.6 Earth's Magnetic Field Influence and Faraday Rotation: Explain the phenomenon of Faraday rotation and derive the rotation angle for a linearly polarized wave traveling through the ionosphere in the presence of the Earth's magnetic field. Assume a 14-MHz signal propagating through a layer with an electron density of $10^{12}\,m^{-3}$ and a magnetic field strength of 50 μT. How does this rotation affect communication systems?

13.7 Multihop Communication Using Skip Distance Calculation: Analyze the relationship between launching angle and single-hop skip distance for high-frequency communication in the ionosphere. Using the derived equations, calculate the launching angles required for achieving skip distances of 1,000 km, 1,500 km, and 2,000 km for a frequency of 12 MHz. Discuss how multihop propagation can extend the communication range and the challenges that it presents.

Selected Bibliography

Barclay, L. W., "Propagation of Radiowaves," *IET*, Vol. 502, 2003.

Barr, R., "The Effect of the Earth's Magnetic Field on the Propagation of ELF and VLF Radio Waves," *Journal of Atmospheric and Terrestrial Physics*, Vol. 33, No. 10, 1971, pp. 1577–1583.

Budden, K. G., *Radio Waves in the Ionosphere: The Mathematical Theory of the Reflection of Radio Waves from Stratified Ionised Layers*, Cambridge, UK: Cambridge University Press, 1961.

David, P., and J. Voge, *Propagation of Waves*, Oxford, UK: Pergamon Press, 1969.

Hitney, H. V., et al., "Tropospheric Radio Propagation Assessment," *Proceedings of the IEEE*, Vol. 73, No. 2, 1985, pp. 265–283.

Rawer, K., *Wave Propagation in the Ionosphere*, Vol. 5, New York: Springer Science & Business Media, 2013.

Reed, H. R., and C. M. Russell, *Ultra High Frequency Propagation*, London: Chapman and Hall, 1966.

Saakian, A., *Radio Wave Propagation Fundamentals*, Norwood, MA: Artech House, 2020.

CHAPTER 14

Advances in Antenna Technology for 5G and Satellite Communication

The rollout of 5G wireless networks marks a major leap in communication technology, offering ultrafast speeds, low latency, and support for a massive number of connected devices. These capabilities are driving significant changes in antenna design, especially as 5G moves into millimeter-wave frequencies and dense urban deployments.

This chapter begins by exploring the key features of 5G and the new challenges that they pose for antenna engineers, such as the need for massive MIMO, precise beamforming, and compact, multiband operation. It introduces a mathematical framework for 5G antenna design and outlines emerging solutions including smart, adaptive, and reconfigurable antennas.

The discussion then transitions to the exciting area of deployable satellite antennas, which are essential for global broadband access, Earth observation, and military missions. It covers their classifications, deployment mechanisms, performance metrics, and future trends, supported by examples of practical implementations across low Earth orbit (LEO) and geostationary Earth orbit (GEO) systems.

Consequently, this chapter now aims for greater narrative cohesion by highlighting the synergy between high-frequency terrestrial networks and evolving space-based antenna architectures. Cross-references are incorporated throughout to improve continuity and enhance comprehension.

A more rigorous examination of antenna types, performance parameters, and terminology is also included, drawing on specialized literature from IEEE Xplore, ScienceDirect, and key industry white papers. This ensures a comprehensive and technically accurate treatment of both terrestrial and satellite-based systems.

The chapter concludes with a forward-looking view on trends in antenna innovation and includes a practical supplement with MATLAB codes for satellite link budget estimation and beam coverage visualization, serving as a hands-on toolkit for students, researchers, and practitioners.

14.1 Introduction to 5G Technology and Its Impact on Antenna Design

5G technology represents a significant advancement in wireless communication, fundamentally changing how data is transmitted, devices are connected, and services are delivered. It addresses the limitations of previous generations (2G, 3G, and

4G) by introducing revolutionary performance benchmarks, enabling a new era of connectivity and communication. The shift to 5G networks introduces key features such as ultralow latency, enhanced mobile broadband, and massive machine-type communications. To meet the demands of this next-generation network, antenna design must also evolve, adapting to the unique challenges presented by 5G's use of high-frequency spectrum, particularly in the millimeter-wave bands.

14.1.1 Key Drivers and Features of 5G

The key drivers behind 5G technology can be categorized into three main areas:

1. Increased capacity and speed:
   Enhanced mobile broadband (eMBB): 5G offers significantly faster download and upload speeds compared to 4G, supporting data-intensive applications such as 4K/8K video streaming, virtual reality (VR), and augmented reality (AR). It ensures seamless connectivity in high-density urban areas and supports the growing number of devices and data demand.
   Network capacity expansion: 5G is designed to handle a much larger number of connected devices per square kilometer, enabling smoother network experiences in crowded environments, such as stadiums or city centers.
2. Ultrareliable low-latency communication (URLLC):
   Real-time applications: 5G technology reduces latency to as low as 1 millisecond, making it suitable for real-time applications such as autonomous driving, remote surgery, industrial automation, and smart grids. These applications demand near-instantaneous data transmission and high reliability.
   Mission-critical communications: For applications where failure is not an option (e.g., safety-critical systems), 5G ensures ultrareliable connectivity that can support continuous and uninterrupted operations.
3. Massive Machine-Type Communications (mMTC):
   IoT: 5G is designed to support billions of IoT devices, enabling communication between a vast array of sensors, devices, and machines. This expands the potential for smart cities, agriculture, healthcare, and industrial applications, where large-scale device connectivity is required with minimal human intervention.
   Energy efficiency: Many IoT devices will be low-power and designed to operate for extended periods without human intervention. 5G supports the power-efficiency demands of these devices.

These areas reflect the broad capabilities of 5G, which aims to meet the evolving needs of consumers, businesses, and industries through faster, more reliable, and highly connected networks.

The key features of 5G technology include the following:

1. *Faster speeds:* 5G offers data transfer rates that can reach up to 10 Gbps, which is significantly faster than 4G. This speed supports applications such as high-definition video streaming, VR, AR, and other data-heavy use cases.

2. *Ultralow latency:* Latency in 5G networks can be as low as 1 millisecond, making it ideal for real-time applications such as autonomous driving, remote surgeries, and industrial automation where instant response times are critical.
3. *Increased capacity:* 5G can support a much larger number of devices connected to the network simultaneously, thanks to its ability to operate in high-frequency millimeter waves. This is essential for the growth of IoT, where billions of devices will require connectivity.
4. *eMBB:* 5G provides enhanced mobile broadband services, enabling faster internet access for users on the go. This supports high-quality streaming, quicker downloads, and better performance in high-traffic areas such as stadiums and airports.
5. *mMTC:* Designed to handle massive numbers of IoT devices, 5G enables devices to communicate efficiently with minimal human intervention. This is key for smart cities, industrial IoT, and agriculture, where many low-power devices need to communicate over long periods.
6. *Network slicing:* 5G networks can be partitioned into multiple virtual networks, each optimized for different use cases. For example, a network slice can be dedicated to low-latency services such as autonomous vehicles, whereas another slice can be optimized for IoT devices with low-power needs.
7. *Improved energy efficiency:* 5G is designed to be more energy-efficient than previous generations, especially for IoT devices. The network can adjust to different types of traffic, conserving energy when data usage is low.
8. *Better coverage with beamforming and MIMO:* Technologies such as massive MIMO and beamforming enhance signal strength and coverage. Beamforming allows for targeted transmission of signals to specific users, reducing interference and improving connection quality.
9. *Reliability and availability:* 5G provides ultrareliable communication, ensuring a stable and consistent network experience. This is essential for mission-critical applications such as healthcare and public safety.
10. *Support for advanced technologies:* 5G integrates and supports emerging technologies such as AR, VR, artificial intelligence (AI), and edge computing, enabling new business models and services across various industries.

These features collectively enable 5G to transform multiple sectors, including healthcare, automotive, entertainment, manufacturing, and public safety, by offering faster, more reliable, and highly scalable connectivity.

14.1.2 Challenges for Antenna Designers

With the introduction of 5G, antennas play a more critical role than ever, as designers face several challenges to meet the demands of the technology. Unlike 4G, which primarily operates in sub-6-GHz frequencies, 5G expands into the millimeter-wave spectrum (24 GHz to 100 GHz), offering larger bandwidths and faster data rates but also presenting challenges such as increased path loss, reduced penetration through obstacles, and shorter propagation distances. To accommodate the massive number of connected devices, antennas must improve spectrum efficiency by supporting multiple users simultaneously without interference, relying on innovations such as massive MIMO and beamforming. Massive MIMO uses a large array of antennas to serve multiple users simultaneously, while beamforming directs energy toward specific users instead of broadcasting signals uniformly. Additionally, multibeam capabilities are essential, especially in millimeter-wave communications, to form and steer multiple beams for several users or devices. Finally, antenna designers must develop compact, high-performance antennas that can be integrated into limited spaces, particularly for MIMO and beamforming applications in increasingly smaller mobile devices.

Figure 14.1 illustrates the architectural differences between 2G, 3G, 4G, and 5G networks. In 4G, data is transmitted from base stations to user devices in a relatively uniform manner, with limited support for advanced techniques such as MIMO. In contrast, 5G networks use more complex architectures with numerous small cells, MIMO arrays, and beamforming technology to improve data rates, reduce latency, and handle more connections.

Here is an illustration of the main acronyms used in the networks:

2G (GSM Network Core):

- BSC: Base Station Controller;
- MSC: Mobile Switching Center;
- VLR: Visitor Location Register;

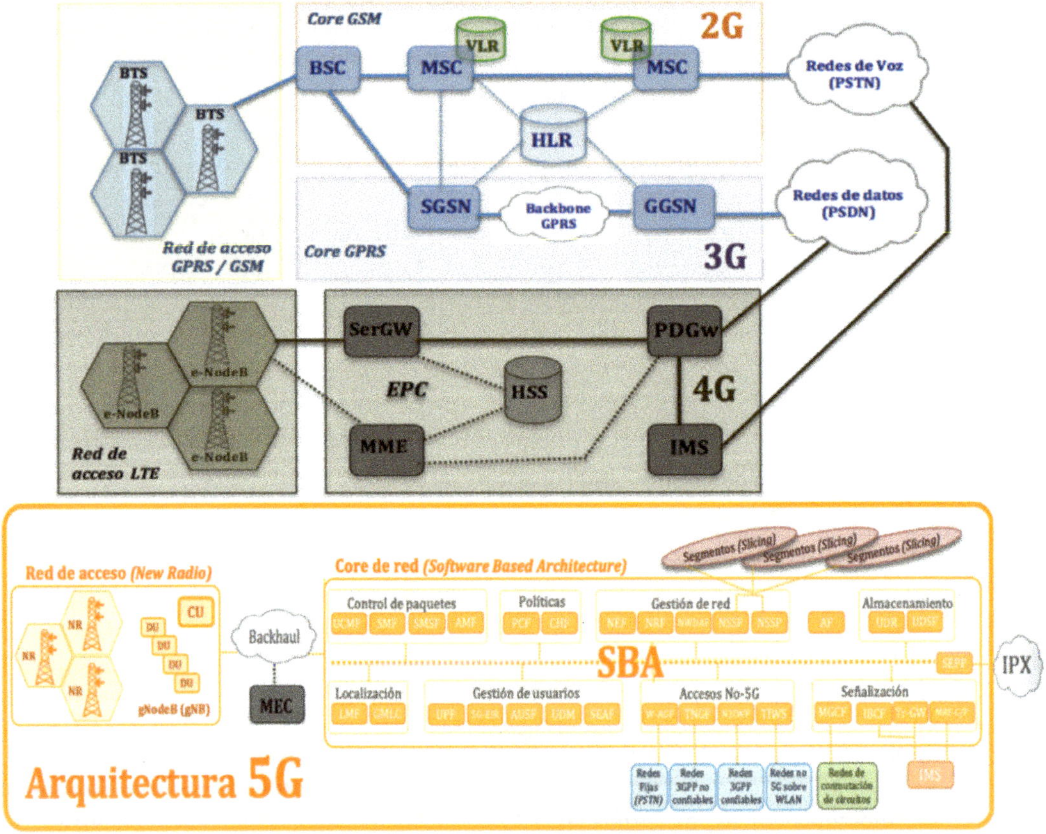

Figure 14.1 Evolution of 2G-3G-4G-5G networks. (Source: Acorletti.)

- HLR: Home Location Register;
- SGSN: Serving GPRS Support Node;
- GGSN: Gateway GPRS Support Node;
- PSTN: Public Switched Telephone Network (Voice Networks);
- PSDN: Packet Switched Data Network (Data Networks).

3G (GPRS Core Network):

- Core GPRS: Backbone for General Packet Radio Services.

4G (LTE Access Network):

- eNodeB: Evolved Node B;
- SerGW: Serving Gateway;
- PDGW: Packet Data Gateway;
- EPC: Evolved Packet Core;
- MME: Mobility Management Entity;
- HSS: Home Subscriber Server;
- IMS: IP Multimedia Subsystem.

5G Architecture:

- Access Network (New Radio);
- NR: New Radio;
- CU: Central Unit;
- DU: Distributed Unit;
- gNodeB (gNB): Next-generation Node B;
- Backhaul: Backhaul Network;
- MEC: Multi-access Edge Computing.

Core Network (Software-Based Architecture (SBA)):

- Package Control: UCMF, SMSF, AMF;
- Policies: PCF, CHF;
- Network Management: NEF, NRF, NWDAF, NSSP;
- User Management: UDM, AUSF, UPF, SMF;
- Storage: UDR, UDSF;
- Access for non-5G: W-AGF, TNGF, N3IWF, TIWS;
- Serialization: MCGF, IBFCF, T-GW, MRF-C/P.

14.1.3 New Antenna Requirements for 5G Networks

As the demands of 5G networks increase, antenna design must meet several new requirements to ensure optimal performance:

1. *MIMO technology:* Massive MIMO is an essential component of 5G technology. By employing dozens or even hundreds of antennas at both the

transmitter and receiver, MIMO systems can significantly increase data throughput and spectral efficiency. This is achieved by transmitting multiple data streams simultaneously across different antennas, taking advantage of the spatial dimension. The use of massive MIMO is critical for overcoming the challenges posed by higher frequencies, where signals attenuate more rapidly.

2. *Beamforming:* Beamforming is another key technique employed in 5G antennas to direct signals towards specific users, rather than broadcasting signals in all directions. This not only improves signal strength but also enhances the signal-to-noise ratio (SNR), making the communication link more robust. In millimeter-wave systems, where signals can be easily blocked by physical obstacles, beamforming becomes even more critical.

Beamforming is accomplished using phased array antennas, which consist of multiple antenna elements that can control the phase of the signal transmitted from each element. By adjusting the phase, the antenna can steer the direction of the beam without physically moving the antenna. Phased arrays are particularly well suited for 5G applications because they offer the flexibility needed to implement beamforming techniques. These antennas can dynamically adjust the direction of their beams, enabling them to track mobile users and compensate for changes in the environment, such as obstacles or interference.

In addition, phased array antennas are capable of supporting multibeam technology, where the antenna can form multiple beams simultaneously, each directed at different users or devices. This is particularly useful in high-density urban environments where multiple users are located within the same area.

In this context, millimeter-wave antennas play a crucial role, as their small size at high frequencies enables the integration of multiple antenna elements necessary for advanced techniques such as beamforming and multibeam operation. Millimeter-wave antennas are designed to operate at the higher frequencies (24–100 GHz) used in 5G networks. At these frequencies, the wavelength of the signal becomes very short, allowing for the design of smaller, more compact antennas. However, millimeter-wave signals are highly susceptible to path loss, meaning that millimeter-wave antennas must be designed with high gain to compensate for signal attenuation. Antenna size is inversely proportional to frequency, meaning that millimeter-wave antennas are much smaller than those used in lower-frequency bands. This presents both challenges and opportunities for integrating multiple antennas into mobile devices.

Table 14.1 shows the frequency bands allocated for 5G, including the sub-6-GHz bands and the millimeter-wave bands (24–100 GHz). It highlights the different regions of the spectrum, showing the trade-offs between lower-frequency bands (greater coverage, lower data rates) and millimeter-wave bands (higher data rates, shorter range).

14.1.4 Mathematical Framework for 5G Antenna Design

The design of 5G antennas relies on several key mathematical equations that form the foundation for understanding performance metrics, system requirements, and

Table 14.1 Global Snapshot of Allocated/Targeted 5G Spectrum*

Country	*<1 GHz*	*2–3 GHz*	*3–4 GHz*	*4–5 GHz*	*5–8 GHz*	*24–30 GHz*	*37–50 GHz*	*60 GHz*	*>95 GHz*
United States	600 MHz (2 × 35 MHz) 900 MHz (2 × 3 MHz)	2.5/2.6 GHz (B41/n41)	3.1–3.45 GHz 3.45–3.55 GHz 3.55–3.7 GHz 3.7–3.98 GHz	4.94–4.99 GHz	5.9–7.1 GHz	24.25–24.45 GHz 24.75–25.25 GHz 27.5–28.35 GHz	37–37.6 GHz 37.6–40 GHz 42–42.5 GHz 47.2–48.2 GHz	57–71 GHz	>95 GHz
Canada	600 MHz (2 × 35 MHz)		3.475–3.65 GHz 3.65–4.0 GHz		5.9–7.1 GHz	26.5–27.5 GHz 27.5–28.35 GHz	37–37.6 GHz 37.6–40 GHz	57–71 GHz	>95 GHz
European Union	700 MHz (2 × 30 MHz)		3.4–3.8 GHz		5.9–6.4 GHz	24.5–27.5 GHz		57–66 GHz	
United Kingdom	700 MHz (2 × 30 MHz)		3.4–3.8 GHz		5.9–6.4 GHz	24.5–27.5 GHz	40.5–43.5 GHz	57–66 GHz	
Germany	700 MHz (2 × 30 MHz)		3.4–3.8 GHz		5.9–6.4 GHz	26 GHz		57–66 GHz	
France	700 MHz (2 × 30 MHz)		3.46–3.8 GHz		5.9–6.4 GHz	26 GHz		57–66 GHz	
Italy	700 MHz (2 × 30 MHz)		3.6–3.8 GHz		5.9–6.4 GHz	26.5–27.5 GHz		57–66 GHz	
China	700 MHz 900 MHz	2 GHz (n1) 2.5/2.6 GHz (B41/n41)	3.3–3.6 GHz	4.8–5.0 GHz		24.75–27.5 GHz	37–43.5 GHz		
South Korea	700/800 MHz	2.3–2.39 GHz	3.4–3.7 GHz 3.7–4.0 GHz	4.72–4.82 GHz	5.9–7.1 GHz	25.7–26.5 GHz 26.5–28.9 GHz 28.9–29.5 GHz	37 GHz	57–64 GHz	
Japan	700/800 MHz	2.3 GHz		4.5–4.9 GHz	5.9–6.4 GHz	27.0–29.5 GHz		57–66 GHz	
India	600 MHz (2 × 40 MHz) 700 MHz (2 × 30 MHz)					24.25–27.5 GHz			
Australia		2.3 GHz	3.4–3.7 GHz		5.9–6.4 GHz	24.25–29.5 GHz	39 GHz	57–66 GHz	

*5G is being designed for diverse spectrum types/bands; Qualcomm.

overall efficiency. Below are the crucial formulas that guide 5G antenna design, especially at higher frequencies such as millimeter waves.

14.1.4.1 Link Budget Equation

The link budget equation is central to calculating the total gain or loss in a communication system. It considers various factors such as transmitted power, antenna gains, and losses due to propagation and environmental conditions. For 5G systems, especially those operating at millimeter-wave frequencies, accurately determining the link budget is crucial, as higher frequencies introduce losses that are more significant. The link budget equation is given by:

$$P_{\text{received}} = P_{\text{transmitted}} + G_{\text{transmitter}} + G_{\text{receiver}} - L_{\text{path}} - L_{\text{other losses}} \quad (14.1)$$

where P_{received} is the power received at the receiver, $P_{\text{transmitted}}$ is the power transmitted by the antenna, $G_{\text{transmitter}}$ and G_{receiver} represent the gains of the transmitting and receiving antennas, respectively, L_{path} is the free-space path loss, accounting for the reduction in signal strength over a distance, and $L_{\text{other losses}}$ include additional losses from factors such as fading, shadowing, interference, or attenuation due to buildings and environmental obstructions.

For 5G at higher frequencies, factors such as path loss, fading, and reflection significantly affect the link budget. Antenna designers must carefully balance transmission power and antenna gain to ensure reliable connectivity while compensating for these additional losses, especially in urban environments or areas with dense obstacles.

14.1.4.2 Free-Space Path Loss Equation

One of the primary concerns in millimeter-wave systems is the significant path loss over distances. Higher frequencies in the millimeter-wave range (24 to 100 GHz) experience greater signal attenuation, requiring more sophisticated antenna technologies such as beamforming and massive MIMO to maintain strong communication links. The free-space path loss equation quantifies this signal attenuation and is expressed as:

$$L_{\text{path}} = 20\log_{10}\left(\frac{4\pi d f}{c}\right) \quad (14.2)$$

where d is the distance between the transmitter and the receiver, f is the frequency of the transmitted signal, and c is the speed of light.

As distance (d) increases or frequency (f) rises, the path loss increases logarithmically. This poses significant challenges for 5G systems operating at millimeter-wave frequencies, where path loss becomes a major limiting factor in communication range. To mitigate this, 5G networks rely on techniques such as:

- Beamforming, which focuses antenna energy in specific directions to enhance signal strength over longer distances.

- High-gain antennas, which compensate for increased path loss by providing directional communication with minimal dispersion.

14.1.4.3 Beamforming and Massive MIMO

Advanced 5G systems employ technologies such as beamforming and massive MIMO to overcome challenges introduced by high path loss and interference. Beamforming allows antennas to concentrate signals toward specific users, rather than broadcasting in all directions, thus improving signal quality and reducing interference. Massive MIMO uses a large array of antennas to serve multiple users simultaneously, maximizing spectrum efficiency and increasing network capacity.

Beamforming can be mathematically described by focusing the antenna array's radiation pattern toward specific angles (θ, ϕ) to enhance gain in those directions. For an array of NNN elements, the total gain is a function of the phase shift between elements and their spatial arrangement, which can be tuned dynamically to steer beams in real time.

14.1.4.4 Friis Transmission Equation

The Friis transmission equation is also fundamental for determining the power received over a line-of-sight communication link, especially in free-space conditions. It is expressed as:

$$P_{\text{received}} = P_{\text{transmitted}} \cdot \frac{G_{\text{transmitter}} G_{\text{receiver}} \lambda^2}{(4\pi d)^2} \tag{14.3}$$

where λ is the wavelength of the signal and d is the distance between the transmitter and the receiver.

This equation highlights how the received power decreases as distance increases and how higher antenna gains or shorter distances can mitigate losses.

14.1.4.5 Antenna Array Factor

The array factor is another critical equation in 5G antenna design, especially for massive MIMO and phased-array systems. The array factor describes how the individual elements of an antenna array combine to produce the total radiation pattern. For an array of N elements, the array factor can be written as:

$$AF(\theta) = \sum_{n=0}^{N} e^{j(knd\cos\theta + \varphi_n)} \tag{14.4}$$

where θ is the angle of observation, $k = 2\pi/\lambda$ is the wave number, d is the distance between adjacent antenna elements, and ϕ_n is the phase shift applied to the nth antenna element.

This equation is critical for optimizing the beamforming capabilities of 5G antenna arrays, enabling precise control over beam direction and enhancing signal strength in desired directions.

In summary, the mathematical framework for 5G antenna design involves calculating link budgets, accounting for path loss, and leveraging advanced technologies such as massive MIMO and beamforming. These calculations ensure efficient, high-performance 5G communication, even in challenging environments where high-frequency signals are subject to greater attenuation and interference.

The introduction of 5G technology marks a significant paradigm shift in wireless communication, demanding a rethinking of traditional antenna design and deployment strategies. Unlike previous generations, 5G operates at much higher frequencies, including the millimeter-wave spectrum, which introduces both opportunities and challenges for antenna engineers. These higher frequencies enable faster data rates and greater network capacity but also come with limitations such as increased path loss, reduced signal penetration through obstacles, and shorter propagation distances. To meet these demands, antenna designers must adopt innovative solutions such as massive MIMO, beamforming, and multibeam technologies, which enhance spectrum efficiency and enable reliable communication despite these challenges. Moreover, 5G antennas must be compact, highly integrated, and capable of supporting a massive number of connected devices in various environments, from dense urban areas to rural regions. The mathematical frameworks, including link budget calculations and path loss equations, are critical tools that guide the design of these advanced antennas, ensuring that they meet the performance requirements of 5G networks. As a result, the evolution of antenna design plays a pivotal role in the success of 5G technology, paving the way for faster, more reliable, and more connected communication systems in the future.

MATLAB Program 28 simulates and plots the radiation pattern of an array factor of a 16-element ULA, steered to 30 degrees, as shown in Figure 14.2.

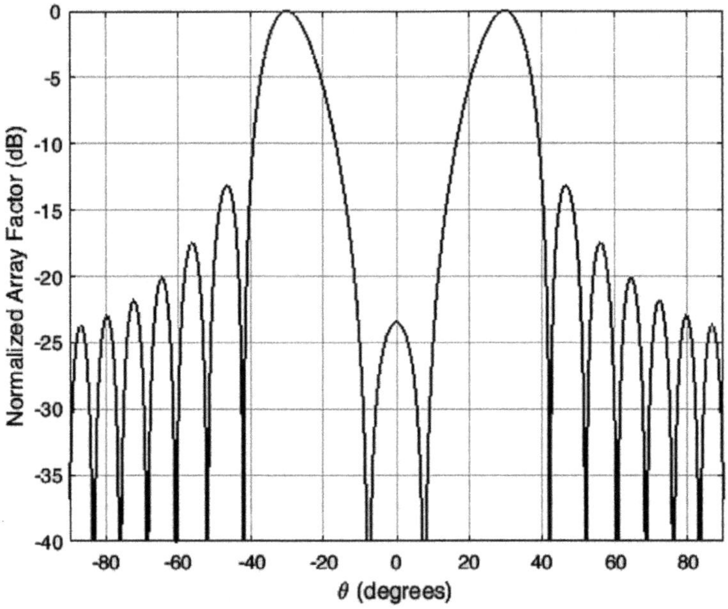

Figure 14.2 Array factor of 16-element ULA steered to 30°.

MATLAB Program 28

```
% MATLAB Simulation: Array Factor of an 16-element Uniform Linear
Array (ULA), steered to 30°
% Parameters
theta = -90:0.1:90;          % Observation angle in degrees
theta_rad = deg2rad(theta);   % Convert to radians
lambda = 1;                  % Normalized wavelength
k = 2 * pi / lambda;          % Wave number
d = lambda / 2;              % Element spacing
N = 15;                   % Number of elements (N+1 total)
n = 0:N;                  % Element indices

% Beam steering direction (in degrees)
theta_0 = 30;
phi_n = -k * d * cosd(theta_0) * n;  % Progressive phase shift
for steering

% Compute array factor
AF = zeros(size(theta_rad));      % Initialize array factor
for i = 1:length(theta_rad)
  AF(i) = sum(exp(1j * (k * d * n * cos(theta_rad(i)) + phi_n)));
end

% Normalize and plot
AF_normalized = abs(AF) / max(abs(AF));  % Normalize to max gain
plot(theta, 20*log10(AF_normalized), 'k', 'LineWidth', 1.5);
xlabel('\theta (degrees)');
ylabel('Normalized Array Factor (dB)');
title('Array Factor of 16-element ULA Steered to 30°');
grid on;
ylim([-40 0]);  % Set dB range
xlim([-90 90]); % Observation range
```

14.2 Antenna Design for 5G Networks

While 5G terrestrial networks introduce high-capacity, low-latency services across dense urban and industrial landscapes, the future of global connectivity increasingly depends on hybrid systems that combine terrestrial and satellite platforms. Deployable satellite antennas, particularly in LEO constellations, are emerging as critical infrastructure for extending 5G services to remote, rural, or mobile environments (e.g., ships, aircraft). These satellite systems utilize phased array antennas, beam-hopping technologies, and electronically steerable arrays, mirroring many of the same beamforming and massive MIMO concepts used in terrestrial 5G. This convergence not only aligns system architectures but also prompts unified design methodologies that span the atmosphere, from ground base stations to spaceborne antenna platforms.

14.2.1 Introduction to 5G Antenna Design Challenges

5G networks represent a significant leap from previous generations, bringing about a new era in wireless communication that promises to revolutionize industries such as healthcare, transportation, and smart cities. However, to achieve these capabilities, the design of antennas plays a pivotal role. The frequency spectrum for 5G is divided into two main segments: sub-6 GHz and millimeter-wave (above 24 GHz). While the sub-6-GHz range is valuable for broad coverage and reliability, the millimeter-wave spectrum holds the key to unleashing the full potential of 5G with ultrahigh data rates and low latency.

However, operating in the millimeter-wave spectrum introduces several design challenges:

1. *Increased path loss:* At higher frequencies, signal attenuation becomes more significant due to free-space path loss and material absorption. The free-space path loss, L_{path}, increases with the square of frequency f and can be described by (14.2).
2. *Reduced penetration through obstacles:* Millimeter-wave signals have poor penetration through walls, buildings, and even trees, requiring a highly directional antenna design to avoid signal blockages and dead zones. The use of beamforming becomes crucial here.
3. *Smaller wavelengths and antenna design:* The wavelength of millimeter-wave signals is much shorter (in the order of millimeters), meaning that antennas can be smaller. This allows for the integration of larger antenna arrays in smaller form factors, but also requires precision in design to ensure efficiency and performance.

In addition to these challenges, 5G antennas must cater to increased traffic loads and diverse applications, such as URLLC for autonomous vehicles and mMTC for the IoT. These applications demand antennas with high reliability, flexibility, and efficiency, which necessitates the exploration of new materials and design methodologies.

For example, the deployment of 5G in urban environments, where the millimeter-wave spectrum will be primarily utilized, has led to the development of dense networks with small cells and high-gain antennas to counteract the propagation challenges. Designing antennas for these scenarios requires careful consideration of both the physical environment and the communication protocol used in 5G.

14.2.1.1 Bridging Terrestrial and Orbital Antenna Design

While the challenges discussed above pertain mainly to terrestrial 5G systems, many of these principles, particularly beamforming, spatial multiplexing, and phased array miniaturization are also central to the design of LEO satellite antennas. The growing convergence between terrestrial 5G and spaceborne communication platforms highlights the need for unified antenna solutions capable of adaptive beam control, interference mitigation, and real-time reconfiguration across highly dynamic environments.

14.2.2 Massive MIMO and Beamforming: Enabling 5G Performance

Massive multiple-input multiple-output (MIMO) is one of the core technologies in 5G that has dramatically changed antenna system architectures. The leap from conventional MIMO systems, with a few antennas, to massive MIMO systems with hundreds or even thousands of antennas, presents both opportunities and challenges. The massive number of antenna elements allows for multiple data streams to be transmitted and received simultaneously, improving spectral efficiency and capacity.

14.2.2.1 Theoretical Foundations of Massive MIMO

The key to massive MIMO's success lies in its ability to spatially multiplex multiple user signals. The channel matrix H in a massive MIMO system is larger and offers a high degree of freedom, allowing the system to serve many users simultaneously. The channel capacity C for a MIMO system is given by the following equation:

$$C = B\log_2\left(\det\left(I + \frac{P}{N_0}HH^H\right)\right) \tag{14.5}$$

where B is the bandwidth, P is the transmit power, N_0 is the noise power, H is the channel matrix, I is the identity matrix, and H^H is the Hermitian transpose of the channel matrix.

In massive MIMO, the dimensionality of the matrix H increases, improving the channel capacity exponentially. Additionally, multi-user MIMO allows multiple users to be served in the same frequency time slot by using different spatial streams, which further enhances the overall capacity of the system.

14.2.2.2 Beamforming in Massive MIMO Systems

Beamforming is integral to massive MIMO, and its implementation can significantly improve 5G network performance. The ability to steer the beam dynamically toward a specific user reduces interference, increases signal strength, and enhances the signal-to-interference-plus-noise ratio (SINR).

There are two main types of beamforming employed in 5G networks:

1. *Analog beamforming:* In analog beamforming, phase shifters are used to adjust the phase of the signal at each antenna element, creating a directed beam. This method is simpler and less costly in terms of hardware, as it requires fewer RF chains. However, analog beamforming is limited in its ability to generate multiple beams.
2. *Digital beamforming:* Digital beamforming provides much greater flexibility by controlling each antenna element individually through digital signal processing. This method enables multiple beams to be formed simultaneously and dynamically steered, but it requires a higher number of RF chains, increasing power consumption and cost.
3. *Hybrid beamforming:* Hybrid beamforming combines the advantages of both analog and digital techniques, reducing the number of RF chains while maintaining the ability to steer multiple beams.

Figure 14.3 illustrates a digital antenna array as a digital beamforming architecture with a synthetic aperture radar (SAR) processor.

14.2.2.3 Challenges in Beamforming for Millimeter-Wave Communications

Beamforming is particularly important in the millimeter-wave bands, where high path loss necessitates the use of directional antennas to extend communication range. The trade-offs in beamforming design are complex. For instance, narrow beams can provide high gain and improve range but at the cost of increased sensitivity to user mobility. Wider beams offer better coverage but suffer from lower gains.

The beamforming gain G_b can be calculated as follows:

$$G_b = N \times G_e \tag{14.6}$$

where N is the number of antenna elements and G_e is the gain of a single element. The larger the array, the greater the beamforming gain, which helps to overcome the high path loss experienced at millimeter-wave frequencies. G_b grows linearly with the number of antenna elements, indicating the substantial enhancement in signal directivity. In decibel scale, this is expressed as: $G_b(\text{dB}) = 10\log10(N) + G_e(\text{dB})$. For example, a 64-element array can offer approximately 18-dB gain over a single element, assuming identical element characteristics. This gain compensates for the high free-space path loss observed in millimeter-wave systems.

For example, a real-world implementation of beamforming can be seen in urban environments with high-rise buildings, where reflections and blockages create challenging propagation scenarios. Beamforming can direct the signal along clear paths, reducing interference and improving signal quality.

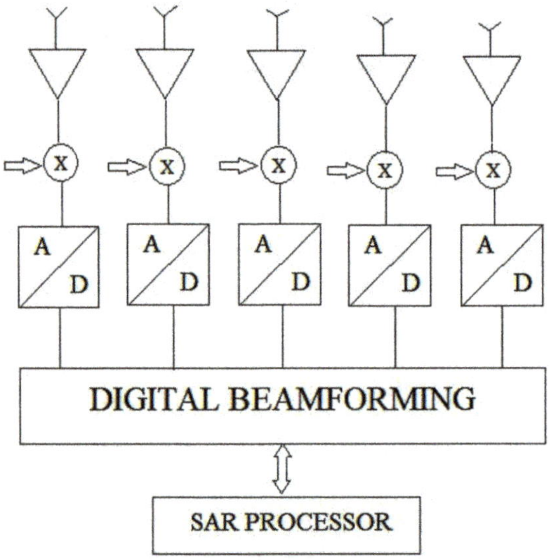

Figure 14.3 Digital antenna array. (Source: Irighter.)

14.2.3 Compact and Multiband Antenna Design for 5G and IoT

As 5G devices become more compact and portable, the demand for compact antenna designs that can operate efficiently in confined spaces has increased. In addition to compactness, 5G antennas must support multiband operation, covering both sub-6-GHz and millimeter-wave frequencies. This requires innovative designs that optimize space while maintaining high performance across multiple frequency bands.

14.2.3.1 Impedance Matching for Multiband Antennas

Designing multiband antennas requires careful consideration of impedance matching across different frequency bands. A good impedance match ensures that maximum power is transferred from the transmission line to the antenna, minimizing reflection and improving efficiency.

The reflection coefficient Γ, which indicates how much of the signal is reflected back due to impedance mismatch, is given by:

$$\Gamma = \frac{Z_L - Z_0}{Z_L + Z_0} \tag{14.7}$$

where Z_L is the load impedance of the antenna and Z_0 is the characteristic impedance of the transmission line (typically 50 ohms). Figure 14.4 demonstrates the reflection coefficient that varies with antenna load impedance, as run by MATLAB Program 29.

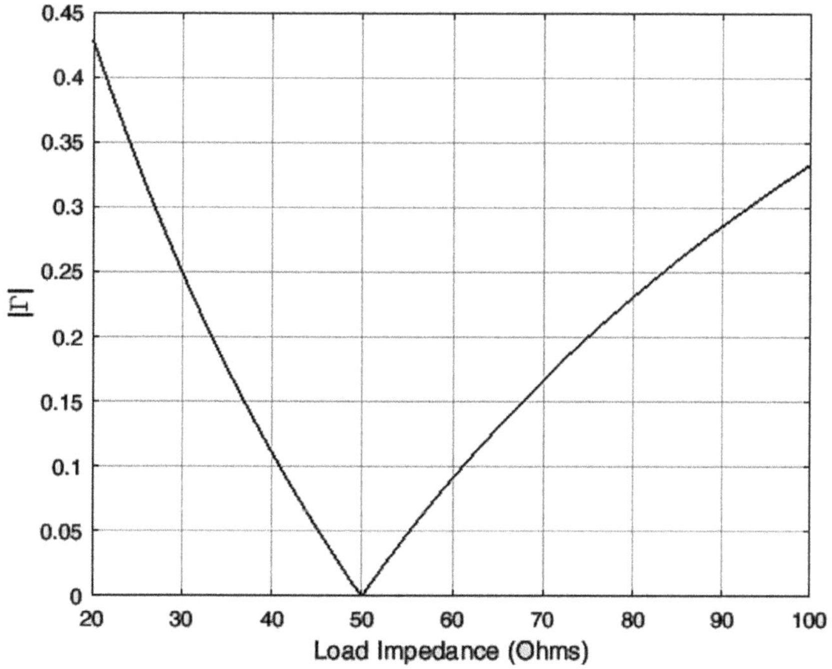

Figure 14.4 Reflection coefficient versus load impedance.

MATLAB Program 29

```
% The following MATLAB code plots how the reflection coefficient
varies with antenna load impedance:

% Plot reflection coefficient vs. load impedance
Z0 = 50;                     % Transmission line impedance (Ohms)
ZL = linspace(20, 100, 500);   % Range of load impedances (Ohms)
Gamma = (ZL - Z0) ./ (ZL + Z0);  % Reflection
coefficient calculation
plot(ZL, abs(Gamma), 'k', 'LineWidth', 1.5);
xlabel('Load Impedance (Ohms)');
ylabel('|\Gamma|');
title('Reflection Coefficient vs. Load Impedance');
grid on;
```

For multiband antennas, the challenge is to maintain a good match across a wide range of frequencies. Techniques such as matching networks and reactive components are often used to achieve this.

14.2.3.2 Miniaturization Techniques in Antenna Design

To address the space constraints in 5G devices, several miniaturization techniques have been developed:

1. *Microstrip patch antennas:* These antennas are characterized as lightweight, low-profile, and easy to fabricate. They are well suited for integration into mobile devices but have limitations in terms of bandwidth and efficiency at higher frequencies.
2. *Fractal antennas:* Fractal geometries, characterized by self-similar patterns, allow for efficient multiband performance while occupying minimal space. These antennas use repeating geometric shapes to create multiple resonant frequencies, making them ideal for 5G devices that need to operate across a wide frequency range.
3. *Metamaterials:* Metamaterials with engineered electromagnetic properties can be used to miniaturize antennas while enhancing their performance. For instance, metamaterials may exhibit a negative refractive index property, which can be used to focus electromagnetic waves more efficiently, allowing for smaller antennas with higher gains.

14.2.3.3 Reconfigurable Antenna Technologies for 5G

The complexity and diversity of 5G communication environments demand antennas that are not only capable of operating across multiple frequency bands but also adaptable to changing network conditions. This is where reconfigurable antennas come into play, offering significant advantages in terms of flexibility, adaptability, and spectrum efficiency.

A reconfigurable antenna can dynamically change its operating parameters, such as frequency, radiation pattern, or polarization, through the use of tunable components such as varactors, PIN diodes, RF microelectromechanical systems (RF MEMS), or even metamaterials. This allows the antenna to optimize performance based on real-time requirements, making it highly suited to the varying demands of 5G networks.

Reconfigurable antennas can dynamically adjust their operating parameters (frequency, radiation pattern, polarization) to optimize performance under varying network conditions. This section is subdivided into:

1. *Frequency reconfigurable antennas:* These antennas can switch between different frequency bands, enabling multiband operation. For 5G, this is particularly useful as the system must handle sub-6-GHz frequencies as well as millimeter-wave frequencies (e.g., 28 GHz, 39 GHz). The ability to reconfigure between these frequency bands allows for efficient use of the available spectrum and reduces the need for multiple antenna structures.

   For example, a frequency-reconfigurable antenna could switch from a lower band (e.g., 3.5 GHz) for wide coverage to a millimeter-wave band (e.g., 28 GHz) for high-speed data transmission, depending on the user's proximity to the base station or the overall network load.
2. *Pattern reconfigurable antennas:* These antennas can alter their radiation pattern in response to changing environmental conditions. By steering the main beam toward a specific user or avoiding interference from nearby devices, pattern reconfigurable antennas improve signal quality and reduce energy consumption. Beamforming techniques are often used in conjunction with pattern reconfigurable antennas to further enhance the directionality of the signal.

   For example, in a crowded urban area, a pattern-reconfigurable antenna might adjust its radiation pattern to avoid interference from other nearby 5G base stations while simultaneously focusing its main beam toward a high-priority user.
3. *Polarization reconfigurable antennas:* These antennas can switch between different polarization modes (e.g., vertical, horizontal, circular) to match the polarization of the incoming or outgoing signal. This is crucial in environments where signal polarization can change due to reflection, diffraction, or scattering, as is often the case in indoor or urban environments.

   For example, in a high-rise building where signals may undergo multiple reflections, a polarization-reconfigurable antenna could adjust its polarization to maximize signal reception, thereby improving overall link reliability.

4.2.3.4 Applications of Reconfigurable Antennas in 5G

1. *Dynamic spectrum access:* Reconfigurable antennas enable more efficient use of the available spectrum by switching between congested and less-used

frequency bands, especially in the millimeter-wave range. This helps to alleviate congestion in densely populated areas and provides users with higher data rates.

2. *Cognitive radio networks:* Reconfigurable antennas are key components of cognitive radio systems, where the network can sense unused spectrum and dynamically adjust its operation to avoid interference with licensed users.
3. *MIMO and beamforming:* In massive MIMO systems, where multiple antennas are used to serve multiple users, reconfigurable antennas can enhance the system's ability to manage beam patterns and frequencies for different users. This improves spatial diversity and overall system capacity.

14.2.3.5 Compact Antennas for IoT Devices

As the IoT expands, the demand for compact, power-efficient antennas that can operate in 5G environments is increasing. IoT devices are often constrained by their small size and limited power supply, making antenna design a critical challenge. These devices need antennas that are not only small in form factor but also capable of operating across multiple frequency bands to support the diverse communication protocols used in IoT ecosystems.

IoT devices, with their stringent size and power requirements, call for innovative antenna solutions. Key characteristics include:

1. *Miniaturization:* IoT devices such as wearable health monitors, smart meters, and environmental sensors are often small, necessitating compact antenna designs. Miniaturization techniques, such as using high-permittivity materials or meandered structures, help to reduce the antenna's physical size without significantly compromising performance.
2. *Multiband operation:* IoT devices typically need to communicate over different networks, such as Wi-Fi, Bluetooth, and cellular (including 5G). Therefore, multiband antennas, which can operate across a range of frequencies, are essential. For instance, a wearable device may need to switch between sub-1-GHz bands for long-range communication (e.g., LoRa) and higher frequencies for high-speed data transmission (e.g., 5G).

   For example, a multiband compact antenna integrated into a smart meter can support both long-range low-power wide-area network communication for remote monitoring and 5G for high-speed data transmission when necessary.
3. *Energy efficiency:* IoT devices are often battery-powered, so the antennas must be designed to minimize power consumption. This can be achieved by optimizing the antenna's impedance matching to reduce power losses and ensuring that the antenna operates efficiently across its intended frequency bands.
4. *Low profile and integration:* For IoT devices such as wearables or medical implants, the antenna needs to be as low profile as possible to maintain user comfort. Planar inverted-F antennas (PIFAs), patch antennas, and loop antennas are commonly used in these applications due to their compact design and ability to be easily integrated into the device's form factor.

14.2.3.6 Future Directions in Antenna Design:

1. *Integration with 5G New Radio (NR):* This means the radio access technology (RAT) that forms the foundation of 5G mobile networks. As IoT devices increasingly rely on 5G for communication, compact antenna designs will need to evolve to handle the stringent requirements of 5G NR, including mMTC and URLLC. New materials and design techniques, such as nanomaterials and 3D printing, may play a role in future antenna designs.
2. *Sustainability and low-cost manufacturing:* With billions of IoT devices expected to be deployed, there is a growing emphasis on sustainable and low-cost antenna manufacturing processes. Techniques such as printed electronics and the use of biodegradable materials for antenna substrates are being explored to reduce the environmental impact of large-scale IoT deployments.

14.2.4 Switched Beam and Adaptive Beam Smart Antenna Systems

Smart antenna systems are an essential innovation in 5G technology, enabling more efficient use of the spectrum and enhancing network reliability. Smart antennas are generally classified into switched beam antennas and adaptive beam antennas. Both types incorporate real-time signal processing, but they differ in how they adjust their radiation patterns to improve communication in complex environments.

Switched beam antennas select from a set of predefined beams, choosing the one that offers the best signal for the user's location. These beams are fixed but can dynamically switch, based on the direction of incoming signals.

Figure 14.5 demonstrates how the switched beam antenna system works by selecting one of several predefined beams that provide coverage in specific directions. Each beam is generated by the phase-shifted signals from the antenna array. The RF switch chooses the beam that offers the best signal quality, based on the measurements from the detector. The control logic guides this selection process. This system is typically used in wireless communication to improve reception by focusing the antenna's radiation pattern in the direction of the desired signal, thereby reducing interference and enhancing the overall communication quality.

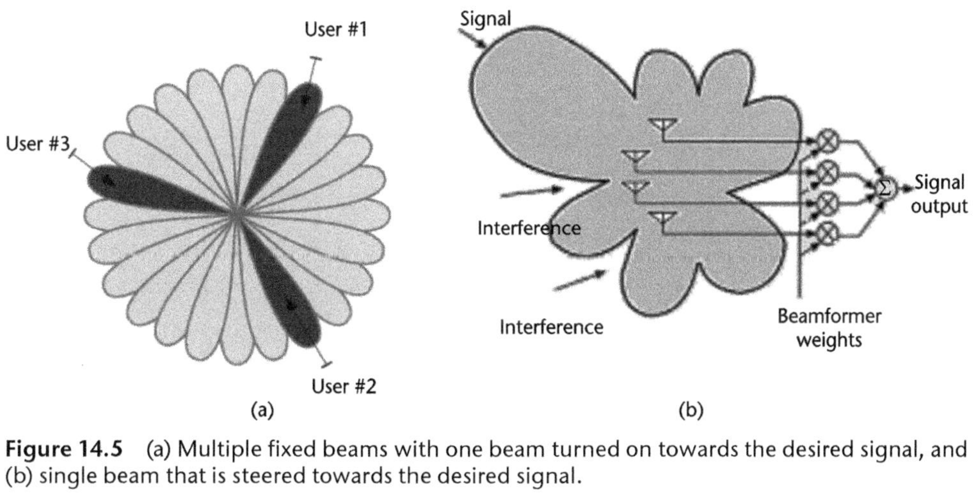

Figure 14.5 (a) Multiple fixed beams with one beam turned on towards the desired signal, and (b) single beam that is steered towards the desired signal.

The block diagram in Figure 14.6 illustrates the working principle of a switched beam antenna system. Here is a breakdown of the different components and their functions:

1. *Antenna array:* This section includes multiple antenna elements, arranged in an array. The angle θ refers to the direction from which the incoming signal is received. These elements receive signals from different directions and feed them into the system for processing.
2. *Phase shifting network:* The phase shifting network applies a phase shift to each signal received from the antenna elements. The purpose of this block is to steer the beams formed by the antenna array toward different directions. By applying the appropriate phase shifts, the system can focus on different angles or directions of incoming signals.
3. *RF switch:* This is an RF switch that controls which beam (or signal from a specific direction) is selected. It chooses one of the predefined beam patterns, depending on the strength or quality of the incoming signals. The switch toggles between beams to select the optimal direction of the desired signal.
4. *Detector:* The detector processes the selected signal. This block detects the strength or quality of the signal received from the RF switch. It is responsible for measuring parameters such as signal power or SNR.
5. *Control logic:* This is the decision-making unit of the system. Based on the input from the detector (signal strength or quality), the control logic decides which beam to select. It commands the RF switch to change to a different beam if necessary. The control logic helps to ensure that the best beam is chosen to maximize signal quality.
6. *Output:* This is the final output signal after the optimal beam has been selected. The output is the best possible signal from the selected beam direction, ready for further processing or transmission.

The following code demonstrates how single and multiple beams, pointing to several targets at several angles, are simulated. MATLAB Program 30 simulates and plots, as shown in Figure 14.7(a), single beam pattern with beam pointing to a target at 45°, and Figure 14.7(b) multibeam pattern with beam pointing to several targets at several angles simultaneously.

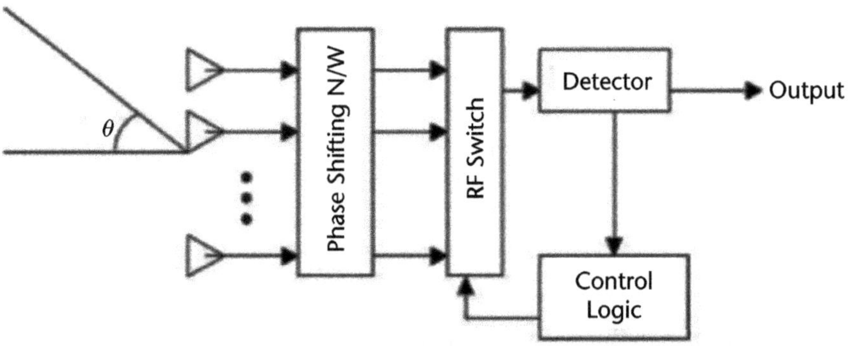

Figure 14.6 Block diagram of switched beam antenna.

(a)

(b)

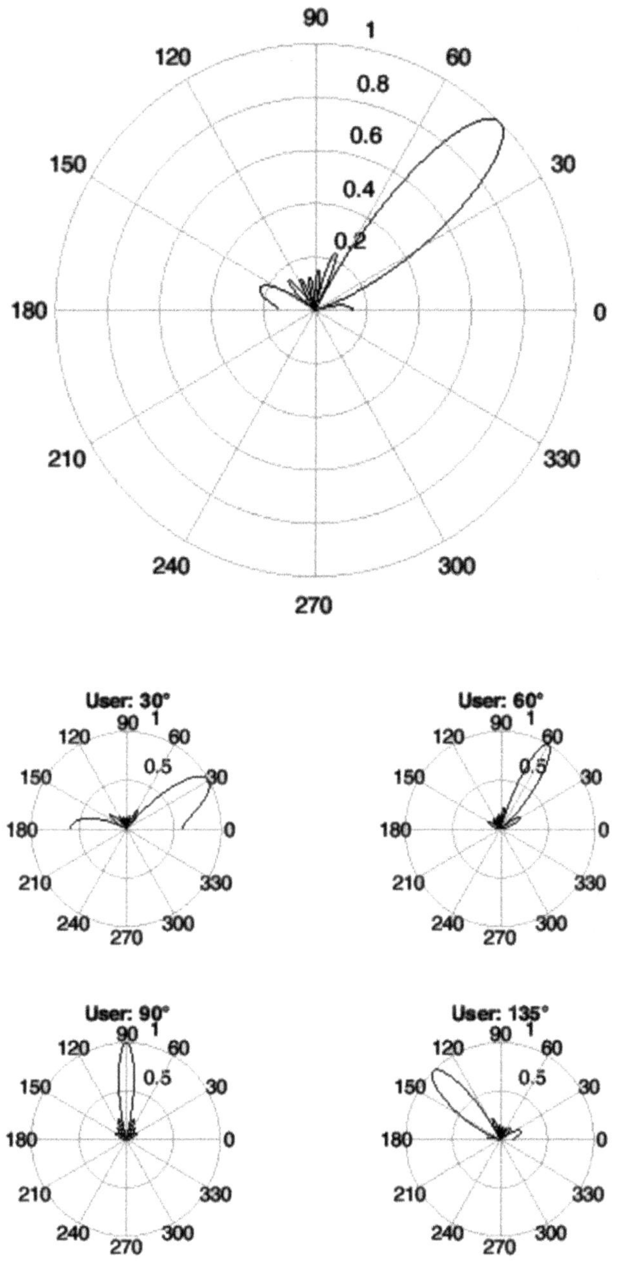

Figure 14.7 (a) Single beam pattern with beam pointing to a target at 45°, and (b) multibeam pattern with beam pointing to several targets at several angles simultaneously.

MATLAB Program 30

```
% Beamforming Simulation for 5G unidirectional uniform linear
Array (ULA) Antenna
% This MATLAB program simulates adaptive beamforming for a
uniform linear array U(LA)
% of antennas in a 5G system. Figure 1 only plots for a single
desired angle, while
```

```
% Figure 2 plots multiple patterns, for several users at various
directions, simultaneously.

clc;     % Clears the command window for a clean output display
clear;   % Clears all variables from the workspace to prevent
conflicts with old data
close all; % Closes all existing figure windows to start with a
fresh plotting environment

% Parameters
fc = 28e9;      % Carrier frequency set to 28 GHz, a common
frequency for 5G systems
c = 3e8;        % Speed of light in meters per second
lambda = c/fc;  % Wavelength of the signal, calculated using
lambda = c/fc
d = lambda/2;   % Distance between array elements, set to half
the wavelength
N = 8;          % Number of antenna elements in the uniform linear
array (ULA)
theta_desired = 45; % Desired beam direction in degrees (0 to 180
for unidirectional)

% Array Steering Vector for Desired Direction
theta = 0:0.1:180; % Define the angle range from 0 to 180 degrees
for plotting
steering_vector = exp(1j*2*pi*d*(0:N-1)'*cosd(theta_desired)/
lambda);

% Compute Beamforming Weights (maximizing gain at desired angle)
weights = conj(steering_vector)/norm(steering_vector);
% The weights are computed by conjugating the steering vector to
maximize gain at the desired angle

% Array Factor (beam pattern calculation)
AF = zeros(1, length(theta));  % Initialize the array factor (AF)
for n = 1:N
  AF = AF + weights(n) * exp(1j*2*pi*(n-1)*d*cosd(theta)/lambda);
  % Sum up the contributions of each element in the array to form
the total beam pattern
end

% Normalize Array Factor
AF = abs(AF)/max(abs(AF));  % Normalize the array factor so that
the maximum value is 1

% Plot the beam pattern for the desired direction
figure;  % Create a new figure for the plot
polar(deg2rad(theta), AF, 'k' );  % Polar plot of the beam
pattern using the polar function
```

```
title(['Beam Pattern with Beam Pointing to ', num2str(theta_
desired), '°']);  % Title of the plot
set(gca, 'FontSize', 12);  % Set font size for clarity

% Adjust beam angles for different users
angles = [30 60 90 135]; % Define the angles for multiple users
(in degrees), ensuring they are in the range of 0 to 180

figure;  % Create a new figure for plotting multiple
beam patterns
title('Beamforming for Different Users with ULA', 'FontSize',
14); % Main title for the figure

for i = 1:length(angles)
  % New steering vector for each user at different angles
  steering_vector = exp(1j*2*pi*d*(0:N-1)'*cosd(angles(i))/
lambda);
  % Recompute the steering vector for the new angle (angles(i))

  % Compute new weights for beamforming toward the current user
  weights = conj(steering_vector)/norm(steering_vector);

  % Reset the array factor for the new beam pattern
  AF = zeros(1, length(theta));  % Reinitialize the array factor
for each user
  for n = 1:N
    AF = AF + weights(n) * exp(1j*2*pi*(n-1)*d*cosd(theta)/
lambda);
    % Sum up the contributions from all antenna elements
  end
  AF = abs(AF)/max(abs(AF)); % Normalize the array factor for
consistent scaling

  % Plot beam pattern for each user angle
  subplot(2, 2, i);  % Create a subplot for each user angle
  polar(deg2rad(theta), AF, 'k'); % Polar plot of the beam
pattern using the polar function
  title(['User: ', num2str(angles(i)), '°'], 'FontSize', 10); %
Simplified title for the subplot with a smaller font size
  set(gca, 'FontSize', 10); % Adjust the font size of the axis
for better readability
end
```

Adaptive beam antennas dynamically adjust their radiation patterns in real time, fine-tuning the beam direction to optimize signal reception and mitigate interference.

Figure 14.8 illustrates the adaptive array system, which dynamically adjusts the beam pattern of the antenna array by altering the weights applied to each element's signal. The adaptive algorithm continuously updates the weights to direct the antenna's focus toward the desired signal and away from interference or noise

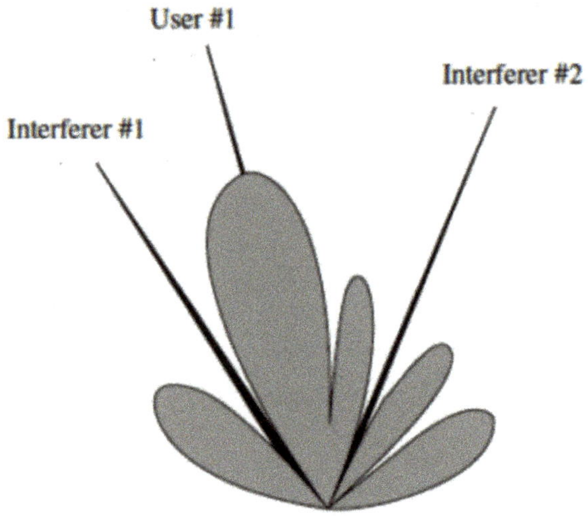

Figure 14.8 Adaptive array antenna radiation pattern where the beam points towards the desired user and nulls undesired interferers.

sources. This system is essential in environments where signal conditions change rapidly, as it allows the antenna to automatically adjust its performance to maintain optimal signal reception. This technology is widely used in wireless communication systems to improve signal quality, reduce interference, and increase capacity.

The block diagram in Figure 14.9 illustrates the working principle of an adaptive array system used in smart antenna technology:

1. *Antenna array:* This section consists of multiple antennas, arranged in an array, designed to receive signals from various directions. The antenna elements capture signals from multiple sources, including desired signals and interference, and feed them into the system.

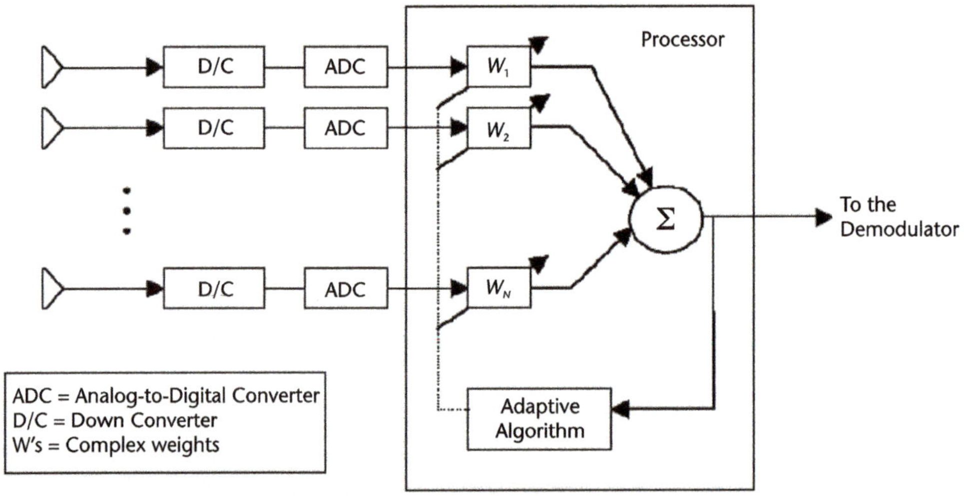

Figure 14.9 Block diagram of adaptive array beam antenna.

2. *Downconverter:* A downconverter reduces the frequency of the incoming signals received by each antenna element. It converts the high-frequency RF signals into lower intermediate frequencies (IF) or baseband signals for easier processing. This process is essential for extracting the relevant information from the carrier signal.
3. *Analog-to-digital converter (ADC):* The ADC is used to convert the analog IF or baseband signals (after downconversion) into digital signals. This conversion is critical because digital signals are easier to process with modern signal processing techniques. The ADC allows the digital processor to manipulate and adapt the signals to improve performance.
4. *Processor:* The processor is the main component that performs signal processing and controls the adaptive beamforming algorithm. The processor adjusts the weights of each antenna element signal to focus the beam in the direction of the desired signal and reduce interference. The weights are represented by W_1, W_2, ..., W_N, and these complex weights can change dynamically based on the adaptive algorithm. The signals from the antenna elements are multiplied by their corresponding weights before being summed.
5. *Adaptive algorithm:* This is the brain of the adaptive array system, responsible for continuously adjusting the complex weights. The adaptive algorithm uses input from the received signals to compute the optimal weights for each antenna element in real time. It updates the weights to maximize the reception of the desired signal while minimizing interference and noise. Common algorithms include the least mean squares (LMS) or recursive least squares (RLS) methods, which minimize errors in signal estimation.
6. *Summation block (Σ):* This block sums the weighted signals from all the antenna elements. After the processor adjusts the weights of each signal, they are combined into a single output signal. This summation process helps to form a focused beam in the desired direction while nullifying or minimizing unwanted signals from other directions.
7. *Output to demodulator:* The output from the summation block is sent to the demodulator. The demodulator extracts the original information from the modulated signal. By this point, the signal has been optimized for quality, and the demodulator can recover the intended message or data efficiently.

MATLAB Program 31 simulates and plots, as shown in Figure 14.10(a), beam pattern with beam pointing to 0°, which illustrates a single beam pointing to the user, and as shown in Figure 14.10(b), beam pattern with nulls at 30° and 150°, which illustrates multiple beams directed simultaneously at different users.

MATLAB Program 31

```
% This MATLAB program simulates a uniform linear array (ULA)
antenna operating at a
% 28 GHz frequency, commonly used in 5G systems. The objective is
to generate an
% adapted beam pattern for a desired direction (0 degrees) and
steer nulls in
```

(a)

(b)

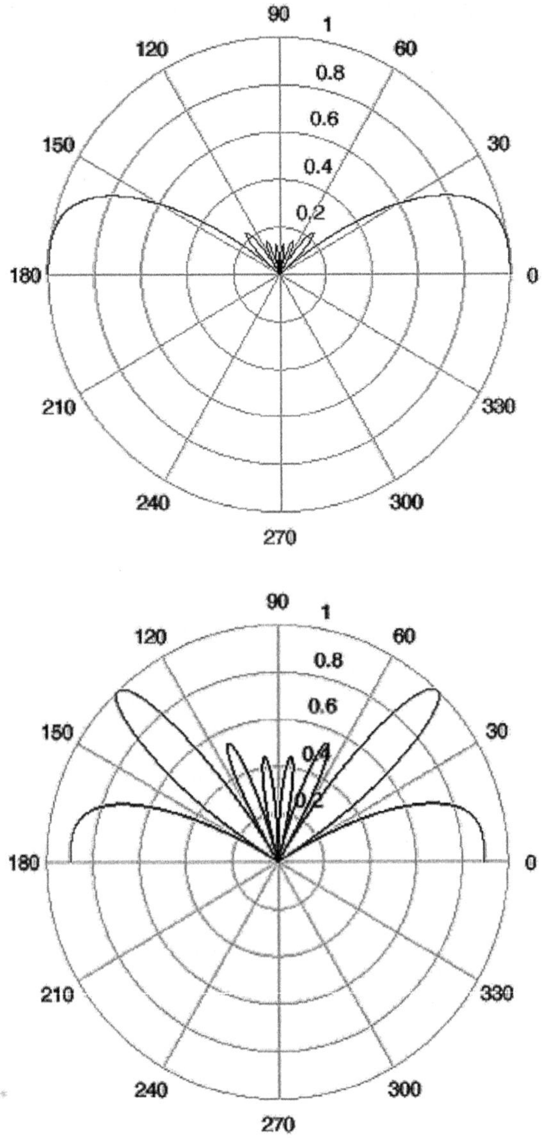

Figure 14.10 (a) Beam pattern with beam pointing to 0°, and (b) beam pattern with nulls at 30° and 150°.

```
% specific directions (30 and 150 degrees) to suppress
interference or undesired signals.

clc;    % Clears the command window for a clean output display
clear;   % Clears all variables from the workspace to prevent
conflicts with old data
close all; % Closes all existing figure windows to start with a
fresh plotting environment
```

```
% Parameters
fc = 28e9;    % Carrier frequency set to 28 GHz, a common
frequency for 5G systems
c = 3e8;      % Speed of light in meters per second
lambda = c/fc;  % Wavelength of the signal, calculated using
lambda = c/fc
d = lambda/2;   % Distance between array elements, set to half
the wavelength
N = 8;        % Number of antenna elements in the uniform linear
array (ULA)
theta_desired = 0; % Desired beam direction in degrees for
Figure 1

% Array Steering Vector for Desired Direction (Figure 1)
theta = 0:0.1:180;  % Define the angle range from 0 to 180
degrees for plotting
steering_vector = exp(1j*2*pi*d*(0:N-1)'*cosd(theta_desired)/
lambda);

% Compute Beamforming Weights for Desired Angle (60 degrees)
weights = conj(steering_vector)/norm(steering_vector);

% Array Factor (beam pattern calculation for Figure 1)
AF = zeros(1, length(theta));  % Initialize the array factor (AF)
for n = 1:N
    AF = AF + weights(n) * exp(1j*2*pi*(n-1)*d*cosd(theta)/
lambda);
end

% Normalize Array Factor
AF = abs(AF)/max(abs(AF));  % Normalize the array factor so that
the maximum value is 1

% Plot the beam pattern for the desired direction (Figure 1)
figure(1);  % Create a new figure for the plot
polar(deg2rad(theta), AF, 'k');  % Polar plot of the beam pattern
using the polar function
title(['Beam Pattern with Beam Pointing to ', num2str(theta_
desired), '°']);  % Title of the plot
set(gca, 'FontSize', 12);  % Set font size for clarity

% Null Steering at 30 degrees and 180 degrees for Figure 2
theta_nulls = [30, 150]; % Directions where we want to
place nulls

% Initialize the null-steered weights with the original weights
null_steered_weights = weights;

% Modify the weights to introduce nulls at the specified angles
for m = 1:length(theta_nulls)
```

```
  % Create the steering vector for the null direction
  null_steering_vector = exp(1j*2*pi*d*(0:N-1)'*cosd(theta_
nulls(m))/lambda);

  % Calculate the projection of the weights onto the null
steering vector
  projection = (null_steering_vector' * weights) /
norm(null_steering_vector)^2;

  % Subtract the projection from the weights to create a null
  null_steered_weights = null_steered_weights - (null_steering_
vector * projection);
end

% Calculate new array factor for the null-steered pattern
(Figure 3)
AF_null = zeros(1, length(theta));
for n = 1:N
  AF_null = AF_null + null_steered_weights(n) * exp(1j*2*pi*(n-
1)*d*cosd(theta)/lambda);
end

% Normalize the array factor for the null-steered pattern
AF_null = abs(AF_null)/max(abs(AF_null));

% Plot the null-steered beam pattern (Figure 2)
figure(2);  % Create a new figure for the null-steered plot
polar(deg2rad(theta), AF_null, 'k');  % Polar plot of the beam
pattern with null steering
title(['Beam Pattern with Nulls at ', num2str(theta_nulls(1)), 'o
and ', num2str(theta_nulls(2)), 'o']);
% Title of the plot
set(gca, 'FontSize', 12);  % Set font size for clarity
```

The tracking behavior of adaptive antennas for both single-user and multi-user scenarios is illustrated as follows:

1. *Single user tracking:* The adaptive antenna system focuses on tracking a single user by adjusting its beam direction towards that user's location.
   - The system uses beamforming to direct the signal towards a single active user.
   - Only one transceiver is needed to communicate with the single user.
   - The beam is continuously adjusted as the user moves, ensuring that the signal strength and quality remain optimal.
2. *Multi-user tracking:* When multiple users are active, the adaptive antenna system creates different beam patterns for each user. This allows for simultaneous communication with multiple users.

- The system generates different beam patterns for each user, so they can all receive signals simultaneously.
- Achieves space division multiple access, which allows multiple users to be served at the same time using the same frequency band by spatial separation of their signals.
- Multiple transceivers work in parallel to support the communication of each user with separate beam pairs.

In summary, switched beam and adaptive array antenna systems differ in their performance and capabilities. Switched beam systems provide less coverage and have limited interference rejection, making them suitable for simpler environments where high precision is not critical. In contrast, adaptive array systems offer several advantages, including greater coverage, improved interference rejection, and reduced transmission power. This makes adaptive arrays more effective in dynamic environments, as they can adjust their beam patterns to focus on specific users or signals, thereby enhancing overall system efficiency and performance.

14.2.4.1 Key Features of Switched Beam and Adaptive Beam Antennas

Both switched beam and adaptive beam antennas offer several key features that enhance the performance of 5G networks. These are:

1. *Interference reduction:* By dynamically adjusting the antenna's radiation pattern to steer nulls toward sources of interference, adaptive antennas can significantly improve signal quality and SNR. This capability is particularly important in high-density urban areas where multiple devices may operate in close proximity.
2. *Optimized coverage:* Adaptive antennas can adjust their beam patterns to focus on users in specific locations, improving both coverage and capacity. For instance, in a scenario where users are concentrated in a particular sector of a cell, the antenna can steer its main lobe toward that sector, reducing power wastage in areas where there are no users.
3. *Multi-user support:* In a 5G network, multiple users can be supported simultaneously using adaptive antennas by employing spatial multiplexing. This involves creating distinct beams for different users, allowing them to share the same frequency resources without causing interference.

Smart antennas offer distinct advantages over traditional antennas due to their ability to dynamically manage signal reception and transmission. By using an array of antenna elements, they can selectively choose which elements to receive signals from or transmit on, optimizing performance. A key feature of smart antennas is their ability to nullify interference from other nodes, enhancing signal clarity and reducing noise. Additionally, they can intelligently control the power distribution across each antenna element, improving overall system efficiency and signal quality, making them well suited for modern communication systems such as 5G.

14.2.4.2 Advantages of Smart Antennas

Smart antennas offer a range of significant advantages that enhance the performance of wireless communication systems. One of the key benefits is range extension, allowing signals to reach further distances with improved reliability. Additionally, smart antennas improve building penetration and effectively address hole filling, ensuring that signals can navigate through obstacles such as walls and other physical barriers, reducing dead zones. They also manage delay spread, minimizing the time difference between received signal paths, which is critical for maintaining signal clarity in complex environments.

Another major advantage is their ability to mitigate cochannel interference, allowing different signals using the same frequency to coexist without significant disruption. Similarly, multiple-access interference, interference from multiple users transmitting at the same time, is reduced, improving overall communication quality. Smart antennas excel in multipath management, using advanced techniques to handle reflections and scattered signals, ensuring stable connections even in challenging environments. Lastly, they contribute to system capacity and power control by optimizing the use of frequency resources and adjusting transmission power intelligently, leading to more efficient power consumption and enhanced network capacity.

14.2.4.3 Applications of Smart Antennas

Smart antennas are pivotal in numerous advanced technologies due to their ability to dynamically manage signal reception and transmission, offering enhanced performance in critical applications. In cellular and wireless networks, smart antennas significantly improve network efficiency by extending coverage, ensuring stronger signal penetration through obstacles, and managing interference in high-density environments. This is particularly crucial in the era of 5G and beyond, where network demands are rapidly increasing, and efficient use of bandwidth and spatial resources is essential. By steering beams toward active users and nullifying interference from unwanted sources, smart antennas enhance both capacity and data transmission quality, leading to a seamless user experience.

In radar systems, smart antennas elevate the capability of detecting, tracking, and imaging objects. Through adaptive beamforming, they allow radar systems to focus on specific areas or targets, improving detection accuracy and range resolution. This makes them indispensable in applications such as air traffic control, weather monitoring, and military surveillance, where precision is paramount.

Smart antennas are also crucial in electronic warfare, particularly as a countermeasure against electronic jamming. In hostile environments, where adversaries use jamming techniques to disrupt communication or radar systems, smart antennas dynamically adjust their radiation patterns to nullify interference. This enhances the survivability and functionality of military communication systems and ensures the reliability of radar and sensor networks, even in contested electromagnetic environments.

In satellite communication systems, the long distances and complex atmospheric conditions that satellite signals must traverse present significant challenges. Smart antennas address these challenges by dynamically adjusting their beams to maintain robust links between satellites and ground stations or between satellites in space. They enhance signal strength, reduce interference, and improve data throughput, ensuring reliable communication even across vast distances. These capabilities are critical for applications such as Global Positioning Systems (GPS), satellite-based internet services, and satellite TV broadcasting, where continuous and reliable communication is essential.

The versatility of smart antennas across these diverse fields—cellular networks, radar systems, electronic warfare, and satellite communications—underscores their importance in modern technological infrastructure. Their ability to enhance system capacity, manage multipath interference, and intelligently control power and signal direction makes them an essential tool in both civilian and defense applications, ensuring efficient, reliable, and high-performance communication and sensing systems.

14.2.4.4 Real-Time Sensing and Adaptation by using AI/Machine Learning Algorithms

The integration of AI and machine learning (ML) into smart antenna systems has opened new possibilities for real-time adaptation. By using AI/ML algorithms, antennas can predict user movement, adapt to changes in traffic demand, and even anticipate interference sources. For instance, an AI-powered antenna system could analyze traffic patterns in real time and adjust its beam to ensure optimal coverage and performance.

In massive deployments such as smart cities or autonomous vehicle networks, AI-enhanced smart antennas can reduce latency, improve spectral efficiency, and ensure seamless connectivity by continuously optimizing their parameters based on the environment.

In a real-world application, smart antennas can be seen in 5G-enabled autonomous vehicle networks, where adaptive antennas are crucial for maintaining low-latency, high-reliability communication. By continuously adjusting beam patterns to track vehicles and avoid interference, these antennas ensure a stable connection for real-time data exchange between vehicles and infrastructure.

14.2.5 Future Trends and Opportunities in 5G Antenna Design

As 5G continues to evolve and pave the way for 6G, antenna technology must rise to meet the increasing demands for higher data rates, ultralow latency, and seamless connectivity across diverse environments. Several emerging trends in antenna design are poised to redefine the landscape of wireless communication, including reconfigurable intelligent surfaces (RIS), metamaterials, and integrated antenna systems for 6G networks. These innovations, along with a focus on sustainability, will form the foundation of the next generation of communication infrastructure.

14.2.5.1 RIS

One of the most transformative technologies in future antenna design is the concept of RIS. RIS consist of arrays of sub-wavelength elements that can dynamically control electromagnetic waves, modifying their phase, amplitude, and polarization. Unlike traditional antennas, RIS do not actively transmit or receive signals. Instead, they manipulate the propagation of incoming waves, making them highly energy-efficient. This opens new possibilities for redirecting signals, enhancing coverage, improving spectral efficiency, and mitigating interference without consuming much power.

RIS can be deployed on building facades, walls, or other surfaces to act as smart reflectors, overcoming the limitations of millimeter-wave signals such as high path loss and poor penetration through obstacles. In urban environments, where buildings and other obstacles create signal blockages, RIS can ensure that 5G networks deliver consistent and reliable coverage.

The phase shift introduced by RIS elements can be expressed as:

$$\varphi_i = 2\pi \frac{d}{\lambda} \sin(\theta_i) \tag{14.8}$$

where φ_i is the phase shift of the ith element, d is the distance between adjacent elements, λ is the wavelength, and θ_i is the angle of incidence.

By carefully adjusting the phase shifts across the RIS elements, the reflected wavefronts can be steered toward the desired direction, creating highly flexible and programmable environments. Future research in RIS aims to improve integration with AI for real-time adaptation and to extend these capabilities to both indoor and outdoor communication networks.

14.2.5.2 Metamaterials for Antenna Design

Another significant advancement lies in metamaterials, which are engineered materials with properties not found in nature. These materials, composed of periodic subwavelength structures, exhibit unique electromagnetic properties, such as negative refraction and superlensing, that can be exploited to enhance antenna performance.

In the context of 5G and 6G antennas, metamaterials offer several key benefits. For example, they enable the design of highly compact antennas without sacrificing gain or efficiency. This is especially important for mobile devices and IoT applications, where space is limited. Metamaterial antennas can also be optimized for better beamforming, allowing for highly directional radiation patterns that reduce interference and enhance communication range.

Moreover, research into metamaterial absorbers and cloaking technologies holds promise for reducing electromagnetic interference and enhancing the security of wireless communication systems. As frequency bands shift towards the terahertz range in 6G, metamaterials will become even more crucial in addressing the challenges of wave propagation at such high frequencies.

14.2.5.3 Integrated Antenna Systems for 6G and Beyond

Looking ahead, integrated antenna systems will be at the forefront of 6G and future wireless technologies. The frequency spectrum for 6G is expected to extend into the terahertz range, which will present new challenges in terms of antenna design, propagation, and hardware integration.

At terahertz frequencies, antennas must be extremely precise and capable of supporting ultrafast data rates with minimal latency. To achieve this, future antenna systems will likely be embedded directly into devices, forming highly integrated modules that can support a wide range of applications, from autonomous vehicles and smart cities to advanced medical diagnostics and quantum communication.

One exciting area of research is the development of graphene-based antennas for terahertz communication. Graphene, known for its exceptional electrical properties, offers the potential for extremely fast, efficient, and compact antennas that could meet the demands of 6G and beyond. Such antennas could enable data transfer rates far exceeding those possible with current technologies, opening new frontiers in wireless communication.

In addition to handling ultrahigh frequencies, integrated antenna systems will need to be energy-efficient, robust, and capable of operating in harsh environmental conditions. This will be particularly important for applications such as satellite-based communication networks, which require antennas to function reliably in space or other extreme settings.

14.2.5.4 Sustainability in Antenna Design

With the increasing deployment of 5G infrastructure and the proliferation of connected devices, the environmental impact of wireless networks is becoming a pressing concern. Future antenna design must prioritize sustainability, focusing on reducing power consumption, minimizing the use of nonrenewable resources, and extending the lifespan of communication systems.

One approach to achieving sustainable antenna systems is through energy-harvesting technologies, where antennas not only transmit and receive signals but also capture ambient energy to power themselves. This could significantly reduce the carbon footprint of dense 5G and 6G networks, particularly in urban environments where energy demands are high.

Additionally, recyclable materials and eco-friendly manufacturing processes will play an increasingly important role in the design of next-generation antennas. As communication networks expand globally, particularly in developing regions, the ability to produce low-cost, sustainable antennas will be essential for bridging the digital divide while minimizing environmental impact.

14.2.5.5 Conclusion and Future Outlook

As the world moves toward 5G, 6G, and beyond, the evolution of antenna technology will be key to unlocking the full potential of wireless communication. RIS, metamaterials, integrated antenna systems, and sustainable design practices are just a few of the innovations that will shape the future of antennas. The journey toward more efficient, flexible, and scalable communication networks is still unfolding, but it is clear that antennas will remain at the forefront of this transformation. By embracing these cutting-edge technologies, the wireless industry will continue to push the boundaries of what is possible, enabling a future of ubiquitous, high-performance, and sustainable communication.

14.3 Deployable Satellite Antennas: An Emerging Frontier

Building upon the foundational concepts of terrestrial smart antenna systems discussed in the previous sections, including adaptive beamforming, AI-enhanced directionality, and compact 5G antenna arrays, we now shift our focus to their extraterrestrial analog: deployable satellite antennas. While smart antennas on Earth are designed to optimize coverage and signal quality in dense, dynamic environments, deployable antennas face a different set of challenges, namely, the constraints of launch volume, mass, and the harshness of space. Yet both domains share a common goal: maximizing performance from compact, efficient hardware through innovative structural and material design. This section explores the types, mechanisms, and materials behind modern deployable antenna systems, highlighting how their ability to unfold and operate autonomously in orbit is revolutionizing satellite communications and deep-space missions.

As satellite communication systems evolve, the advent of deployable antennas has revolutionized the field. These antennas can be compactly stowed during launch and subsequently deployed in space, forming large aperture antennas that enhance communication capabilities. This section delves into the technology, types, materials, mechanisms, and critical equations associated with deployable antennas.

14.3.1 Introduction to Deployable Antenna Technology

14.3.1.1 Overview of Deployable Antennas

Deployable antennas are sophisticated structures designed to be launched in a compact form and then expanded or deployed once they reach their operational environment. The need for such technology has become increasingly important due to the limitations of traditional antenna designs, which often compromise on size, weight, and performance. With the growing demand for high-bandwidth communications, the development of deployable antennas presents a viable solution to accommodate larger apertures without increasing the overall launch payload.

Historically, satellite antennas were fixed structures, limiting the performance of communication systems. As satellite technology advanced, the realization that larger apertures could significantly enhance performance drove innovation. Deployable antennas allow for the integration of larger surfaces that can capture more signal energy, thus improving gain and directivity.

14.3.1.2 Advantages of Deployable Antennas

The advantages of deployable antennas are multifaceted:

1. *Increased performance:* Larger aperture sizes contribute to improved gain and directivity, critical parameters in satellite communications. This performance enhancement allows satellites to transmit and receive signals over greater distances with higher clarity and reliability.
2. *Versatility:* Deployable antennas are suitable for various applications, including telecommunications, Earth observation, and scientific research. Their

ability to adapt to different missions makes them invaluable in the ever-evolving landscape of satellite applications.
3. *Cost-effectiveness:* By reducing the volume and weight of satellite payloads, deployable antennas can lead to significant cost savings during launch. The ability to pack more functionality into smaller spaces is particularly appealing for commercial and governmental space agencies, which are constantly seeking to optimize launch costs.
4. *Space utilization:* In space, every cubic inch matters. Deployable antennas maximize the available space on a satellite, allowing for more efficient designs that can accommodate additional payloads or instruments.
5. *Scalability:* The technology behind deployable antennas allows for scalability in design, accommodating different missions with varying requirements without a complete redesign of the satellite.

14.3.2 Types of Deployable Antennas

The design of deployable antennas can vary significantly depending on their intended use, deployment mechanisms, and performance requirements. Here we describe some of the prominent types of deployable antennas.

14.3.2.1 Umbrella Antennas

1. *Structure and deployment:* Umbrella antennas are deployable antennas that feature a parabolic or similar reflective surface held by a radial structure. They fold into a compact shape for launch and deploy in space like an umbrella, expanding outward to form a large, high-gain reflective surface. This design allows for significant size expansion, which is crucial for long-distance communications. Figure 14.11 demonstrates an umbrella paraboloid antenna.
2. *Deployment mechanism:* The deployment typically involves torsion or spring-loaded hinges that initiate and control the unfolding. As the structure opens, the radial "spokes" extend outward, pulling the antenna's surface into a precise shape.
3. *Applications:* Umbrella antennas are particularly suited for deep-space missions and high-gain communications, as demonstrated by the Chang'e-4 relay satellite's 4.2-m umbrella antenna. Positioned in lunar orbit, it facilitates a reliable link between Earth and lunar landers and rovers, especially where long-distance signal integrity is essential.
4. *Advantages:* They offer a high storage-to-deployment ratio, allowing larger apertures without adding excessive weight or volume. The robust yet simple deployment process provides reliability, while the parabolic shape enables excellent directivity and signal focus.
5. *Challenges:* Structural stability is essential in ensuring that the antenna remains rigid and precise after deployment. Vibrations during launch or the stresses of deep-space environments may affect this type of antenna, so advanced materials and reliable locking mechanisms are used to maintain shape integrity.

Figure 14.11 Umbrella paraboloid antenna. (Source: Wdwdbot. Wikimedia Commons.)

14.3.2.2 Mesh Antennas

1. *Structure and design:* Mesh antennas consist of a flexible, woven mesh that serves as the antenna's reflective surface. This mesh is stretched over a support frame or truss system, forming a parabolic shape essential for focused signal reflection. The design is extremely lightweight compared to traditional solid reflectors and folds compactly for stowage. Figure 14.12 demonstrates a pair of 29-foot mesh dish antennas.
2. *Deployment mechanism:* Mesh antennas deploy by unfolding a frame or truss that stretches the mesh into a precise parabolic shape. This is often done using rigid or semi-rigid struts and hinges that lock into place once the mesh is fully expanded.
3. *Applications:* These antennas are widely used in Earth observation and navigation systems, such as the Beidou-3 (BD-3) navigation satellite. Large mesh antennas are particularly useful in SAR systems, where multibeam capabilities and high gain are required for detailed, high-resolution imaging.
4. *Advantages:* Mesh antennas provide excellent flexibility, light weight, and high storage efficiency, allowing them to be packed tightly for launch. Their large apertures improve signal gain and provide multibeam functionality, making them suitable for applications needing wide-area coverage, multi-frequency operation, or high resolution.

Figure 14.12 Pair of 29-foot mesh dish antennas at the Praline Mountain Signal Site. (Source: Glen Vehring. Wikimedia Commons.)

5. *Challenges:* As mesh reflectors are flexible, maintaining precise surface accuracy can be challenging, especially when exposed to thermal variations or radiation. High-precision manufacturing, material resilience, and dynamic deployment systems are crucial for mesh antennas to meet the demands of high-resolution communication and imaging.

14.3.2.3 Inflatable Antennas

1. *Structure and design:* Inflatable antennas consist of a membrane that inflates to form the desired shape. The membrane, usually made of durable, lightweight materials such as polyimide, is compactly stowed and inflates in orbit to create a large reflective surface. Figure 14.13 demonstrates an inflatable antenna which was deployed in LEO from the Space Shuttle Endeavour. The Inflatable Antenna Experiment (IAE) was carried on the SPARTAN-207 free-flyer, which the Shuttle's robotic arm released; the mylar dish was inflated on-orbit (May 19–20, 1996).
2. *Deployment mechanism:* These antennas are stowed as deflated structures and expand upon reaching orbit through gas inflation or, in some cases, pressure-induced hardening. Once deployed, the structure remains in place, either through internal pressure or by a subsequent rigidizing process that solidifies the membrane.

Figure 14.13 Inflatable antenna experiment. (Image courtesy of NASA.)

3. *Applications:* Inflatable antennas are highly suitable for short-duration missions or lower-Earth orbit applications, such as disaster monitoring, rapid-deployment communications, and Earth observation, where lightweight construction is prioritized. Inflatable structures have been explored by agencies such as NASA for emergency satellite communications and temporary observation needs.
4. *Advantages:* Their lightweight and compact stowage make inflatable antennas ideal for payload-limited missions. They allow very large apertures to be deployed quickly with minimal mass, which can be especially advantageous for short-term or emergency deployments.
5. *Challenges:* Due to their dependence on inflation and membrane stability, inflatable antennas may not be as robust or durable over long-term missions compared to other deployable types. Factors such as micrometeoroid impacts, UV radiation, and prolonged thermal cycling can degrade the material over time, so they are often better suited for temporary missions.

14.3.3 Materials and Mechanisms

The effectiveness of deployable antennas largely depends on the materials and mechanisms used in their design. This section explores the critical materials and mechanical elements that contribute to the performance and reliability of these antennas.

14.3.3.1 Material Selection

The choice of materials is crucial for the success of deployable antennas. The following materials are commonly used:

1. *Shape memory alloys (SMAs):* SMAs are metallic materials that can return to a predefined shape when heated. This property is particularly useful in deployable antennas, where temperature changes in space can trigger deployment mechanisms. The main benefits of SMAs include:
   - *Reliability:* SMAs provide a reliable method for actuating deployment mechanisms, ensuring that antennas unfold smoothly in space.
   - *Minimal power requirement:* The ability to deploy using heat reduces the need for complex and power-hungry motor systems.
2. *Carbon fiber composites:* Carbon fiber composites are widely used due to their excellent strength-to-weight ratio and resistance to thermal fluctuations. Key advantages include:
   - *Lightweight:* These materials significantly reduce the overall weight of the antenna, allowing for larger apertures without compromising the satellite's launch capabilities.
   - *Durability:* Carbon fiber composites are resistant to environmental degradation, ensuring the longevity and reliability of antennas in the harsh space environment.
3. *Polymeric materials:* In addition to SMAs and carbon fiber, various polymers are employed in inflatable antennas and other deployable structures due to their flexibility and lightweight properties. These materials can provide effective barriers against moisture and other environmental factors.

14.3.3.2 Mechanical Mechanisms

The deployment of antennas involves intricate mechanical systems designed to ensure accurate and reliable unfolding. The following elements are essential:

1. *Actuators and hinges:* Actuators play a critical role in the deployment process. They can be electric, hydraulic, or thermal, depending on the design. The hinges must be robust yet lightweight, allowing for smooth movement without added friction. Common types include:
   - *Rotary hinges:* Used to allow sections of the antenna to rotate into place during deployment.
   - *Sliding mechanisms:* Enable components to slide past each other, facilitating compact stowage and smooth deployment.
2. *Control systems:* The control systems manage the deployment process, ensuring that antennas unfold correctly and reach their operational position without issues. Modern control systems utilize:
   - *Sensors:* Sensors monitor the position and state of the antenna during deployment.
   - *Feedback mechanisms:* These systems provide real-time data to adjust the deployment process as needed, enhancing reliability.

3. *Deployment strategies:* Various strategies exist for deploying antennas, including sequential deployment, where segments unfold one after another, and simultaneous deployment, where all parts deploy at once. Each strategy has its advantages and is selected based on the mission requirements.

14.3.3.3 Performance Optimization

To ensure the best performance, optimization techniques are employed during the design phase. These include computational modeling and simulations to predict how the antenna will behave once deployed. Factors such as wind resistance, thermal expansion, and material fatigue are considered to enhance the stability and precision of deployable antennas in space.

14.3.4 Performance Calculation

To quantify the performance of deployable antennas, several key equations are essential:

1. *Aperture efficiency equation for satellite antennas:* The aperture efficiency of an antenna is a critical measure of its effectiveness in capturing and utilizing incoming signals. The equation for aperture efficiency is given by:

$$\eta = \frac{A_e}{A_p} \tag{14.9}$$

where η = aperture efficiency, A_e = effective aperture area (m^2), and A_p = physical aperture area (m^2).

The effective aperture area considers the losses due to surface imperfections, dielectric losses, and other factors. A higher aperture efficiency indicates that a larger fraction of the input power is effectively captured and utilized by the antenna.

2. *Gain calculation for parabolic reflectors in satellite communication:* The gain of a parabolic reflector antenna is an essential parameter, providing insight into how effectively the antenna can transmit or receive signals. The gain can be calculated using the following formula:

$$G = \frac{4\pi A_e}{\lambda^2} \tag{14.10}$$

where G = gain (dimensionless), A_e = effective aperture area (m^2), and λ = wavelength of the operating frequency (m).

This equation highlights the direct relationship between aperture size and gain, illustrating the advantages of deployable antennas that can achieve larger effective areas in orbit.

3. *Radiation pattern calculation:* The radiation pattern of an antenna can be characterized using the following equation, which defines the power radiated per unit solid angle:

$$P(\theta,\varphi) = P_0 \cdot \frac{G(\theta,\vartheta)}{4\pi} \tag{14.11}$$

where $P(\theta, \phi)$ = power density in the direction of angles θ and ϕ (W/m^2), P_0 = total power transmitted (W), and $G(\theta, \phi)$ = gain as a function of angles (dimensionless).

This equation is fundamental in assessing the performance of deployable antennas, particularly when analyzing how effectively they can communicate with ground stations or other satellites.

4. *Thermal expansion effects:* Thermal expansion can significantly impact the performance of deployable antennas in space. The change in length (ΔL) due to temperature variations can be calculated using:

$$\Delta L = L_0 \cdot \alpha \cdot \Delta T \tag{14.12}$$

where ΔL = change in length (m), L_0 = original length (m), α = coefficient of linear expansion (°C^{-1}), and ΔT = change in temperature (°C).

Understanding thermal expansion is critical in the design of deployable antennas to ensure that they can withstand the temperature fluctuations experienced in space without compromising their structural integrity.

14.3.5 Future Trends in Deployable Antenna Technology

The landscape of deployable satellite antennas is continuously evolving, driven by advances in materials science, engineering, and space technology. The following trends are anticipated to shape the future of deployable antennas:

14.3.5.1 Integration with CubeSats and Small Satellites

As the demand for small satellite missions increases, deployable antennas designed for CubeSats are gaining attention. The compact nature of CubeSats presents unique challenges for antenna design, but advancements in deployable technology can enable these small satellites to achieve effective communication capabilities comparable to larger counterparts. Miniaturized deployable antennas are being developed to optimize space and enhance performance, allowing for the successful integration of multiple missions into a single satellite platform.

14.3.5.2 Smart Materials and Adaptive Systems

The use of smart materials, such as self-healing polymers and adaptive structures, promises to enhance the functionality of deployable antennas. These materials can respond to environmental changes, such as temperature variations or mechanical stresses, improving the resilience and longevity of antennas in space. Future deployable antennas may feature systems that adjust their shapes or configurations based on real-time data, maximizing performance in dynamic environments.

14.3.5.3 Collaborative and Swarm Missions

As satellite constellations and swarm missions become more prevalent, deployable antennas will play a crucial role in facilitating communication between multiple satellites. The development of antenna systems that can dynamically adjust their configurations for optimal communication links within constellations is an emerging area of research. These collaborative systems will enhance the efficiency and effectiveness of satellite networks, enabling seamless data transfer and improved coverage.

14.3.5.4 Advancements in Simulation and Modeling

The continued refinement of computational modeling and simulation tools will further enhance the design and optimization of deployable antennas. Advanced simulation techniques allow engineers to predict performance under various conditions, enabling the development of more efficient and reliable designs. This trend will lead to quicker iterations in the design process and the successful deployment of innovative antenna solutions in future satellite missions.

To summarize, deployable satellite antennas represent a transformative advancement in satellite communication technology. Their ability to provide enhanced performance, versatility, and cost-effectiveness makes them indispensable in modern space missions. As the demand for more sophisticated communication capabilities grows, ongoing research and development in deployable antenna technology will drive innovation, paving the way for new applications and expanded capabilities in the ever-evolving realm of space exploration and communication.

14.4 Applications of Deployable Satellite Antennas

Deployable satellite antennas are fundamental to modern satellite technology, offering a compact and efficient solution for achieving large aperture antennas in space-constrained environments. These antennas are critical for the success of various satellite missions in both commercial and defense sectors. By allowing satellites to launch with smaller stowed dimensions and expand to full operational size in orbit, deployable antennas improve performance capabilities across several applications. In this section, we delve into the vital roles of deployable antennas in communication satellites (COMSATs), Earth observation satellites, and military applications, providing a comprehensive analysis of their functions, benefits, and technological advancements.

14.4.1 COMSATs

Deployable antennas have become indispensable in communication satellites, particularly in GEO and LEO. As demand for high-speed data services continues to increase, the need for robust, high-gain antennas that can be launched compactly and expanded in orbit has risen.

The primary mission of the CubeSat is to provide access to space for small payloads. It provides a standard for the design of nanosatellites to reduce cost and development time, increase accessibility to space, and sustain frequent launches. Furthermore, the CubeSat Standard is being used to develop picosatellites by high schools, universities, government agencies, and private companies all over the world.

CubeSats, as small satellite platforms, often utilize deployable antennas to maximize their communication capabilities despite limited size (Figure 14.14). The ability of these antennas to unfold and extend in space allow CubeSats to achieve performance levels comparable to larger satellites while maintaining a compact launch profile. This design is particularly beneficial for missions requiring low-cost and rapidly deployable satellite solutions, such as disaster monitoring and scientific research.

14.4.1.1 Enhanced Coverage and Data Rates

In GEO, communication satellites equipped with deployable antennas provide reliable, high-quality data services to vast regions, including remote areas that lack terrestrial infrastructure. GEO satellites operate at approximately 35,786 km above the equator, remaining in a fixed position relative to the Earth. This altitude demands high-gain antennas to achieve effective data transmission across such long distances. Deployable antennas meet this requirement by providing a larger aperture in space, which improves the antenna's ability to focus and receive signals over extensive areas, effectively increasing data rates and coverage zones.

In LEO, where satellites orbit much closer to Earth (between 500 and 2,000 km), deployable antennas enable the miniaturization of satellite structures while

Figure 14.14 Illustration of a deployable antenna on a CubeSat. (Source: ESA–G. Porter. European Space Agency.)

ensuring sufficient antenna performance for high-speed data downlinks. LEO satellites are increasingly used in constellations such as Starlink, OneWeb, and Project Kuiper, due to their low latency and capacity to deliver broadband internet. These constellations rely on deployable antennas to maintain efficient data relay across the network, ensuring continuity in service as each satellite rapidly moves across the Earth's surface.

14.4.1.2 Compact Launch Configurations and Cost-Efficiency

The ability of deployable antennas to be compactly stowed and expanded once in orbit significantly reduces launch mass and volume. For communication satellites, this translates into reduced launch costs, as launch providers typically charge based on the size and weight of the payload. Deployable antennas enable the launch of larger, more capable antennas without occupying excessive payload space, facilitating cost-effective missions that meet stringent communication demands.

In addition, many deployable antennas are designed with modularity in mind, allowing for customization based on specific mission requirements. This flexibility enables satellite operators to choose antennas that best balance performance, size, and cost constraints, further enhancing mission adaptability and cost-effectiveness.

14.4.1.3 Applications in LEO and Mega-Constellations

Mega-constellations in LEO are designed to provide global coverage by deploying thousands of small satellites. Each satellite must be able to communicate effectively with ground stations and other satellites in the constellation. Deployable antennas play a crucial role here, as they allow each satellite to transmit high volumes of data at high speeds, enabling broadband internet and other communication services in underserved regions.

As these constellations expand, deployable antennas facilitate scalability, allowing each new satellite to seamlessly integrate into the network. They also enhance resilience, as the loss or degradation of one satellite can be compensated for by the surrounding network, maintaining consistent service coverage.

14.4.2 Earth Observation Satellites

Deployable antennas are equally valuable for Earth observation satellites, which require high-resolution data collection for a range of scientific and practical applications. These satellites are critical for climate monitoring, disaster management, and resource management, providing the data necessary for informed decision-making.

14.4.2.1 High-Resolution Data Collection

Deployable antennas on Earth observation satellites enable high data throughput, essential for transmitting the large volumes of information generated by onboard sensors back to Earth. For example, SAR and multispectral imaging sensors produce vast amounts of data that must be transmitted in near real time to support monitoring activities. Deployable antennas provide the high gain necessary for such high-resolution data transmission, enabling clear and detailed observations of the Earth's surface.

Applications range from tracking deforestation and urban expansion to mapping changes in polar ice caps and monitoring marine environments. By providing real-time data on ecological changes, these satellites support conservation efforts, resource management, and environmental policy development.

14.4.2.2 Adaptability in Response to Environmental Conditions

Earth observation satellites often need to adjust their data transmission rates to accommodate varying environmental conditions or mission demands. Deployable antennas enable this adaptability by offering a larger gain-to-size ratio compared to fixed antennas, making it easier to modulate transmission rates and ensure data is relayed efficiently even during periods of increased demand.

For instance, during natural disasters such as hurricanes or wildfires, satellites with deployable antennas can increase their data transmission rates to deliver vital information to emergency responders and government agencies. This flexibility is crucial in supporting rapid response efforts, allowing agencies to better coordinate disaster relief, assess damage, and allocate resources where they are most needed.

14.4.2.3 Resource and Ecosystem Monitoring

Deployable antennas also support satellites tasked with monitoring natural resources and ecosystems. These satellites are used extensively in agriculture, forestry, and coastal management, providing data on soil health, crop yields, forest density, and coastline changes. With a deployable antenna, an Earth observation satellite can transmit detailed imagery and data with minimal delay, supporting sustainable practices and informed resource management.

For example, SAR-equipped satellites with deployable antennas can detect changes in biomass, soil moisture, and land deformation. Such data is invaluable for monitoring environmental health, tracking agricultural productivity, and predicting natural disasters.

14.4.3 Military Applications

Deployable antennas are essential for military satellites, providing secure and reliable communication links, surveillance capabilities, and intelligence-gathering support. The defense sector relies heavily on these antennas for their versatility, durability, and ability to enhance situational awareness across diverse operational environments.

14.4.3.1 Surveillance and Reconnaissance

Military satellites equipped with deployable antennas play a vital role in surveillance and reconnaissance, gathering intelligence on potential threats and monitoring conflict zones. High-resolution imaging, radar, and signal interception capabilities allow these satellites to capture detailed information on enemy movements, troop deployments, and infrastructure. Deployable antennas support the real-time transmission of this data to ground stations, enabling timely decision-making and strategic planning.

For instance, surveillance satellites in low and medium Earth orbits can benefit from deployable antennas with synthetic aperture radar, which provides high-resolution images even in adverse weather conditions. This capability is critical for monitoring adversarial activities and assessing potential security risks, offering a significant advantage in the domain of space-based intelligence.

14.4.3.2 Secure and Resilient Communications

Military communications depend on secure and resilient links to maintain operational effectiveness. Deployable antennas provide high-gain, narrow-beam transmissions that reduce the risk of interception and interference. These antennas also enhance the satellite's ability to communicate with ground stations and other assets in challenging environments, including mountainous or densely forested areas.

The modularity of deployable antennas allows for rapid adjustments in response to mission needs, enhancing security by enabling frequent reconfiguration of communication links. This flexibility supports covert operations, as military assets can be quickly reoriented to maintain secure, reliable links with command centers.

14.4.3.3 Covert Operations and Signal Intelligence

Deployable antennas support signal intelligence (SIGINT) and electronic intelligence (ELINT) operations by intercepting and analyzing electronic signals from potential adversaries. These operations are essential for gathering intelligence on enemy radar and communication systems, identifying patterns, and assessing technological capabilities.

In covert operations, deployable antennas enhance the satellite's ability to collect data without detection, minimizing the chance of compromise. Their ability to switch frequencies and modulation techniques makes it harder for adversaries to detect and jam signals, increasing the effectiveness of intelligence-gathering efforts. Additionally, deployable antennas can be stowed or deployed as needed, reducing the likelihood of detection during sensitive missions.

Table 14.2 illustrates how deployable antennas outperform fixed antennas across key metrics such as gain, coverage area, and data transmission rate. The chart compares parameters such as stowed volume, weight, and overall mission cost-effectiveness, demonstrating the practical benefits of deployable technology.

To summarize, deployable satellite antennas are revolutionizing the way that satellites operate by providing versatile, high-performance solutions across communication, Earth observation, and military applications. Their ability to combine compact launch configurations with extensive in-orbit capabilities makes them a preferred choice for modern space missions. As satellite networks grow in scale and complexity, the role of deployable antennas will continue to expand, driving innovations in design and materials to meet the evolving demands of both commercial and defense sectors. This adaptability ensures that deployable antennas will remain central to the future of satellite communications, Earth monitoring, and secure military operations.

Table 14.2 Performance Comparison Between Fixed and Deployable Satellite Antennas

Performance Metric	*Fixed Satellite Antennas*	*Deployable Satellite Antennas*
Gain	20–30 dBi	30–50 dBi
Coverage area	Up to 500 km^2 (depends on frequency and design)	Adjustable; can cover multiple regions simultaneously
Data transmission rate	Up to 1 Gbps	Up to 10 Gbps
Stowed volume	1–3 m^3 (varies by size)	0.1–1 m^3 (compact when stowed)
Weight	100–500 kg (varies by design)	10–200 kg (lighter due to materials used)
Deployment complexity	Low (fixed installation)	Moderate to high (mechanisms required for deployment)
Overall mission cost-effectiveness	Low initial cost, limited capabilities (e.g., approximately $1 million to $5 million per unit)	Higher initial cost (approximately $5 million to $20 million per unit), greater flexibility and capability
Operational flexibility	Fixed, less adaptable	Highly adaptable to mission needs
Lifespan	10–15 years (varies by environmental conditions)	10–20 years (with potential for enhancement via upgrades)

14.5 Conclusion and Future Directions

14.5.1 Key Takeaways

As we conclude this chapter on advanced antenna technologies, several pivotal themes and insights emerge that underscore the transformative evolution-taking place in this field. The rapid advancements in antenna design, driven by the increasing demands of modern communication systems, are reshaping our understanding of electromagnetic performance, operational strategies, and deployment methodologies.

One of the most significant advancements is the introduction of RIS. This innovative technology signifies a paradigm shift in how antennas can be utilized to enhance wireless communication. By enabling dynamic control over the wireless environment, RIS can significantly enhance signal strength and improve overall communication efficiency, especially in densely populated urban settings. The adaptability of RIS is particularly valuable in scenarios where conventional signal propagation techniques face limitations. These surfaces can manipulate the propagation of electromagnetic waves by reflecting, refracting, or absorbing signals, allowing for optimized channel conditions and a reduction in interference. This adaptability not only enhances user experience but also paves the way for more efficient spectrum usage, which is becoming increasingly crucial as the demand for bandwidth escalates.

Furthermore, the integration of metamaterials into antenna design has emerged as a game changer. Metamaterials exhibit unique electromagnetic properties that are not found in naturally occurring materials. These properties include negative refractive indices and the ability to manipulate electromagnetic waves in novel ways. The incorporation of metamaterials in antenna design allows for the development of compact, high-performance antennas that can meet the rigorous demands of high-frequency communications, including those required for 5G and future generations of wireless technology. The ability to create antennas with enhanced bandwidth and directivity while maintaining a small form factor is particularly valuable in applications ranging from mobile communications to satellite systems.

In addition to RIS and metamaterials, advancements in deployable satellite antennas are redefining the landscape of satellite communication. The increasing deployment of mega-constellations—networks of satellites that work together to provide global coverage—has heightened the need for innovative antenna solutions that can ensure reliable connectivity. Deployable satellite antennas allow for rapid deployment and reconfiguration, which is crucial in responding to the dynamic requirements of communication networks. These antennas are designed to be lightweight and compact, making them ideal for small satellites and enabling them to function efficiently even in the challenging conditions of space.

Overall, the key takeaways from this chapter highlight that the future of antenna technology is being shaped by a convergence of innovative materials, advanced design techniques, and the integration of new functionalities that enhance performance and adaptability. This evolution not only addresses current communication challenges but also anticipates future needs, ensuring that antenna technologies remain at the forefront of wireless communication advancements.

14.5.2 Future Trends

Looking ahead, several exciting trends are set to shape the future of antenna technology. One prominent trend is the integration of ML and AI into antenna design and operation. The adoption of ML and AI holds significant potential for optimizing antenna performance in real time. Smart antennas equipped with advanced algorithms can analyze and respond to changing environmental conditions, user demands, and interference patterns. This capability allows for dynamic adjustments in antenna configurations, improving throughput and reliability in varying conditions. The application of these technologies not only enhances user experience but also contributes to more efficient resource utilization, which is increasingly important in the context of limited spectrum availability.

In parallel, there is a growing emphasis on sustainable antenna technologies. As environmental concerns gain prominence across industries, the development of eco-friendly antenna solutions is becoming a priority. Researchers are exploring biodegradable materials and energy-efficient designs that minimize the environmental footprint of antenna production and deployment. Innovations in this area not only will contribute to sustainability but will also enhance the adaptability of antennas in various applications, from urban environments to remote sensing scenarios. As society continues to prioritize sustainable practices, antenna technologies must evolve to meet these expectations, ensuring a balance between performance and environmental responsibility.

Another critical trend is the exploration of terahertz frequencies for communication applications. Terahertz communication is emerging as a promising frontier, particularly for high-data-rate applications in wireless networks. Operating at frequencies between microwave and infrared, terahertz communication has the potential to achieve data rates that far exceed those of current technologies. However, developing antennas that operate efficiently in this frequency range presents unique challenges, including propagation losses and the need for precise material properties. Ongoing research in this area aims to develop antennas capable of harnessing the benefits of terahertz frequencies, unlocking new possibilities for wireless communication.

The deployment of heterogeneous networks is also a significant trend that will shape the future of antenna technology. Heterogeneous networks combine various types of access nodes—such as small cells, macro cells, and satellite systems—into a cohesive network designed to provide seamless connectivity. This integration requires sophisticated antenna designs that ensure efficient spectrum utilization and robust performance across different nodes. Future research will focus on developing multifunction antennas that can operate across diverse frequency bands, optimizing performance in these complex network architectures. This approach not only enhances coverage and capacity but also addresses the challenges posed by the increasing number of devices connected to the network.

Finally, the rapid growth of the IoT presents both challenges and opportunities for antenna design. As IoT devices proliferate across various sectors, the demand for antennas that can handle diverse connectivity requirements is increasing. Future antenna designs must focus on low-power, compact solutions capable of supporting a multitude of devices across applications, from smart cities and healthcare to industrial automation. The ability to develop antennas that are energy-efficient and can operate effectively in diverse environments will be crucial for facilitating the widespread adoption of IoT technologies.

14.5.3 Conclusion

The landscape of antenna technology is poised for dramatic evolution, driven by advancements in materials, design philosophies, and integration with emerging technologies. As we embrace these innovations, the future of communication systems will hinge on our ability to adapt and innovate in antenna design. The convergence of RIS, metamaterials, deployable technologies, and intelligent algorithms will not only enhance the performance of existing systems but also pave the way for new applications and capabilities in an increasingly connected world.

As we stand on the brink of these advancements, it is imperative that researchers, practitioners, and industry leaders collaborate to explore the frontiers of antenna technology. The integration of interdisciplinary approaches, combining insights from physics, engineering, and computer science, will be essential for addressing the complexities of next-generation communication systems. The journey ahead promises to be exciting, filled with opportunities to leverage the latest technological advancements to improve connectivity, enhance user experiences, and ultimately transform the way that we communicate.

In conclusion, the evolution of antenna technology reflects the dynamic nature of our communication landscape. By remaining attuned to emerging trends and prioritizing innovation, we can ensure that antenna systems continue to meet the ever-evolving demands of society. The future is bright for antenna technology, with the potential to drive significant advancements in global communication networks, thereby enhancing connectivity and fostering innovation in an increasingly interconnected world.

Problems for Students

14.1 (a) Explain the key differences between 4G and 5G communication systems. How has the evolution from 4G to 5G impacted antenna design requirements?
(b) Discuss the role of millimeter-wave frequencies in 5G systems. What challenges do these frequencies pose for antenna design?

14.2 (a) Calculate the required antenna array gain for a 5G base station designed to support a 1-km cell radius with a minimum SNR of 20 dB.
(b) Discuss the challenges in designing small-form-factor antennas for dense urban environments, where multiple antennas are deployed in close proximity.

14.3 (a) Define antenna diversity in the context of 5G networks. How can MIMO technology improve communication reliability in 5G systems?
(b) Design a simple 2 × 2 MIMO system for a 5G base station and calculate the expected capacity enhancement compared to a single-antenna system.

14.4 (a) Explain the concept of beamforming and how it is utilized in 5G networks to enhance signal quality and coverage.
(b) Using an array of 8 elements in a uniform linear configuration, calculate the beamforming gain at an angle of 30°. Assume that the array is working at 28 GHz.

14.5 (a) What is massive MIMO, and how does it address the growing demand for data rates in 5G systems?
(b) Calculate the increase in capacity when a 5G base station with 64 antennas (massive MIMO) operates at a frequency of 3.5 GHz, compared to a system with a single antenna.

14.6 (a) Explain the concept of deployable antennas in satellite communication. How do deployable antennas contribute to the efficiency of satellite systems?
(b) Design a deployable satellite antenna system that can support multiple frequency bands (e.g., L-band, C-band, and Ku-band). Consider the size and material constraints of the deployable structure.

14.7 (a) Discuss the role of deployable satellite antennas in LEO satellite constellations for global broadband communication.
(b) Calculate the aperture efficiency of a deployable satellite antenna with a diameter of 3m operating at a frequency of 14 GHz. Assume an efficiency factor of 0.65.

14.8 (a) Explain the importance of antenna pointing accuracy in satellite communication systems, particularly in LEO and geostationary satellite systems.
(b) Calculate the required pointing accuracy for a satellite antenna in an LEO system to maintain continuous communication with a ground station located 1,000 km away.

14.9 (a) Derive the link budget for a satellite communication system using a deployable antenna. Assume that a satellite transmits at 30W and the ground station has a receiver gain of 50 dB.
(b) Given that the satellite antenna has a 3-m diameter and the frequency is 12 GHz, calculate the received power at the ground station 1,500 km away.

14.10 (a) Discuss the emerging antenna technologies that are expected to drive the evolution of 5G networks in the coming decade, including reconfigurable and intelligent surfaces.

(b) Based on the current trends in antenna miniaturization and energy efficiency, predict how the integration of 5G and satellite communication could evolve to support IoT applications.

Problems for Instructors

14.1 Antenna Design Requirements for 5G: Discuss the key differences in antenna design requirements for sub-6-GHz 5G systems versus millimeter-wave 5G systems. Include considerations such as size, beamforming capabilities, and propagation characteristics. Propose a design approach for a 5G millimeter-wave antenna array suitable for use in urban areas.

14.2 Beamforming and Massive MIMO in 5G Networks: Using the concept of beamforming, design a 4×4 massive MIMO antenna array for a 5G base station operating at 3.5 GHz. Determine the interelement spacing, radiation pattern, and beam-steering angles to achieve optimal coverage in a 120° azimuth sector.

14.3 Link Budget Analysis for 5G and Satellite Communication: Perform a link budget analysis for a 5G communication link operating at 28 GHz with a base station antenna gain of 15 dBi and a user equipment gain of 5 dBi. Compare this with a satellite link operating at 12 GHz with a parabolic reflector gain of 35 dBi and a receiver antenna gain of 25 dBi. Assume similar path losses and system parameters, and explain the observed differences in received power.

14.4 Deployable Satellite Antenna Design Considerations: Propose a deployable satellite antenna design for a small satellite mission aimed at providing broadband coverage. Discuss the trade-offs between ability of mechanical deployment, gain, and structural rigidity. Calculate the aperture size required for a gain of 40 dBi at 14 GHz.

14.5 Applications of Deployable Antennas: Identify three real-world applications of deployable satellite antennas in modern communication systems. For each application, describe how the design features of deployable antennas address specific mission requirements such as frequency range, coverage, or mobility.

14.6 Future Trends in Antenna Technology for 5G and Satellites: Critically analyze how emerging technologies such as RIS or meta-surfaces can transform antenna design for 5G and satellite systems. Provide examples of their potential impact on reducing cost, enhancing coverage, or improving spectral efficiency.

14.7 Integration of 5G and Satellite Communication: Design a hybrid communication system integrating 5G terrestrial networks with satellite communication. Discuss how antennas can be optimized to support dual functionality in such a system, addressing challenges such as operating in multiple frequency bands and maintaining high performance under different propagation conditions.

Selected Bibliography

Abulgasem, S., et al., "Antenna Designs for CubeSats: A Review," *IEEE Access*, Vol. 9, 2021, pp. 45289–45324.

Ahmed, I., et al., "A Survey on Hybrid Beamforming Techniques in 5G: Architecture and System Model Perspectives," *IEEE Communications Surveys & Tutorials*, Vol. 20, No. 4, 2018, pp. 3060–3097.

Alkhateeb, A., et al., "Channel Estimation and Hybrid Precoding for Millimeter Wave Cellular Systems," *IEEE Journal of Selected Topics in Signal Processing*, Vol. 8, No. 5, 2014, pp. 831–846.

Björnson, E., L. Sanguinetti, and M. Debbah, "Massive MIMO with Imperfect Channel Covariance Information," *2016 50th IEEE Asilomar Conference on Signals, Systems and Computers*, November 2016, pp. 974–978.

Brodecki, M., and Z. D. Groot, "SIGINT: The Mission Cubesats Are Made For," *32 Annual AIAA/USU Conference on Small Satellites*, 2018.

Du, H., et al., "Performance and Optimization of Reconfigurable Intelligent Surface Aided THz Communications," *IEEE Transactions on Communications*, Vol. 70, No. 5, 2022, pp. 3575–3593.

Fakoukakis, F. E., et al., "Development of an Adaptive and a Switched Beam Smart Antenna System for Wireless Communications," *Journal of Electromagnetic Waves and Applications*, Vol. 20, No. 3, 2006, pp. 399–408.

Freeland, R. E., et al., "Large Inflatable Deployable Antenna Flight Experiment Results," *Acta Astronautica*, Vol. 41, No. 4-10, 1997, pp. 267–277.

Gaspar, J., et al., "Structural Test and Analysis of a Hybrid Inflatable Antenna," *48th AIAA/ASME/ASCE/AHS/ASC Structures, Structural Dynamics, and Materials Conference*, January 2007, p. 1832.

Hartl, D. J., and D. C. Lagoudas, "Aerospace Applications of Shape Memory Alloys," *Proceedings of the Institution of Mechanical Engineers, Part G: Journal of Aerospace Engineering*, Vol. 221, No. 4, 2007, pp. 535–552.

Im, C., et al., "Design of a Deployable Broadband Mesh Reflector Antenna for a SIGINT Satellite System Considering Surface Shape Deformation," *Sensors*, Vol. 24, No. 2, 2024, p. 384.

Ingerson, P., and W. Wong, "The Analysis of Deployable Umbrella Parabolic Reflectors," *IEEE Transactions on Antennas and Propagation*, Vol. 20, No. 4, 1972, pp. 409–414.

Kodheli, O., et al., "Satellite Communications in the New Space Era: A Survey and Future Challenges," *IEEE Communications Surveys & Tutorials*, Vol. 23, No. 1, 2020, pp. 70–109.

Lu, Y., M. Hao, and R. Mackenzie, "Reconfigurable Intelligent Surface Based Hybrid Precoding for THz Communications," *Intelligent and Converged Networks*, Vol. 3, No. 1, 2022, pp. 103–118.

Mailloux, R. J., *Phased Array Antenna Handbook*, Norwood, MA: Artech House, 2017.

Marzetta, T. L., "Noncooperative Cellular Wireless with Unlimited Numbers of Base Station Antennas," *IEEE Transactions on Wireless Communications*, Vol. 9, No. 11, 2010, pp. 3590–3600.

Moraitis, N., and P. Constantinou, "Indoor Channel Measurements and Characterization at 60 GHz for Wireless Local Area Network Applications," *IEEE Transactions on Antennas and Propagation*, Vol. 52, No. 12, 2004, pp. 3180–3189.

Qualcomm, "Global 5G Spectrum Status and Innovations for Future Wireless Systems," May 31, 2023, https://www.qualcomm.com/content/dam/qcomm-martech/dm-assets/documents/global-5g-spectrum-status-and-innovations-for-future-wireless-systems.pdf.

Rao, G. A., and S. P. Mahulikar, "Integrated Review of Stealth Technology and Its Role in Airpower," *The Aeronautical Journal*, Vol. 106, No. 1066, 2002, pp. 629–642.

Rappaport, T. S., et al., "Wideband Millimeter-Wave Propagation Measurements and Channel Models for Future Wireless Communication System Design," *IEEE Transactions on Communications*, Vol. 63, No. 9, 2015, pp. 3029–3056.

Shah, B. M., M. Murtaza, and M. Raza, "Comparison of 4G and 5G Cellular Network Architecture and Proposing of 6G, a New Era of AI," *2020 5th IEEE International Conference on Innovative Technologies in Intelligent Systems and Industrial Applications (CITISIA)*, November 2020, pp. 1–10.

Sonawane, N. S., and J. Thangavelautham, "Precision Pointing of Antennas in Space Using

Arrays of Shape Memory Alloy Based Linear Actuators," 2017, *arXiv preprint arXiv:1701.07561*.

Tan, L., D. Wang, and K. D. Xu, "Terahertz Metamaterials for Spectrum Modulation: Structural Design, Materials and Applications," *Materials & Design*, Vol. 113217, 2024.

Xu, W., L. Xie, and Y. Ying, "Mechanisms and Applications of Terahertz Metamaterial Sensing: A Review," *Nanoscale*, Vol. 9, No. 37, 2017, pp. 13864–13878.

Zohuri, B., *Radar Energy Warfare and the Challenges of Stealth Technology*, Berlin: Springer, 2020, pp. 1–110.

APPENDIX 14A: MATLAB Codes for Satellite Link Budget Calculation, Beam Coverage Simulation, and Ionospheric Absorption Modeling

MATLAB Program 32 calculates the power budget for a satellite link and displays the results of the received power, noise power, and signal-to-noise ratio.

MATLAB Program 32

```
% MATLAB Code for Satellite Link Power Budget Calculation.
% This script calculates the power budget for a satellite
communication link.
% It aims to compute the following key parameters:
% (1) Received Power (Pr): The power received at the receiver,
calculated in both dBm and Watts, taking into account the
transmitter power, gains, and losses.
% (2) Noise Power (Pn): The thermal noise power at the receiver,
calculated in dBm, considering the bandwidth and noise figure.
% (3) Signal-to-Noise Ratio (SNR): The ratio of the received
signal power to the noise power, expressed in dB, which indicates
the quality of the communication link.

% Define Constants
Pt_dBm = 20;            % Transmitter power (dBm)
Gt_dB = 10;            % Transmitter gain (dB)
Gr_dB = 7;             % Receiver gain (dB) - adjusted down to
lower SNR
L_dB = 120;             % Free space loss (dB)
L_other_dB = 3;          % Other losses (e.g., cable, connectors)
(dB)
NF_dB = 12;              % Noise Figure (dB) - increased to
reduce SNR
k = 1.38e-23;           % Boltzmann's constant (J/K)
T0 = 290;               % Standard temperature (K)

% Convert dBm to Watts
Pt_W = 10^((Pt_dBm - 30) / 10); % Transmitter power in Watts

% Calculate Received Power in dBm
Pr_dBm = Pt_dBm + Gt_dB + Gr_dB - L_dB - L_other_dB;
```

```
% Calculate Received Power in Watts
Pr_W = 10^((Pr_dBm - 30) / 10); % Received power in Watts

% Calculate Noise Power (Pn) in dBm
B = 1e6;              % Bandwidth (Hz) - 1 MHz
Pn_dBm = 10*log10(k * T0 * B) + 30 + NF_dB; % Noise Power in dBm

% Calculate Signal-to-Noise Ratio (SNR) in dB
SNR_dB = Pr_dBm - Pn_dBm;

% Display Results
fprintf('Transmitter Power (dBm): %.2f\n', Pt_dBm);
fprintf('Transmitter Gain (dB): %.2f\n', Gt_dB);
fprintf('Receiver Gain (dB): %.2f\n', Gr_dB);
fprintf('Free Space Loss (dB): %.2f\n', L_dB);
fprintf('Other Losses (dB): %.2f\n', L_other_dB);
fprintf('Received Power (dBm): %.2f\n', Pr_dBm);
fprintf('Noise Power (dBm): %.2f\n', Pn_dBm);
fprintf('Signal-to-Noise Ratio (SNR) (dB): %.2f\n', SNR_dB);
```

Display Results

Received Power (dBm): -86.00
Noise Power (dBm): -101.98
Signal-to-Noise Ratio (SNR) (dB): 15.98

MATLAB Program 33 calculates and plots the satellite antenna beam coverage patterns of a parabolic dish for various diameters with f/D=0.5, as shown in Figure 14A.1.

MATLAB Program 33

```
%  MATLAB Code for Simulating Satellite Antenna Beam Coverage.
This script calculates and plots the beam coverage patterns of a
parabolic dish with a fixed f/D ratio of 0.5 and diameters: 0.5,
1 and 2 meters at 8 GHz and 12 GHz.

% Clear previous data and figures
clear; clc; close all;

% Parameters
frequencies = [8e9, 12e9]; % Operating frequencies in Hz (e.g.,
X-band, Ku-band)
c = 3e8;              % Speed of light in m/s
f_D_ratio = 0.5;         % Fixed f/D ratio for this simulation

% Define dish diameters, line styles, and markers for
higher frequency
diameters = [0.5, 1, 2]; % Dish diameters in meters
```

```
line_styles = {'-', '--', ':'}; % Line styles for each diameter
markers = {'o', 's', 'd'};     % Marker styles for higher
frequency (circle, square, diamond)

% Beamwidth Calculation and Plotting
figure;
hold on;
for f = 1:length(frequencies)
  wavelength = c / frequencies(f); % Calculate wavelength
  for d = 1:length(diameters)
    D = diameters(d);

    % Half-power beamwidth (HPBW) approximation for a
parabolic dish
    HPBW = 70 * wavelength / D; % HPBW in degrees

    % Plotting beam pattern
    theta = linspace(-HPBW, HPBW, 100); % Angular range
    beam_pattern = cosd(theta / HPBW * 90); % Simple cosine
beam pattern

    % Plotting beam pattern
    % Create the plot for the main line
    plot(theta, beam_pattern, 'k', 'LineStyle', line_styles{d},
...
      'DisplayName', sprintf('Freq: %.1f GHz, D: %.1f m, f/D:
%.2f, HPBW: %.2f°', ...
      frequencies(f)/1e9, D, f_D_ratio, HPBW));

    % If it's the higher frequency, add markers selectively
    if f == 2
      % Add markers at specified intervals to get 11 markers
      marker_indices = round(linspace(1, length(theta), 11)); %
11 markers
      plot(theta(marker_indices), beam_pattern(marker_indices),
'k', ...
        'Marker', markers{d}, 'LineStyle', 'none'); % No line,
just markers
    end
  end
end

% Customize plot
title('Satellite Antenna Beam Coverage Patterns for Various
Diameters with f/D=0.5');
xlabel('Angle (degrees)');
ylabel('Normalized Beam Intensity');
legend show;
grid on;
hold off;
```

MATLAB Program 34 models ionospheric absorption loss (in dB/km) versus frequency (in MHz) and altitude (in km) for daytime and nighttime ionospheric conditions, as shown in Figure 14A.2.

MATLAB Program 34

```
% IONOSPHERIC ABSORPTION MODELING
%
% Purpose:
%  This MATLAB program models ionospheric absorption of high-
frequency (HF)
%  radio waves (1-30 MHz) due to collisions between electrons
and neutral
%  particles, primarily in the D-layer (60-120 km altitude).
%
%  It computes and visualizes the absorption loss (in dB/km) as
a function
%  of frequency and altitude for both daytime and
nighttime ionospheric
%  conditions using a Chapman-like profile for electron density.
%
% Motivation:
%  - Complements Faraday rotation and effective dielectric
constant analysis.
%  - Demonstrates real-world attenuation effects for
HF communications.
%  - Useful for understanding day vs. night radio
wave propagation.
```

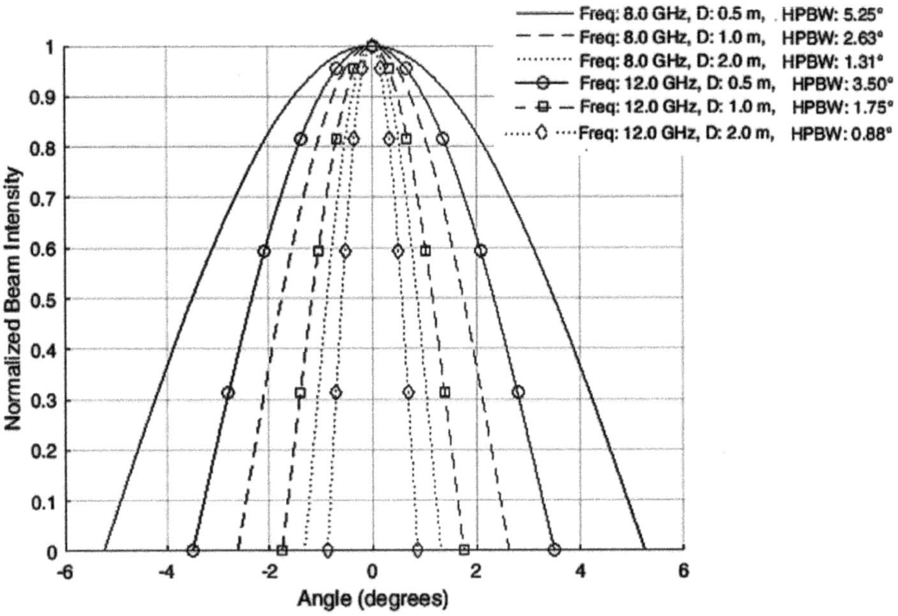

Figure 14A.1 Satellite Antenna Beam Coverage Patterns for Various Diameters with $f/D = 0.5$.

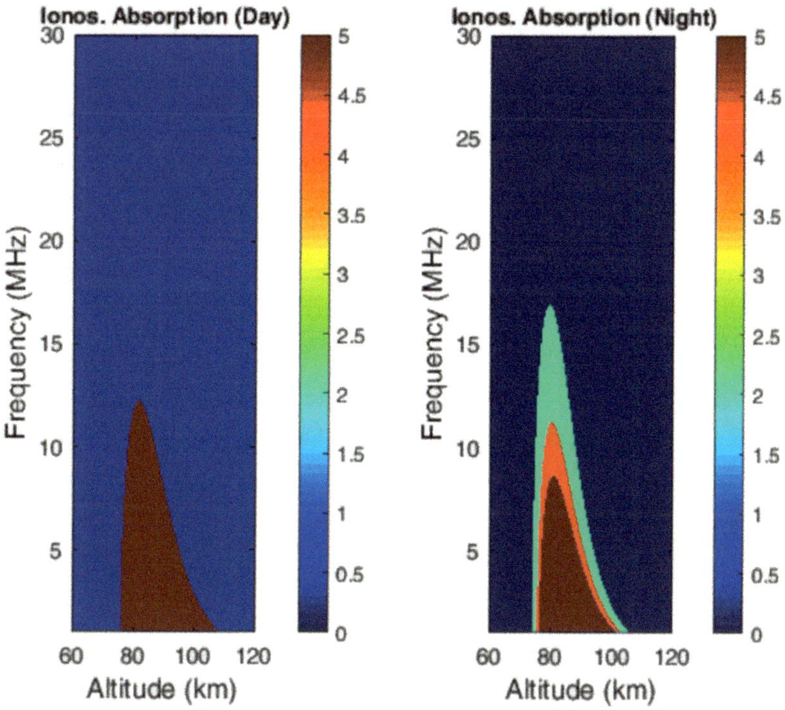

Figure 14A.2 Ionospheric absorption loss (dB/km) versus frequency (MHz) and altitude (km).

```
%
% Outputs:
%   - Two contour plots:
%     1. Daytime absorption (dB/km) vs. frequency and altitude
%     2. Nighttime absorption (dB/km) vs. frequency and altitude
%
%-------------------------------------------------------------------

clc; clear; close all;

% ------------------ CONSTANTS ------------------
e = 1.602e-19;    % Elementary charge (C)
eps0 = 8.854e-12; % Vacuum permittivity (F/m)
me = 9.109e-31;    % Electron mass (kg)
c = 3e8;          % Speed of light (m/s)

% ------------------ GRID SETUP ------------------
h = linspace(60e3, 120e3, 200); % Altitude (m)
f = linspace(1e6, 30e6, 200); % Frequency (Hz)
[H, F] = meshgrid(h, f);      % 2D grid for altitude and frequency

% ------------------ IONOSPHERIC PROFILES ------------------

% Daytime parameters
Nm_day = 1e11; hm_day = 90e3; H_day = 10e3;
```

```
Ne_day = Nm_day .* exp(1 - (H - hm_day)./H_day - exp(-(H
- hm_day)./H_day));

% Nighttime parameters
Nm_night = 1e10; hm_night = 85e3; H_night = 8e3;
Ne_night = Nm_night .* exp(1 - (H - hm_night)./H_night - exp(-(H
- hm_night)./H_night));

% Collision frequency profile ν(h)
nu = 1e9 .* exp(-(H - 60e3)./6000);  % Hz

% ------------------ ABSORPTION CALCULATION ------------------

% Plasma frequency ω_p = sqrt(Ne * e^2 / (me * eps0))
wp_day = sqrt(Ne_day .* e.^2 ./ (me .* eps0));      % Day
wp_night = sqrt(Ne_night .* e.^2 ./ (me .* eps0));  % Night

% Angular wave frequency ω = 2πf
w = 2 * pi * F;

% Absorption coefficient α ≈ (ω_p^2 * ν) / (c * (ω^2 + ν^2)) in
Np/m
alpha_day = (wp_day.^2 .* nu) ./ (c * (w.^2 + nu.^2));
alpha_night = (wp_night.^2 .* nu) ./ (c * (w.^2 + nu.^2));

% Convert to dB/km: (1 Np = 8.686 dB), and m to km
alpha_day_db = alpha_day * 8.686 * 1e3;
alpha_night_db = alpha_night * 8.686 * 1e3;

% ------------------ PLOTTING ------------------

figure;

% Daytime absorption
subplot(1,2,1);
contourf(h/1e3, f/1e6, alpha_day_db, 30, 'LineColor', 'none');
xlabel('Altitude (km)', 'FontSize', 12);
ylabel('Frequency (MHz)', 'FontSize', 12);
title('Ionos. Absorption (Day)', 'FontSize', 10);
colorbar; colormap(jet);
caxis([0 5]);  % Set color scale range

% Nighttime absorption
subplot(1,2,2);
contourf(h/1e3, f/1e6, alpha_night_db, 30, 'LineColor', 'none');
xlabel('Altitude (km)', 'FontSize', 12);
ylabel('Frequency (MHz)', 'FontSize', 12);
title('Ionos. Absorption (Night)', 'FontSize', 10);
colorbar; colormap(jet);
caxis([0 5]);  % Keep consistent color range
```

CHAPTER 15

Advanced Antenna Applications and Simulations Using MATLAB

This chapter presents a unified exploration of several emerging and application-driven areas in antenna and electromagnetic engineering. Specifically, it focuses on reconfigurable intelligent surfaces (RIS), reflect arrays, biomedical antennas, and wave propagation in lossy media, each representing a frontier in modern wireless communication, sensing, or healthcare systems.

While these topics may appear diverse in their end-use applications, they share a common foundation in electromagnetic wave manipulation, antenna design principles, and the need for efficient simulation tools to model performance in complex environments. By grouping these topics together, this chapter aims to offer a comprehensive yet cohesive framework for understanding how advanced antennas can be designed, optimized, and deployed across both terrestrial and biomedical domains.

A distinguishing feature of this chapter is its consistent use of MATLAB-based simulations to reinforce theoretical concepts. Each section integrates customized code examples, allowing readers to visualize radiation patterns, beam steering, field attenuation, or safety metrics such as specific absorption rate (SAR). These simulations serve not only to validate analytical models but also to equip the reader with hands-on tools for real-world design and analysis.

Through a blend of foundational theory, application-specific design insights, and computational modeling, this chapter is intended to support both academic learning and practical engineering development. Whether addressing the programmable surfaces of future 6G networks or the miniaturized antennas inside the human body, the emphasis remains on cultivating simulation-based intuition and system-level understanding.

15.1 RIS for 6G Networks

With the unprecedented growth in connected devices and high-demand applications, the need for advanced wireless capabilities has led to research on sixth-generation (6G) networks. Unlike the fifth-generation (5G) networks, which rely heavily on high-power transmission technologies like millimeter-wave communications and massive MIMO arrays, 6G introduces a novel approach called Smart Radio Environments (SRE). Central to SREs are RIS, also referred to as intelligent reflecting surfaces (IRS). RIS technology transforms passive surfaces into programmable metasurfaces that can adjust signal pathways dynamically and autonomously, creating efficient, enhanced connectivity with minimal power.

RIS leverages its passive structure to reflect and direct wireless signals through phase adjustments, offering several advantages, such as mitigating signal blockages, reducing latency, and increasing energy efficiency. This chapter provides an in-depth look at the fundamentals, control mechanisms, deployment strategies, emerging applications, channel acquisition methods, and AI-powered optimization techniques relevant to RIS technology, with a focus on its integration into the 6G landscape.

Therefore, RIS offers a complementary solution to the high-power, infrastructure-heavy paradigm of 5G satellite and terrestrial networks discussed in Chapter 14. While satellite antennas extend global coverage, RIS augments the quality of service in complex urban and indoor environments through passive beam control. This convergence of passive reflectors and active antenna systems is foundational to the intelligent connectivity that 6G envisions.

15.1.1 Fundamentals of RIS

Figure 15.1 illustrates depicting an RIS with downlink and uplink channels in a mobile network.

15.1.1.1 RIS Structure and Operation

RIS is typically composed of three layers:

- *Reflective element layer:* This top layer consists of numerous small, programmable elements that can adjust the phase of incoming signals.

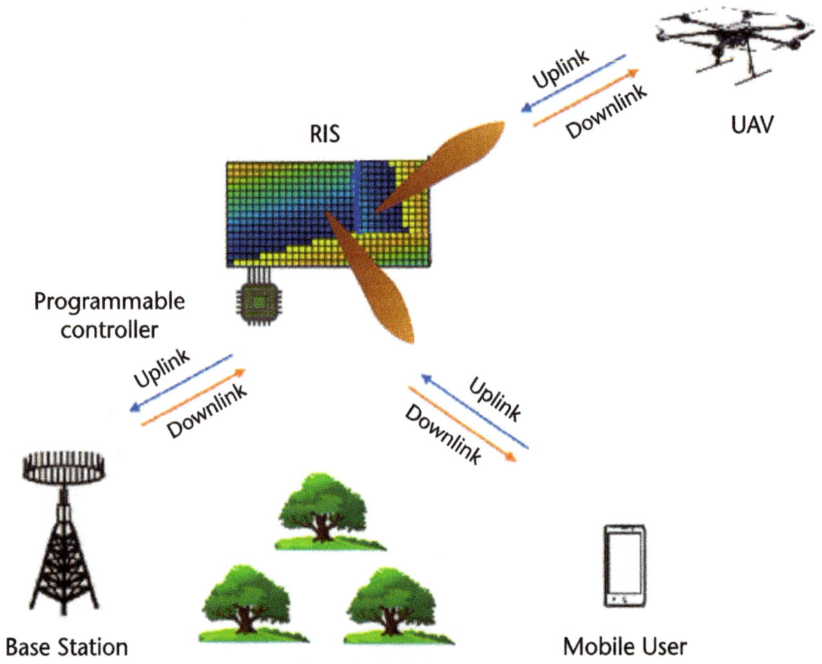

Figure 15.1 An illustration depicting an RIS with downlink and uplink channels in a mobile network.

- *Copper plate layer:* Positioned below the reflective elements, this layer prevents signal leakage and ensures that energy is directed efficiently.
- *Control circuit layer:* This layer includes a control unit, often a microcontroller or field-programmable gate array (FPGA), which governs the phase shifts based on real-time channel state information, allowing precise signal manipulation.

15.1.1.2 Phase Shift and Signal Propagation

The total reflected field $E_{\text{reflected}}$ from an RIS with N unit elements is modeled by the superposition principle:

$$E_{\text{reflected}} = \sum_{n=1}^{N} A_n \cdot e^{j(\varphi_{In}+\theta_n)} \tag{15.1}$$

where A_n is the amplitude, ϕ_n is the incident signal phase, and θ_n is the RIS-induced phase shift for the nth element. This model allows simulation of directional reflection and beamforming, depending on the desired phase profile.

MATLAB Program 35

```
% MATLAB Code: Simulation of Single RIS Element Response to an
Incident Signal
% Description: This code simulates the response of a single RIS
element to an incident signal
% with user-defined parameters.
% Clear workspace and command window
clc; clear;
% Parameters
disp('Enter the following parameters for the simulation:');
A = input('Amplitude of the incident signal (A): ');  %
Initial amplitude
phi = input('Phase of the incident signal in radians (phi): '); %
Initial phase
theta = input('Phase shift introduced by the RIS in radians
(theta): '); % Phase shift
% Compute Reflected Signal
A_reflected = A * exp(1j * (phi + theta));
% Extract amplitude and phase of the reflected signal
amplitude_reflected = abs(A_reflected);
phase_reflected = angle(A_reflected);
% Display Results
disp('----------------------------------------');
disp('Results of the RIS Simulation:');
disp(['Incident Signal Amplitude: ', num2str(A)]);
disp(['Incident Signal Phase (radians): ', num2str(phi)]);
disp(['RIS Phase Shift (radians): ', num2str(theta)]);
disp('----------------------------------------');
disp(['Reflected Signal Amplitude: ',
num2str(amplitude_reflected)]);
disp(['Reflected Signal Phase (radians): ',
num2str(phase_reflected)]);
```

```
disp(['Real Part of Reflected Signal: ',
num2str(real(A_reflected))]);
disp(['Imaginary Part of Reflected Signal: ',
num2str(imag(A_reflected))]);
disp('----------------------------------------');

Display Results
----------------------------------------
Results of the RIS Simulation:
Incident Signal Amplitude: 1
Incident Signal Phase (radians): 0.7854
RIS Phase Shift (radians): 1.0472
----------------------------------------
Reflected Signal Amplitude: 1
Reflected Signal Phase (radians): 1.8326
Real Part of Reflected Signal: -0.25882
Imaginary Part of Reflected Signal: 0.96593
----------------------------------------
```

Understanding the structural components and electromagnetic operation of RIS sets the stage for analyzing how real-time phase control is achieved through advanced materials and circuit-level design. The next section focuses on the technological enablers that translate theoretical beam manipulation into practical deployment.

15.1.2 Control Mechanisms of RIS

RIS can use various control mechanisms to enable flexible and dynamic phase shifts. Key techniques include:

- *PIN diodes:* PIN describes the physical layer structure of the diode (a p-type layer, an intrinsic high-resistivity layer, and an n-type layer, in that order). The intrinsic layer is the key feature, giving PIN diodes their wide depletion region, which improves performance for RF switching, photodetection, and high-voltage applications. These semiconductor devices are widely used in RIS for achieving quick binary phase shifts. They operate by switching states between on and off, altering the impedance and enabling phase control with minimal power consumption. This approach is often chosen for applications needing rapid phase adjustments, such as in real-time beamforming.
- *Varactor-tuned resonators:* Varactors are voltage-controlled capacitors that provide a continuous range of capacitance adjustments. By tuning this capacitance, varactor-based RIS structures can achieve finer phase control over a broader range, allowing more flexibility than binary switching. This characteristic is particularly useful in environments where gradual phase shifts are beneficial for achieving higher resolution in signal steering and shaping.
- *Liquid crystal-based RIS:* This method leverages liquid crystals' variable refractive index, which can be modulated by an external voltage. Liquid-crystal-based RIS offers substantial reconfigurability, allowing for a highly adjustable phase shift. However, the response time is generally slower compared

to PIN diodes and varactors, making it more suitable for applications where reconfiguration speed is less critical.

- *Microelectromechanical systems (MEMS):* MEMS technology provides precise, multilevel phase shifts with low power requirements. MEMS devices are known for their fine-tuned control over phase, which enables highly adaptable beamforming. While they generally have a moderate response speed, faster than liquid crystals but slower than PIN diodes, MEMS are robust and efficient, making them ideal for scenarios where power efficiency and precision are essential, such as in cellular networks and satellite communications.

Each method presents unique advantages in terms of granularity and energy efficiency. The mathematical model governing phase shift in a PIN diode-based RIS can be given as:

$$\theta_{\mathrm{PIN}} = f\left(C_{\mathrm{var}}\right) \tag{15.2}$$

where C_{var} is the variable capacitance controlling the phase.

The following MATLAB code simulates the relationship between varying capacitance values and the corresponding phase shift, specifically focusing on a control mechanism using a PIN diode. The code first defines a range of capacitance values from 0.1 pF to 1 pF. It then calculates the phase shift for each capacitance value using the arctangent function, which serves as an example of how capacitance affects phase shift. MATLAB Program 36 simulates a phase shift control using a PIN diode to visualize the effect of varying capacitance on phase, as shown in Figure 15.2.

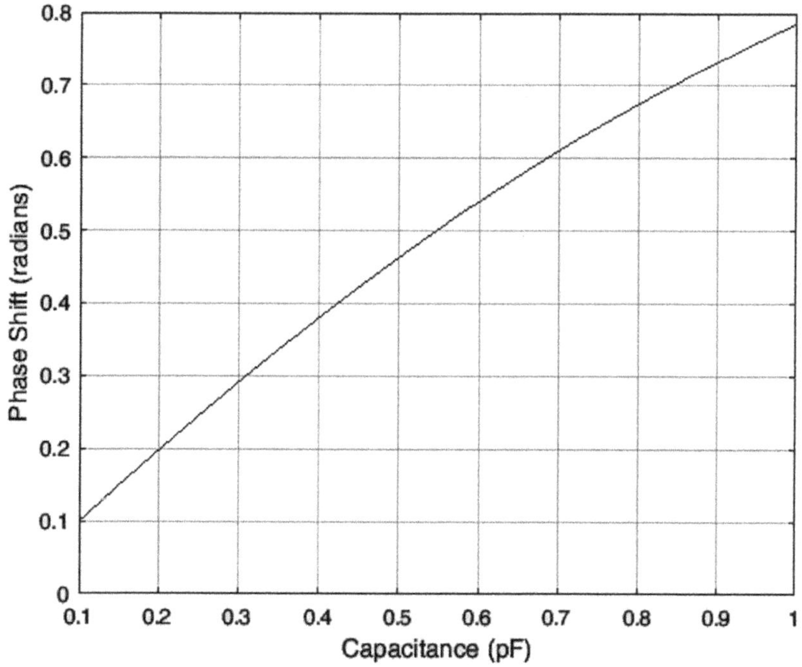

Figure 15.2 Phase shift control using PIN diode.

MATLAB Program 36

```
% A MATLAB code to simulate phase shifting using a control
mechanism, to visualize the effect of varying capacitance
on phase.
C_var = linspace(0.1, 1, 100); % Simulated capacitance values
theta_PIN = atan(C_var); % Example relationship for phase shift

plot(C_var, theta_PIN, 'k-');
xlabel('Capacitance (pF)');
ylabel('Phase Shift (radians)');
grid;
title('Phase Shift Control Using PIN Diode');
```

The above simulation provides insight into how a nonlinear capacitance-to-phase relationship affects the control precision of an RIS. In practical implementations, designers must select tuning mechanisms based on resolution (e.g., 1-bit, 2-bit phase quantization) versus power and response-time trade-offs.

MATLAB Program 37 simulates the reconfiguration speed of RIS by comparing different control mechanisms, including PIN diodes, varactor diodes, MEMS, and liquid crystals. Each mechanism is characterized by its average fast response time and phase shift range, which influence how quickly the RIS can adapt to changing communication conditions. The code generates exponential decay curves that represent the response behavior of each control mechanism over time, visualizing how the phase shift varies with respect to time. MATLAB Program 37 simulates and plots the impacts of different control mechanisims on RIS, as shown in Figure 15.3.

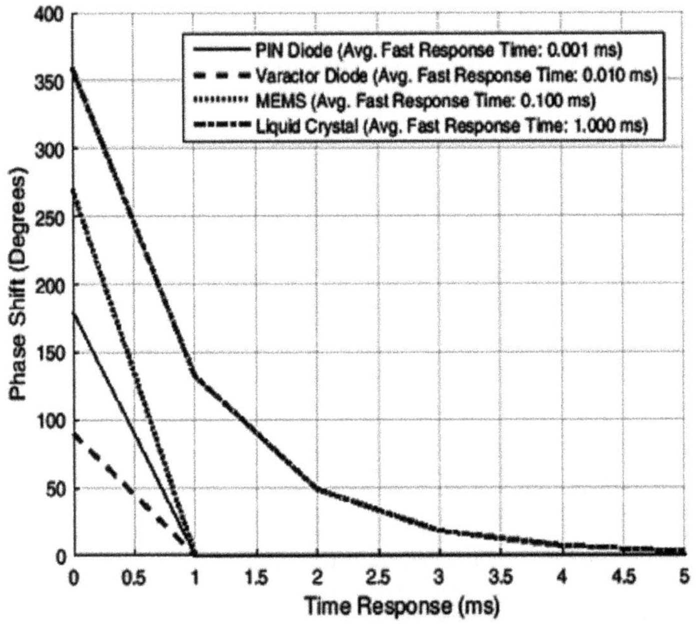

Figure 15.3 RIS average fast response time for different control mechanisms.

MATLAB Program 37

```
% MATLAB Code to Illustrate Different Control Mechanisms for
RIS Reconfiguration
% This code simulates the impact of different control mechanisms
on RIS (Reconfigurable Intelligent Surface)
reconfiguration speed.
% Each mechanism has unique properties such as average response
time and phase shift range.

% Control Mechanism Data
control_mechanisms = {'PIN Diode', 'Varactor Diode', 'MEMS',
'Liquid Crystal'};
response_times = [0.001, 0.01, 0.1, 1];  % Average response times
in milliseconds
phase_shift_range = [180, 90, 270, 360];  % Phase shift range
in degrees

% Initialize figure for plotting
figure;
hold on;
lineStyles = {'-', '--', ':', '-.'}; % Line styles for each
control mechanism

% Loop through each control mechanism to simulate and plot the
response time effect on reconfiguration speed
for i = 1:length(control_mechanisms)
  % Generate time data for simulation (in seconds)
  time = linspace(0, 1, 1000);  % 0 to 1 second, 1000 points for
smooth curves
  response_curve = exp(-time / (response_times(i) / 1000));  %
Exponential decay to simulate response

  % Plot each control mechanism's response curve
  plot(time * 1000, response_curve * phase_shift_range(i), ... %
Convert time to milliseconds for the x-axis
    'Color', 'k', 'LineStyle', lineStyles{i}, 'LineWidth', 1.5,
... % Set line color to black
    'DisplayName', [control_mechanisms{i} ' (Avg. Fast Response
Time: ' num2str(response_times(i), '%.3f') ' ms)']);
end

% Plot formatting
title('RIS Average Fast Response Time for Different
Control Mechanisms');
xlabel('Fast Time Response (ms)'); % Updated x-axis title
ylabel('Phase Shift (Degrees)');
axis([0 5 0 400]);  % Set axis limits
legend('show');
grid on;
hold off;
```

```
% Displaying key information
disp('RIS Reconfiguration Speed Simulation for Different Control
Mechanisms (Using Average Fast Response Time)');
disp('-----------------------------------------------');
for i = 1:length(control_mechanisms)
  fprintf('%s:\n', control_mechanisms{i});
  fprintf('  - Fast Response Time: %.4f ms\n', response_
times(i)); % Updated display in milliseconds
  fprintf('  - Phase Shift Range: %d degrees\n',
phase_shift_range(i));
  disp('-----------------------------------------------');
end
```

As illustrated, the choice of control mechanism directly affects the dynamic adaptability of RIS in fast-changing environments. For instance, in vehicular networks or real-time holographic communications, mechanisms such as PIN diodes are better suited due to their rapid phase shift capabilities, despite offering limited resolution. This highlights the importance of application-driven hardware selection in RIS design.

MATLAB Progr am 38

```
% Program: Beam Steering Using a Linear RIS Array
% This program simulates the far-field beam pattern of a linear
% Reconfigurable Intelligent Surface (RIS) with uniformly spaced
% elements that reflect incoming waves toward a
desired direction.
% The program computes and plots the normalized array factor
% (beam pattern) by applying a linear phase shift
across elements.

clc; clear;  % Clear command window and workspace variables

%% --- Define Physical Parameters of the RIS ---
lambda = 1;              % Wavelength (normalized to 1 unit)
k = 2 * pi / lambda;      % Wave number (k = 2π/λ)
d = lambda / 2;          % Inter-element spacing (half-wavelength)
N = 16;                 % Number of RIS elements in the
linear array
theta0_deg = 30;          % Desired beam steering angle
(in degrees)
theta0 = deg2rad(theta0_deg); % Convert steering angle to radians

%% --- Define Observation Angles ---
theta_scan = linspace(-pi, pi, 1000);  % Observation angle range:
-180° to 180° in radians

% Preallocate space for array factor calculation
```

```
AF = zeros(size(theta_scan));  % Initialize array factor (field
strength) vector

%% --- Phase Shift Profile for Beam Steering ---
% Compute the progressive phase shift across RIS elements
to steer
% the beam in the desired direction (theta0)
beta_n = k * d * (0:N-1)' * sin(theta0);  % Linear phase shifts

%% --- Calculate Array Factor Over All Scan Angles ---
% Loop through each observation angle and compute total
reflected field
for idx = 1:length(theta_scan)
  theta = theta_scan(idx);  % Current observation angle
(in radians)

  % Compute total complex field by summing phasors from
all elements
  AF(idx) = sum(exp(1j * (k * d * (0:N-1)' * sin(theta) +
beta_n)));
end

%% --- Normalize and Convert Array Factor to dB ---
AF_dB = 20 * log10(abs(AF) / max(abs(AF)));  % Convert magnitude
to decibels (normalized)

%% --- Plot Beam Pattern ---
figure;
plot(rad2deg(theta_scan), AF_dB, 'k', 'LineWidth', 1.5);  % Plot
in degrees
xlabel('Angle (degrees)');
ylabel('Normalized Array Factor (dB)');
title(['Beam Steering of RIS Array to ' num2str(theta0_deg)
'°']);
grid on;
axis([-90 90 -40 0]);  % Display ±90° with side lobes truncated
at -40 dB
```

The beam pattern in Figure 15.4 confirms that, by applying a progressive phase shift across a linear RIS array, reflected waves can be steered directionally—here, toward 30°. This passive beamforming approach demonstrates the programmable control that RIS offers, without requiring active RF chains, making it highly suited for dynamic 6G applications.

15.1.3 RIS Deployment Strategies

RIS can be deployed in various configurations to optimize wireless communication performance. Each deployment strategy serves different coverage, capacity, and

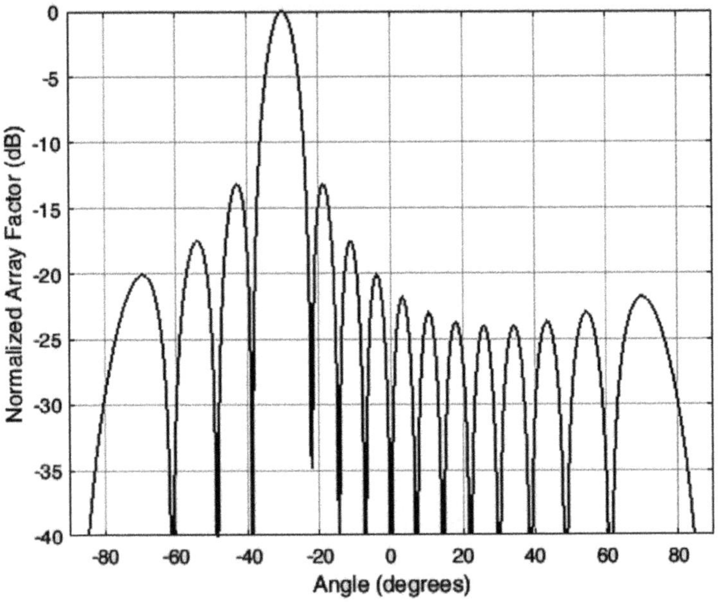

Figure 15.4 Far-field beam pattern of a linear RIS array with 16 elements steering to 30°.

quality objectives in diverse environmental scenarios. The primary RIS deployment configurations are:

- *Centralized RIS (base station side):* In this configuration, RIS units are placed near the base station (BS). This positioning allows the base station to exert maximum control over the RIS, enabling efficient beamforming and directing signals to cover a larger area. By placing RIS near the BS, network operators can achieve improved beamforming gains, making it ideal for extending coverage in open environments or areas where long-range signal control is required.
- *Distributed RIS (user-side):* Here, RIS units are strategically located closer to end-user devices, particularly in densely populated regions, to enhance signal quality and reduce interference. This approach is useful in urban or indoor environments with multiple obstacles, where proximity to users helps to increase the received signal strength and improves user experience, even in high-traffic zones.
- *Hybrid RIS deployment:* This strategy combines centralized and distributed RIS deployments, providing a balanced solution that maximizes both coverage and network capacity. By deploying RIS near both the BS and closer to users, hybrid setups leverage the advantages of wide coverage and high-quality, localized signal enhancement. This dual deployment structure supports more stable connectivity in both open and dense environments.
- *Aerial RIS on unmanned aerial vehicles (UAVs):* Deploying RIS on UAVs adds flexibility by providing nonline-of-sight signal coverage in areas where direct communication paths may be blocked. UAV-mounted RIS can adjust their position dynamically, optimizing signal reflection and providing coverage in otherwise unreachable locations. This approach is particularly useful

for emergency response or temporary events, where fast deployment of communication infrastructure is critical.

- *Centralized RIS coverage and signal strength:* The effectiveness of each deployment method is illustrated through coverage and beamforming gain equations. For centralized RIS, the signal strength S can be expressed as:

$$S \propto \frac{G_{\mathrm{BS}} G_{\mathrm{RIS}}}{d^2} \tag{15.3}$$

where G_{BS} and G_{RIS} are the gains from the base station and RIS, respectively, and d is the distance from RIS to the receiver.

Equation (15.3) highlights that centralized RIS configurations are effective in extending the reach of signals over a wide area by leveraging higher gains from both the BS and RIS. As d increases, the received signal strength decreases, so placement near the BS helps to optimize performance across larger distances.

A more comprehensive path loss model incorporating RIS can be written as:

$$P_r = P_t + G_{\mathrm{BS}} + G_{\mathrm{RIS}} + G_{\mathrm{UE}} - 20\log_{10}(d_1 + d_2) - L_{\mathrm{RIS}} \tag{15.4}$$

where d_1 and d_2 represent the distances from BS to RIS and RIS to user, respectively, and L_{RIS} is the loss due to RIS element imperfections. Furthermore, G_{BS} is the base station antenna gain, G_{RIS} is the RIS aperture gain, and G_{UE} is the user equipment antenna gain. This model aids in determining RIS placement and expected gains under varying topologies.

Each deployment strategy offers unique benefits, making it possible to tailor RIS deployment to specific environmental needs, user densities, and coverage requirements, thereby optimizing both the quality and reach of the wireless network.

15.1.4 Emerging Applications of RIS in 6G

RIS technology is anticipated to be a key enabler of various advanced applications in 6G, addressing unique challenges associated with high-frequency communications and network demands. RIS technology offers multiple advantages across different applications, including:

- *Millimeter-wave and terahertz communications:* Millimeter-wave and terahertz frequencies are promising for 6G due to their high bandwidth, but they suffer from short-range limitations and vulnerability to blockages (e.g., buildings, vehicles). RIS helps to mitigate these issues by intelligently redirecting or reflecting signals, reducing path loss and improving link reliability. By leveraging RIS for beamforming and path redirection, 6G networks can overcome the inherent propagation challenges of millimeter-wave and terahertz signals, thereby enabling high-speed data transmission over greater distances without requiring additional base stations.
- *Mobile edge computing (MEC):* MEC brings computation closer to the user, enabling faster data processing and reduced network latency. RIS technology

further enhances MEC by dynamically optimizing offload paths, which leads to more efficient data routing and reduces latency. RIS can redirect signals toward MEC nodes or improve signal strength in congested areas, making data transfer between edge devices and servers more efficient. This capability is especially useful in supporting low-latency applications such as augmented reality (AR), virtual reality (VR), and real-time analytics in smart cities, where edge computing is crucial.
- *Vehicular networks (VANETs):* In vehicular networks, ensuring stable, reliable connectivity in high-mobility environments is essential. RIS can support consistent signal coverage and reduce handover frequency by reflecting signals around obstacles and optimizing paths in real time, which is critical for autonomous vehicles and connected transportation systems. By enabling seamless signal transition, RIS minimizes disruptions caused by vehicle movements and ensures reliable, low-latency communication for applications such as vehicle-to-everything (V2X) services, traffic management, and safety systems.
- *IoT:* The IoT ecosystem, with its vast number of devices, presents power and connectivity challenges. RIS can enhance IoT network performance by improving power efficiency and offering simultaneous wireless information and power transfer (SWIPT), which allows IoT devices to receive data and recharge wirelessly from the same signal. In dense IoT environments, such as smart homes and factories, RIS helps to reduce energy consumption by intelligently managing signal paths, which is critical for battery-powered or passive devices. This ensures that IoT devices can operate for extended periods without frequent battery replacements, contributing to more sustainable and resilient IoT networks.

These emerging applications signify a paradigm shift in wireless communication, bridging gaps left by traditional 5G satellite antennas covered in Chapter 14. Where deployable satellite reflectors enable global-scale communication, RIS provides localized intelligence, especially in environments characterized by severe multipath fading, latency sensitivity, or infrastructure constraints. The synergy between satellite MIMO and ground-based RIS systems is expected to shape hybrid 6G architectures.

In summary, RIS offers transformative benefits across 6G applications by enhancing coverage, improving energy efficiency, and supporting high-speed, reliable communication in various challenging environments. As 6G networks evolve, RIS is set to play an essential role in meeting the demands of next-generation applications, from ultrafast mobile connectivity to sustainable IoT solutions.

15.2 Reflect Arrays and Their Simulation

Reflect arrays represent an innovative and powerful approach in antenna design that marries the benefits of phased arrays with the simplicity of reflector antennas. Comprising an array of passive elements, reflect arrays can manipulate incoming electromagnetic waves, providing tailored phase shifts that enhance signal strength in desired directions. This technology finds applications in diverse areas such as satellite communications, radar systems, and wireless networks, making it a crucial area of study in antenna engineering.

15.2.1 Principles of Reflect Array Operation

Reflect arrays operate on the principle of phase manipulation. When an incident electromagnetic wave strikes the array, each element reflects the wave, introducing a specific phase shift based on the element's geometry and orientation. The overall radiation pattern is determined by the constructive and destructive interference of the waves reflected from the individual elements.

The design goal is to achieve a desired beam directionality by carefully calculating the necessary phase shifts for each element. This can be expressed mathematically as:

$$\phi_n = k \cdot d_n + \phi_{\text{feed}} \tag{15.5}$$

where ϕ_n is the phase shift introduced by the nth element, k is the wave number (related to the operating frequency), d_n is the distance from the feed point to the nth element, and ϕ_{feed} is the phase imparted by the feeding mechanism.

By adjusting the design of each element, engineers can steer the reflected beam to the desired angle, enhancing performance across various communication systems.

The total reflected field E_r of a reflect array with $N \times M$ elements is typically modeled using the array factor:

$$E_r(\theta,\varphi) = \sum_{n=1}^{N}\sum_{m=1}^{M} A_{nm} \cdot e^{j(kd(n\sin\theta\cos\varphi + m\sin\theta\sin\varphi) + \varphi_{mn})} \tag{15.6}$$

where A_{nm} is the amplitude, d is the interelement spacing, and φ_{mn} is the phase shift applied to the (n, m)th element. By designing φ_{mn} appropriately, reflect arrays can steer beams in specific directions.

15.2.2 Design Considerations for Reflect Arrays

Having understood the phase manipulation principle that governs reflect arrays, we now turn to the design considerations that determine their performance in real-world systems.

The successful design of a reflect array requires careful consideration of multiple factors:

1. *Element geometry:* The choice of geometry for the individual reflective elements (e.g., patch antennas, dipoles, slots, or metasurfaces) is critical. Each design impacts the reflection coefficient and overall performance.
2. *Spacing and arrangement:* The physical arrangement and spacing between elements are vital for ensuring that the desired phase shifts are achieved

while minimizing undesirable coupling effects that may result in performance degradation.

3. *Feed network design:* The feeding mechanism must be designed to distribute the input signal evenly across the array, providing the necessary phase adjustments to facilitate the intended beamforming effect. Techniques such as corporate feeds, series feeds, or sequential rotation can be employed depending on the array's configuration.
4. *Bandwidth and efficiency:* Reflect arrays should be designed to operate efficiently over a wide frequency range. This involves optimizing the element design and layout to minimize losses and ensure consistent performance across the operational bandwidth.
5. *Material selection:* The choice of materials influences the reflectivity, durability, and weight of the array. Materials should be selected based on the application's environmental conditions and mechanical requirements.

15.2.3 Simulation of Reflect Arrays Using MATLAB

With the foundational design factors established, the next logical step is to simulate and visualize the radiation characteristics of reflect arrays. MATLAB offers a practical platform to model these effects before fabrication.

Simulation is a crucial aspect of the design and optimization process for reflect arrays. MATLAB, with its powerful computational capabilities and specialized toolboxes, provides an excellent platform for modeling electromagnetic behavior and predicting the performance of reflect arrays before fabrication.

15.2.3.1 MATLAB Environment Setup

To effectively simulate a reflect array in MATLAB, follow these general steps:

1. *Define element properties:* Specify the characteristics of the individual reflect array elements, including their geometry, dimensions, material properties, and reflection coefficients.
2. *Construct the array layout:* Use MATLAB functions to create the spatial arrangement of the elements, ensuring that the desired spacing and alignment are achieved. Functions such as meshgrid can be helpful for establishing coordinates.
3. *Implement the feeding mechanism:* Model the feed network and its interaction with the reflect array. This may involve defining feed point locations and calculating the phase shifts introduced at each point.
4. *Conduct electromagnetic simulations:* Utilize tools such as the method of moments (MoM), finite-difference time domain (FDTD), or finite element methods (FEM) to analyze the reflect array's performance. These methods allow for the calculation of radiation patterns, gain, and efficiency under varying operational conditions.
5. *Analyze simulation results:* Utilize MATLAB's plotting capabilities to visualize results, including radiation pattern plots, gain versus frequency graphs,

and 3D surface plots of the array's performance. Functions such as polarplot, surf, and plot3 can provide insightful visualizations.

15.2.3.2 Example Simulation Code

Here is an example MATLAB code that demonstrates how to create a simple reflect array model. This code sets up a rectangular reflect array and simulates its radiation pattern. MATLAB Program 39 simulates and plots the radiation pattern of a 5 × 5 reflect array at 10 GHz with elements spaced at half wavelength, as shown in Figure 15.5.

MATLAB Program 39

```
% MATLAB Code for Reflect Array Simulation
% This code simulates the radiation pattern of a 5x5
reflect array.
% The array operates at a frequency of 10 GHz, with elements
spaced at
% half the wavelength (lambda/2). Each element's position
is computed
% relative to the array's center to ensure proper spacing
and alignment.
% Using the array factor method, the code calculates the
azimuthal radiation
% pattern for broadside radiation and normalizes it for plotting
in dB scale.
```

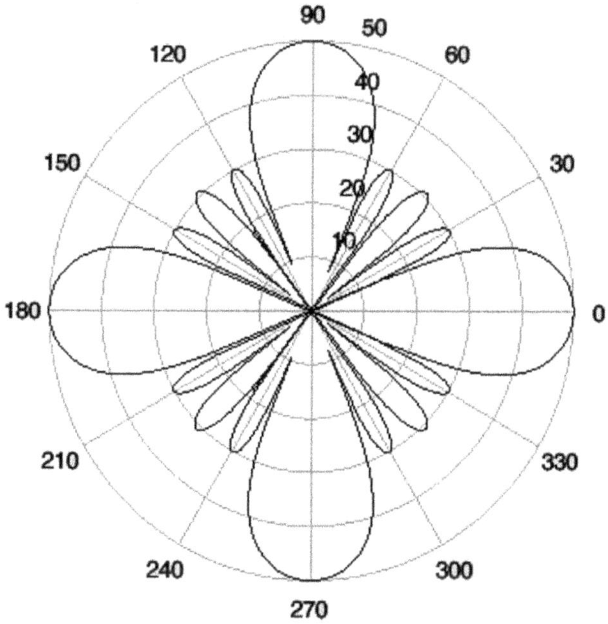

Figure 15.5 Reflect array radiation pattern (dB).

```
% Finally, the radiation pattern is visualized in
polar coordinates.

% Define parameters
frequency = 10e9; % Frequency in Hz
lambda = 3e8 / frequency; % Wavelength
element_size = lambda / 2; % Size of each element
array_size = [5, 5]; % Array dimensions (5x5)
d = lambda / 2; % Distance between elements

% Create reflect array element positions
elements = []; % Array to store element positions
for m = 1:array_size(1)
  for n = 1:array_size(2)
    % Positioning each element
    pos_x = (n - (array_size(2)+1) / 2) * d; % Centering
X position
    pos_y = (m - (array_size(1)+1) / 2) * d; % Centering
Y position
    elements = [elements; pos_x, pos_y]; % Store position
  end
end

% Simulate radiation pattern
theta = linspace(0, 2*pi, 360); % Azimuth angle for
pattern plotting
phi = pi/2; % Elevation angle (broadside)
pattern = zeros(size(theta));

% Calculate the array factor
for i = 1:length(theta)
  af = 0; % Array factor initialization
  for elem = 1:size(elements, 1)
    % Calculate phase shift due to each element position
    phase_shift = 2 * pi / lambda * (elements(elem, 1) *
cos(theta(i)) + elements(elem, 2) * sin(theta(i)));
    af = af + exp(1j * phase_shift); % Sum the contributions
  end
  pattern(i) = abs(af); % Store absolute value of array factor
end

% Normalize and convert to dB
Pattern= abs(pattern)/max(abs(pattern));
pattern_dB = 20 * log10(Pattern);
pattern_dB= pattern_dB+50;

% Plot results
figure;
polar(theta, pattern_dB, 'k');
title('Reflect Array Radiation Pattern (dB)');
grid on;
```

The above simulation demonstrates the azimuthal radiation behavior of a simple 5 × 5 reflect array. As expected, a strong main lobe emerges in the broadside direction, confirming constructive interference from uniformly spaced elements. This visualization provides a baseline for understanding how larger or phase-tuned reflect arrays can steer beams in different directions, similar to the RIS structures introduced earlier.

15.2.4 Applications of Reflect Arrays

Reflect arrays have become a vital technology in modern communication and sensing systems. Their design flexibility and capability to steer beams make them suitable for numerous applications across different sectors. Below are some key areas where reflect arrays play a significant role:

- *Satellite communications:* Reflect arrays are extensively used to enhance satellite communication performance. They provide focused, high-gain beams, which are essential for maintaining reliable connections over large distances and in challenging environments. Reflect arrays are also cost-effective alternatives to traditional parabolic antennas, offering lightweight and compact designs without compromising signal quality. By enabling controlled beam shaping and redirection, reflect arrays help to maximize satellite coverage and link robustness.
- *Wireless networks:* In wireless network applications, especially in dense urban environments, reflect arrays enable efficient beamforming to direct signals toward specific users. This capability minimizes interference among users and improves network efficiency by directing energy where it is most needed. In Wi-Fi, cellular, and other wireless technologies, reflect arrays enhance capacity and connectivity by dynamically adjusting beam patterns to adapt to the user's location, thus optimizing network performance.
- *Radar systems:* Reflect arrays are widely employed in radar applications due to their ability to steer beams and focus reflections for high-resolution imaging and precise target tracking. In systems such as automotive radar, weather radar, and defense applications, reflect arrays enhance detection accuracy by controlling the phase of reflected waves, enabling accurate distance and speed measurement of objects over long ranges. Their adaptability makes them suitable for various radar configurations, providing flexibility for different operational requirements.
- *5G and beyond:* The evolution of mobile networks to 5G and beyond requires increased data rates, connectivity, and efficient spectrum usage. Reflect arrays are instrumental in meeting these demands by supporting advanced beamforming and MIMO configurations. They can provide directional beams that adapt to user locations, resulting in stronger connections and broader coverage areas. In future 5G+ and 6G systems, reflect arrays will help to address connectivity challenges in high-frequency bands, such as millimeter-wave, where direct line of sight is critical.

While reflect arrays and RIS share the concept of manipulating reflected waves, they differ in control complexity and operational design. Reflect arrays are typically fixed or manually tuned, whereas RIS allows for real-time, programmable beam control via embedded electronics. However, both approaches offer cost-effective and power-efficient alternatives to conventional active arrays, particularly in millimeter-wave and satellite applications.

15.2.5 Conclusion

Reflect arrays represent a groundbreaking solution in the field of antenna design, offering a blend of passive elements' simplicity and phased arrays' flexibility. The ability to manipulate reflected waves enables enhanced performance across various communication systems. With the aid of MATLAB simulation, engineers can optimize reflect array designs, ensuring efficient performance before physical deployment. Continued advancements in reflect array technology promise to expand their applications and capabilities in an increasingly connected world.

15.3 Antennas for Biomedical Applications

15.3.1 Introduction to Biomedical Antenna Applications

Antennas are vital components in the rapidly evolving field of healthcare, facilitating innovations that enhance diagnosis, monitoring, and therapeutic treatments. The integration of antenna technology in medical devices has revolutionized patient care, enabling real-time data collection and improved treatment outcomes. Antennas serve as the backbone of various healthcare applications, providing reliable wireless communication essential for both patient and provider.

The transformative impact of antenna technology can be observed across multiple biomedical fields:

- *Diagnostics:* Antennas enable the transmission of critical data from diagnostic devices, allowing for noninvasive assessments of health conditions. For example, portable electrocardiogram (ECG) monitors utilize antennas to relay heart activity data to healthcare providers in real time.
- *Implantable devices:* Devices such as pacemakers and insulin pumps rely on antennas to communicate data wirelessly, facilitating continuous monitoring and control of physiological conditions without the need for invasive procedures.
- *Therapeutic applications:* Antennas are integral to therapeutic devices that require precise targeting and control, such as those used in microwave ablation treatments for cancer.

15.3.2 Wearable and Implantable Antenna

The following sections explore two key biomedical antenna categories: wearable and implantable. Each category poses distinct engineering challenges, particularly in balancing performance, size, safety, and patient comfort.

15.3.2.1 Wearable Antennas

Wearable antennas have revolutionized the healthcare sector by enabling the development of a range of portable devices designed for continuous health monitoring. For instance:

1. ECG monitors utilize wearable antennas to provide real-time heart activity data, allowing for the early detection of cardiovascular issues. Similarly, pulse oximeters measure blood oxygen levels continuously, which is vital for patients with respiratory conditions.
2. Wearable biosensors further expand this capability by integrating multiple physiological measurements, such as glucose levels in diabetic patients, heart rate variability, and physical activity, all in one device. The continuous data that these devices collect fosters preventive care strategies and promotes patient engagement in their own health management, ultimately reducing the frequency of in-person medical visits and hospital admissions.

Design Constraints

The design of wearable antennas presents unique challenges that must be addressed to ensure optimal performance and user satisfaction. First, low-profile designs are critical; the antennas must be unobtrusive and lightweight to maximize comfort for users, who may wear these devices for extended periods. This often necessitates innovative materials and configurations that can conform to the human body while maintaining effective radiation patterns.

Flexibility

Flexibility is another essential design constraint. Wearable antennas must be able to bend and stretch without compromising their operational efficiency. This requirement often leads to the use of flexible substrates, such as conductive textiles or polymer materials, which allow the antenna to adapt to the contours of the body while remaining functional during physical activities.

Additionally, wearable antennas typically operate within industrial, scientific, and medical (ISM) bands, notably at frequencies such as 2.4 GHz. This frequency range is chosen not only for its balance of good propagation characteristics and relatively low absorption by biological tissues but also for its compatibility with established wireless communication protocols, such as Bluetooth and Wi-Fi. These protocols are essential for seamless data transmission to smartphones or healthcare providers, ensuring that patient data is accessible in real time.

In wearable applications, the antenna is typically simulated using an equivalent multilayer model, where layers represent air, fabric, skin, and tissue. The antenna's efficiency depends on its ability to maintain resonance and match impedance in proximity to lossy biological tissue.

The antenna gain and return loss S_{11} can be analyzed across frequency to ensure stable performance in the ISM band. A typical threshold for acceptable matching is $S_{11} < -10$ dB, indicating minimal signal reflection.

To illustrate these concepts, consider the design and simulation of a simple wearable patch antenna. MATLAB can be used to model this antenna, providing insights into its performance metrics and helping to visualize how design choices affect its functionality. MATLAB Program 40 is an example for designing a wearable patch antenna operating at 2.4 GHz. It simulates and plots the return loss of a wearable patch antenna throughout a frequency range 2.3 to 2.5 GHz, as shown in Figure 15.6.

MATLAB Program 40

```
% MATLAB Code: Wearable Patch Antenna Simulation at 2.4 GHz
% This code designs and simulates a wearable patch antenna
operating at
% the ISM band frequency of 2.4 GHz. It focuses on
calculating the
% necessary patch dimensions, simulating the return loss (S11)
to evaluate
% impedance matching, and estimating the specific absorption rate
(SAR)
% to assess safety in wearable applications.
%
%   1. Defining Antenna Parameters: The code calculates
the dimensions
%    of the patch antenna based on the substrate properties
and desired
%    operating frequency.
%   2. Return Loss Simulation: It generates a mock return
loss curve
%    to visualize how well the antenna is impedance matched
at the
%    operating frequency.
```

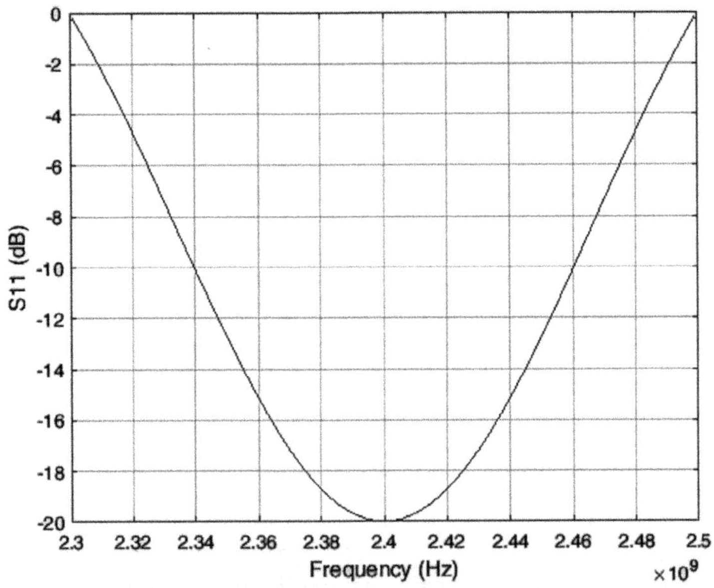

Figure 15.6 Approximate return loss (S_{11}) of wearable patch antenna.

```
%   3. SAR Calculation: It provides an estimate of the specific
%    absorption rate for a typical human tissue model,
helping evaluate
%    the safety of the wearable antenna design.
%
% Users can modify frequency, substrate properties, and
electric field
% values to explore different configurations and applications
for wearable
% antennas.

clc;
clear;

% Constants
frequency = 2.4e9; % Operating frequency in Hz (2.4 GHz for
ISM band)
c = 3e8; % Speed of light in vacuum (m/s)
lambda = c / frequency; % Wavelength calculation

% Substrate Parameters (FR4 for wearable applications)
substrateThickness = 0.0016; % Thickness in meters (1.6 mm)
substratePermittivity = 4.4; % Relative permittivity (FR4)

% Patch Dimensions (Approximate calculations for
resonant frequency)
% Width and Length of the patch
patchWidth = lambda / (2 * sqrt((substratePermittivity + 1) /
2));
effectivePermittivity = (substratePermittivity + 1) / 2 + ...
               ((substratePermittivity - 1) / 2) * ...
               (1 + 12 * substrateThickness / patchWidth)^(-0.5);
deltaL = 0.412 * substrateThickness * ...
     (effectivePermittivity + 0.3) * ...
     (patchWidth / substrateThickness + 0.264) / ...
     ((effectivePermittivity - 0.258) * ...
     (patchWidth / substrateThickness + 0.8));
patchLength = lambda / (2 * sqrt(effectivePermittivity)) - 2 *
deltaL;

% Display calculated patch dimensions
disp(['Patch Width: ', num2str(patchWidth), ' meters']);
disp(['Patch Length: ', num2str(patchLength), ' meters']);

% Create a Simple Return Loss (S11) Plot Simulation
% using an approximate resonance curve
freqRange = linspace(2.2e9, 2.6e9, 100); % Frequency range around
2.4 GHz
S11_approx = -20 * abs(sinc((freqRange - frequency)/1e8)); % Mock
S11 plot
```

```
% Plot Return Loss (S11)
figure;
plot(freqRange, S11_approx, 'k-');
axis([2.3*10^9  2.5 *10^9  -20  0])
title('Approximate Return Loss (S11) of Wearable Patch Antenna');
xlabel('Frequency (Hz)');
ylabel('S11 (dB)');
grid on;

% Radiation Pattern Placeholder (no toolbox function)
% Instead, describe radiation properties theoretically or based
on literature.

% SAR Calculation (Specific Absorption Rate) Example
tissuePermittivity = 38; % Approximate permittivity of skin at
2.4 GHz
tissueConductivity = 1.2; % Conductivity of skin at 2.4 GHz in
S/m
tissueDensity = 1000; % Density of human tissue (kg/m^3)
E_field = 10; % Electric field magnitude in V/m (sample value
for SAR)
SAR = (tissueConductivity * E_field^2) / tissueDensity;
disp(['Estimated SAR: ', num2str(SAR), ' W/kg']);
% Check if SAR is below the general public safety limit
if SAR < 2
  disp('SAR is within safe limits for wearable use.');
else
  disp('Warning: SAR exceeds general public exposure limits.');
end
Display Results
Patch Width: 0.038036 meters
Patch Length: 0.029442 meters
Estimated SAR: 0.12 W/kg
```

The simulated return loss curve demonstrates effective impedance matching around 2.4 GHz, consistent with ISM band requirements. Calculated SAR values remain below safety limits, confirming the design's viability for continuous skin contact in wearable scenarios.

While wearable antennas are externally mounted and benefit from easier access and lower absorption, implantable antennas must function deep within body tissues, making miniaturization, biocompatibility, and electromagnetic safety critically important.

15.3.2.2 Implantable Antennas

Implantable antennas play a crucial role in a variety of medical devices designed for internal use within the human body. These devices, including pacemakers, insulin pumps, and cochlear implants, rely on implantable antennas to function effectively.

1. *Pacemakers:* These devices continuously monitor heart rhythms and deliver electrical impulses to ensure proper heart function. Implantable antennas enable the transmission of real-time cardiac data to external monitors or healthcare providers, allowing for timely interventions if abnormalities are detected.
2. *Insulin pumps:* These devices automatically deliver insulin to diabetic patients based on their blood glucose levels. Implantable antennas facilitate communication between the pump and glucose monitors, ensuring that patients receive precise insulin dosages without manual intervention.
3. *Cochlear implants:* These electronic devices provide a sense of sound to individuals with hearing loss by directly stimulating the auditory nerve. Implantable antennas transmit data between the cochlear implant and external sound processors, enhancing sound quality and user experience.

The efficiency of these antennas in transmitting vital information is critical, as they must operate seamlessly within the complex and dynamic environment of the human body while ensuring patient safety.

Biocompatibility and SAR

When designing implantable antennas, biocompatibility and SAR are paramount considerations. The materials used for the antennas must be biocompatible to prevent adverse reactions within the body, and the design must ensure that the SAR remains low. SAR is a measure of the rate at which energy is absorbed by the body when exposed to an electromagnetic field, expressed in watts per kilogram.

To ensure patient safety, the SAR must be minimized. This can be achieved through:

- *Careful selection of dielectric materials:* The choice of materials affects how electromagnetic fields interact with biological tissues. Materials with lower dielectric losses and suitable thermal properties can help reduce SAR.
- *Miniaturization of antenna components:* Smaller antennas can reduce the amount of electromagnetic energy absorbed by surrounding tissues. Advanced fabrication techniques, such as 3D printing and the use of flexible materials, allow for the creation of compact and efficient implantable antennas.

The SAR can be calculated using the following equation:

$$SAR = \frac{\sigma E^2}{\rho} \tag{15.7}$$

where σ is the conductivity of the tissue (in S/m), E is the electric field (in V/m), and ρ is the tissue density (in kg/m^3).

This equation shows how SAR scales quadratically with electric field strength and linearly with conductivity. To meet IEEE C95.1 standards, localized SAR in head or trunk regions should remain below 1.6 W/kg (averaged over 1g) or 2.0 W/kg (averaged over 10g) for safety.

MATLAB Program 41 is a code example for SAR visualization. It calculates and visualizes the SAR distribution in human tissue based on varying electric field values, shown in Figure 15.7. This example allows engineers and researchers to assess the impact of different electric field strengths on SAR, aiding in the design of safer implantable antennas.

MATLAB Program 41

```
% SAR Distribution Calculation and Visualization
% This code calculates and plots the Specific Absorption Rate
(SAR)
% distribution in human tissue as a function of electric
field strength.

% Parameters
sigma = 1.5; % Conductivity of human tissue in S/m
rho = 1000; % Density of tissue in kg/m^3
E = linspace(1, 100, 100); % Electric field range from 1 V/m to
100 V/m

% Calculate SAR using the provided formula
SAR = (sigma .* E.^2) ./ rho;

% Create a figure for SAR distribution
figure;
plot(E, SAR, 'k-');
xlabel('Electric Field (V/m)', 'FontSize', 12);
ylabel('SAR (W/kg)', 'FontSize', 12);
```

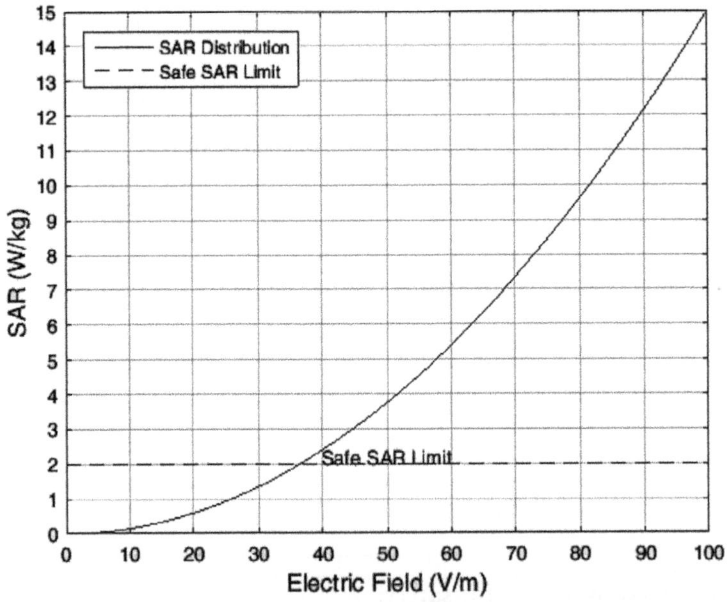

Figure 15.7 SAR distribution versus electric field.

```
title('SAR Distribution vs Electric Field', 'FontSize', 14);
grid on; % Add a grid for better visualization

% Setting y-axis interval to 1
set(gca, 'YTick', 0:1:max(SAR));

% Adding a horizontal line to indicate safe SAR limits
hold on;
safe_limit = 2; % Example safe limit for SAR in W/kg
plot(xlim, [safe_limit safe_limit], 'k--', 'DisplayName', 'Safe
SAR Limit'); % Black dashed line for safe limit

% Adjusting the position of the 'Safe SAR Limit' label above
the line
text(mean(xlim), safe_limit + 0.2, 'Safe SAR Limit',
'HorizontalAlignment', 'center', 'FontSize', 10, 'Color', 'k'); %
Position label above the line

% Displaying the plot
legend('SAR Distribution', 'Safe SAR Limit', 'Location',
'Northwest');
hold off;
```

The SAR distribution plot provides a practical tool for visualizing exposure risk in implantable antenna design. As electric field strength increases, SAR rises rapidly, underscoring the importance of low-power, efficient antennas with tightly confined radiation to minimize tissue heating. This visualization helps designers to assess electromagnetic safety and optimize antenna configurations to remain within established regulatory limits. By balancing performance and safety, engineers can create implantable devices that support advanced biomedical applications while protecting patient health.

Unlike RIS and reflect arrays discussed earlier, biomedical antennas operate at lower frequencies (e.g., ISM band) and in lossy environments (human tissue), which significantly impact design. However, the simulation principles using MATLAB—array factors, SAR calculation, and S-parameters—remain fundamentally aligned across applications.

15.3.3 Other Examples of Biomedical Antenna Applications

15.3.3.1 RFID for Patient Identification and Tracking

RFID technology is increasingly being used in healthcare for patient identification and tracking. By embedding RFID tags on patient wristbands, healthcare facilities can streamline operational efficiency, minimize errors, and enhance patient safety through accurate and reliable tracking. RFID systems in hospitals facilitate:

- Quick patient identification, especially in emergency scenarios;
- Real-time location tracking for monitoring patient movements and ensuring they are in the correct wards or receiving the appropriate treatments;

- Inventory management of medical supplies and equipment, allowing staff to track high-value assets across facilities.

MATLAB Program 42 is a code example for RFID antenna simulation. It simulates a simple dipole antenna designed for RFID applications, operating at the standard RFID frequency of 915 MHz. This frequency is commonly used for passive RFID systems in healthcare for applications such as patient and asset tracking. Figure 15.8 is a simulated image of a dipole antenna for RFID and Figure 15.9 is a 3D radiation pattern of an RFID dipole antenna at 915 MHz. Furthermore, Figure 15.10 demonstrates the reflection coefficient variation with frequency of the dipole antenna, while Figure 15.11 shows the gain pattern of the antenna at an elevation angle of 90 degrees, at 915 MHz.

MATLAB Program 42

```
% RFID Dipole Antenna Simulation at 915 MHz
% This MATLAB script simulates a half-wave dipole antenna,
commonly used in RFID applications,
% operating at 915 MHz. The code includes:
% 1- Visualization of the antenna structure in 3D space.
% 2- Calculation and plotting of the ideal dipole
radiation pattern.
% 3- Simulation of impedance matching by plotting the
S11 parameter.
% 4- Visualization of the gain pattern in the azimuth plane at an
elevation angle of 90°.
```

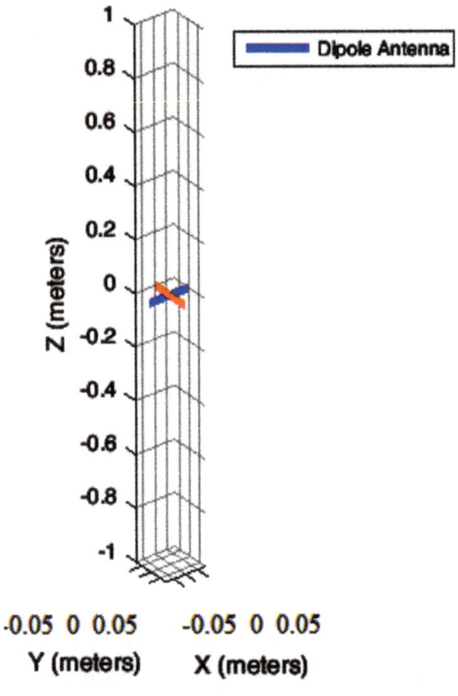

Figure 15.8 Dipole antenna for RFID at 915 MHz.

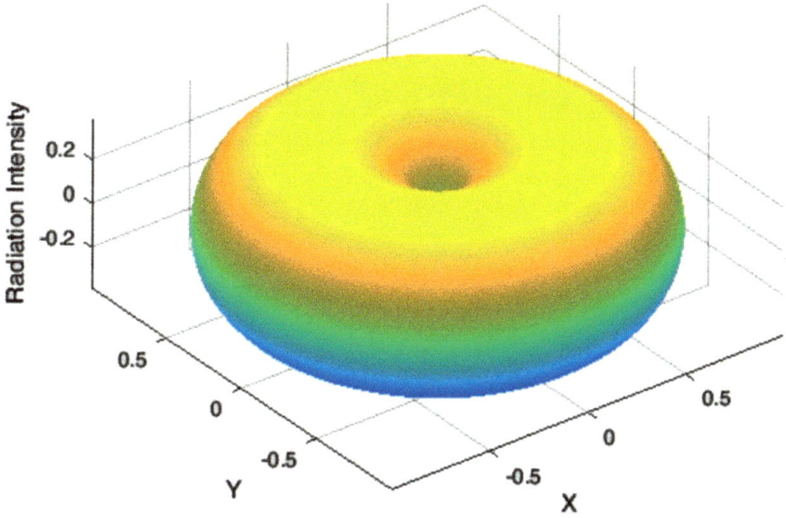

Figure 15.9 The 3D radiation pattern of dipole antenna at 915 MHz.

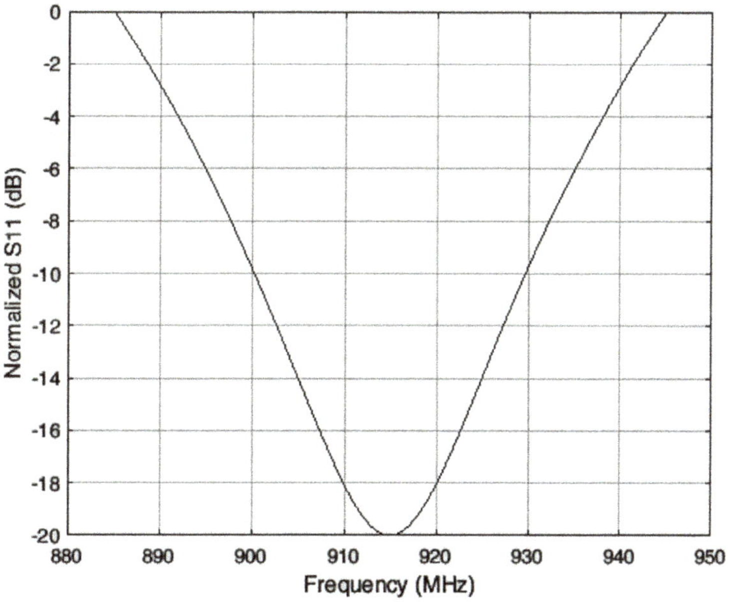

Figure 15.10 Normalized reflection coefficient of dipole antenna at 915 MHz.

```
% Parameters
frequency = 915e6; % RFID frequency in Hz
c = 3e8; % Speed of light in m/s
lambda = c / frequency; % Wavelength (in meters)

% Create Dipole Antenna
L = 0.5 * lambda; % Length of dipole antenna (half wavelength)

% Visualize the Dipole Antenna Structure
```

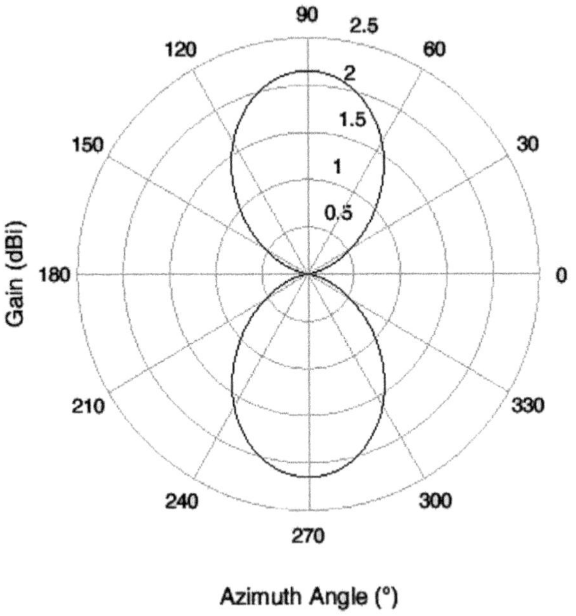

Figure 15.11 Gain pattern at elevation angle of 90°.

```
% Plotting the dipole antenna as a simple line in 3D space
figure;
plot3([-L/2, L/2], [0, 0], [0, 0], 'LineWidth', 4, 'Color', 'b');
hold on;
plot3([0, 0], [-L/2, L/2], [0, 0], 'LineWidth', 4, 'Color', 'r');
title('Dipole Antenna for RFID at 915 MHz');
xlabel('X (meters)');
ylabel('Y (meters)');
zlabel('Z (meters)');
grid on;
axis equal;
view(3); % 3D view
legend('Dipole Antenna');

% Radiation Pattern Simulation
theta = linspace(0, pi, 180); % Elevation angles (from 0 to
180 degrees)
phi = linspace(0, 2*pi, 360); % Azimuth angles (from 0 to
360 degrees)
[thetaGrid, phiGrid] = meshgrid(theta, phi);

% Compute a simplified radiation pattern (ideal dipole)
% Corrected radiation pattern: sin(theta)^2, representing an
ideal dipole's broadside pattern.
radiationPattern = sin(thetaGrid).^2;

% Convert spherical coordinates to Cartesian coordinates
for plotting
```

```
[x, y, z] = sph2cart(phiGrid, pi/2 - thetaGrid,
radiationPattern);

% Polar plot of radiation pattern using surf
figure;
surf(x, y, z);
shading interp;
title('3D Radiation Pattern of Dipole Antenna at 915 MHz');
xlabel('X');
ylabel('Y');
zlabel('Radiation Intensity');
axis equal;
grid on;

% Impedance Matching and S-Parameter Simulation
% Typically a dipole antenna has an impedance of about 73 ohms
at resonance.
% Let's simulate the S11 parameter (reflection coefficient) to
check impedance matching.

% Parameters for S11 calculation
centerFrequency = 915e6; % Resonant frequency at 915 MHz
deltaF = 10e6; % Bandwidth around resonant frequency (in Hz)
frequencyRange = linspace(centerFrequency - 30e6, centerFrequency
+ 30e6, 200); % Frequency range around 915 MHz

% S11 Calculation using a Lorentzian model (ignoring the
minus sign)
S11 = 20 * log10(1 + ((frequencyRange - centerFrequency) /
deltaF).^2);

% Normalize the S11 by subtracting the maximum S11 value to set
the maximum to 0 dB
S11_normalized = S11 - max(S11);

% Plot normalized S11 (Reflection Coefficient) across
frequency range
figure;
plot(frequencyRange / 1e6, S11_normalized, 'k-');
title('Normalized S11 Parameter (Reflection Coefficient) for
Dipole Antenna at 915 MHz');
xlabel('Frequency (MHz)');
ylabel('Normalized S11 (dB)');
grid on;

% Gain Simulation
% The gain of the dipole antenna is approximately 2.15 dBi in
free space.
gain = 2.15; % Gain of dipole antenna in dBi
disp(['Dipole Antenna Gain at ', num2str(frequency/1e6), ' MHz:
', num2str(gain), ' dBi']);
```

```
% Visualizing Gain Pattern in the Azimuth Plane (phi = 0 to 2*pi)
at elevation angle 90 degrees
phiAngles = linspace(0, 2*pi, 360); % Azimuth angles
gainPattern = gain * sin(pi/2).^2 * sin(phiAngles).^2; % Gain
varies with azimuth angle in figure-eight pattern

% Plot gain vs azimuth angle at elevation angle 90 degrees
figure;
polar(phiAngles, gainPattern, 'k-' );
title(['Gain Pattern at Elevation Angle of ', num2str(90), '°']);
xlabel('Azimuth Angle (°)');
ylabel('Gain (dBi)');
grid on;
```

The simulated 3D radiation pattern reveals the characteristic toroidal (doughnut-shaped) distribution of a half-wave dipole antenna, with peak radiation occurring broadside to the antenna axis. This radiation behavior supports efficient tag detection in passive RFID systems by ensuring strong signal coverage in the horizontal plane.

The normalized S_{11} plot shows a resonance dip near 915 MHz, indicating efficient impedance matching and minimal signal reflection—an essential criterion for effective RFID operation.

The gain pattern at 90° elevation reveals directional symmetry, reinforcing the dipole's suitability for omnidirectional tagging environments in clinical spaces.

15.3.3.2 Telemedicine and Remote Patient Monitoring

As patient monitoring expands beyond identification and tracking, telemedicine applications require antennas that balance compactness, power efficiency, and continuous data transmission, key concerns addressed in the next section.

In telemedicine applications, wearable devices equipped with antennas enable real-time monitoring of patients' vital signs. These devices use Bluetooth or Wi-Fi protocols to transmit data to healthcare providers. The design of antennas for such devices requires:

- Compact size to integrate within wearable devices without discomfort;
- Low power consumption to preserve battery life;
- Reliable signal transmission in varying body-worn configurations.

MATLAB Program 43 is a code example for Bluetooth antenna design. This example simulates a monopole antenna designed for Bluetooth, commonly used in telemedicine devices operating at 2.45 GHz, as shown in Figure 15.12.

MATLAB Program 43

```
% Bluetooth Antenna Design for Wearable Monitoring
% This code simulates the design of a quarter-wave monopole
antenna for Bluetooth applications operating at 2.45 GHz.
% It includes the antenna's structural design, visualizes its
radiation pattern, and calculates its gain in free space.
```

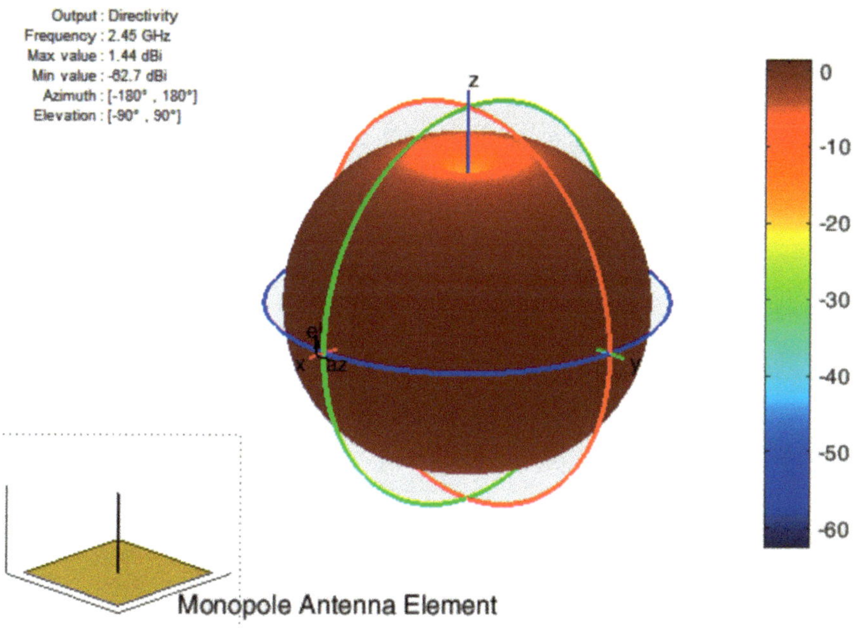

Figure 15.12 The 3D radiation pattern of Bluetooth monopole antenna at 2.45 GHz.

```
% Parameters
bluetooth_freq = 2.45e9;    % Bluetooth operating frequency in Hz
(2.45 GHz)
c = 3e8;                   % Speed of light in meters per second (m/s)
wavelength = c / bluetooth_freq; % Wavelength corresponding to
the Bluetooth frequency in meters
quarter_wavelength = 0.25 * wavelength; % Height of the monopole
antenna, set to a quarter of the wavelength

% Create Monopole Antenna
% - The antenna height is set to one-quarter of the wavelength to
match the monopole design for optimal resonance at the
operating frequency.
% - The width is kept small for simulation purposes, in
accordance with MATLAB's antenna modeling requirements.
% - The ground plane is also small, which is typical for wearable
device applications where space is limited.

bluetoothAnt = monopole('Height', quarter_wavelength, ...
               'Width', 0.001, ...           % Small antenna width
for simulation, typical for practical applications
               'GroundPlaneLength', 0.05, ...  % Ground plane
length suitable for wearable devices
               'GroundPlaneWidth', 0.05);      % Ground plane
width set for small-scale applications
```

```
% Radiation Pattern Visualization in 3D (excluding the
antenna structure)
% - This section computes and displays the radiation pattern of
the monopole antenna at 2.45 GHz.
% - The 3D plot shows the radiation pattern without including the
antenna structure in the visualization.

figure;
% title('3D Radiation Pattern of Bluetooth Monopole Antenna at
2.45 GHz'); % Optional title for the radiation pattern
pattern(bluetoothAnt, bluetooth_freq); % Display the 3D radiation
pattern of the monopole antenna at the Bluetooth frequency
grid on; % Enable grid for better visualization of the pattern

% Add title to the bottom left corner of the radiation
pattern figure
annotation('textbox', [0.15, 0.05, 0.1, 0.1], 'String', '
Monopole Antenna Element', ...
      'EdgeColor', 'none', 'FontSize', 12, 'HorizontalAlignment',
'left');

% Calculate and Display Antenna Gain
% The gain of a monopole antenna in free space is typically
around 2 dBi, but here we use an approximation of 2.15 dBi.
% The gain represents the antenna's ability to focus energy in a
specific direction, compared to an isotropic radiator.

gain = 2.15; % Approximate gain of the monopole antenna in dBi
(decibels relative to an isotropic radiator)
disp(['Monopole Antenna Gain at ', num2str(bluetooth_freq / 1e6),
' MHz: ', num2str(gain), ' dBi'])
```

15.3.3.3 Therapeutic Applications: Microwave Ablation

Beyond communication, antennas play a critical role in therapeutic applications, where energy delivery, not just data transmission, is central. The next section explores microwave ablation antennas that focus RF power into specific tissue regions.

Microwave ablation is a minimally invasive therapeutic technique designed to treat cancerous tissues by applying focused RF energy to targeted regions. Through precise energy delivery, this method selectively heats and destroys tumor cells while minimizing harm to surrounding healthy tissue. Microwave ablation is particularly effective for treating tumors in organs such as the liver, lungs, and kidneys, where precision and control are essential.

In microwave ablation, specialized antennas are positioned strategically near the tumor site to deliver RF energy efficiently. As the RF waves penetrate the tissue, they cause molecules to oscillate, generating heat. This heat propagates through the tissue, raising the temperature of the targeted area to a lethal range that induces apoptosis and necrosis in cancer cells. The result is a localized, controlled ablation, sparing adjacent healthy tissues from significant damage.

Thermal Energy Distribution Equation
The thermal energy absorbed in the tissue is quantified by the following relationship:

$$Q = SAR \cdot \rho \cdot C \cdot \Delta T \tag{15.8}$$

where

Q represents the thermal energy delivered to the tissue, which is a function of the power absorbed per mass.

SAR denotes the rate at which RF energy is absorbed by the tissue per unit mass (W/kg). SAR is directly related to the power of the RF source and the antenna design and configuration. Higher SAR values lead to greater localized heating, making this parameter critical for achieving therapeutic temperatures.

ρ is the tissue density (kg/m^3), an intrinsic property that impacts the rate of temperature increase in tissue. Tissue types with varying densities (e.g., fat versus muscle) absorb and dissipate energy differently, impacting the efficiency of ablation.

C is the specific heat capacity of the tissue (J/kg °C), representing the amount of energy required to raise the temperature of 1 kg of tissue by 1°C. This parameter dictates how quickly the tissue can reach therapeutic temperatures, with lower heat capacity values resulting in faster temperature increases.

ΔT is the temperature increase in the targeted tissue region (°C), which determines the efficacy of the ablation. Typical therapeutic ablation ranges aim to reach temperatures of 50°C and above within the tumor to ensure complete cellular death.

By adjusting these parameters, especially SAR, which can be controlled by antenna power and proximity to the tumor, clinicians can fine-tune the thermal energy delivered to achieve effective tumor ablation without causing significant collateral damage.

Role of Antennas in Energy Delivery
In microwave ablation, antennas are not only the energy delivery system but also the focus point for shaping the energy distribution in tissue. Different types of antennas, such as monopole and dipole configurations, can be used depending on the size and location of the tumor. Key considerations in antenna design and positioning include:

- *Antenna type and geometry:* The shape and length of the antenna affect the SAR distribution and the penetration depth of RF energy. For example, monopole antennas typically provide deeper penetration, suitable for larger or deeper tumors.
- *Frequency and power settings:* The frequency of the RF source, typically between 915 MHz and 2.45 GHz for ablation, determines the wavelength of

the energy and the depth of penetration. Power levels influence SAR and temperature rise rates, which are crucial for achieving the therapeutic threshold.

- *Antenna placement and orientation:* Correctly positioning the antenna within or adjacent to the tumor ensures that the highest energy density is concentrated within the tumor volume, sparing the surrounding healthy tissue.

The following code models the thermal distribution in the tumor region based on the SAR, tissue density, and specific heat capacity. This code is crucial for simulating and visualizing how microwave ablation parameters influence the temperature profile, ensuring efficient tumor ablation with minimal impact on surrounding tissues. Figure 15.13 demonstrates a 2D cross-section of the thermal energy distribution in a tumor during microwave ablation.

MATLAB Program 44

```
% Microwave Ablation Simulation: Modeling Thermal Energy
Distribution for a 2 cm Tumor
% This simulation models the thermal energy distribution in a
tumor during microwave ablation
% to achieve effective cell death by reaching temperatures within
a lethal range (50°C - 100°C).
% Key features:
% 1- Specific Absorption Rate (SAR) defines the microwave energy
absorption per unit mass.
% 2- Thermal Energy Delivery is modeled to stay within 200-500 J/
gram, essential for tumor cell ablation.
```

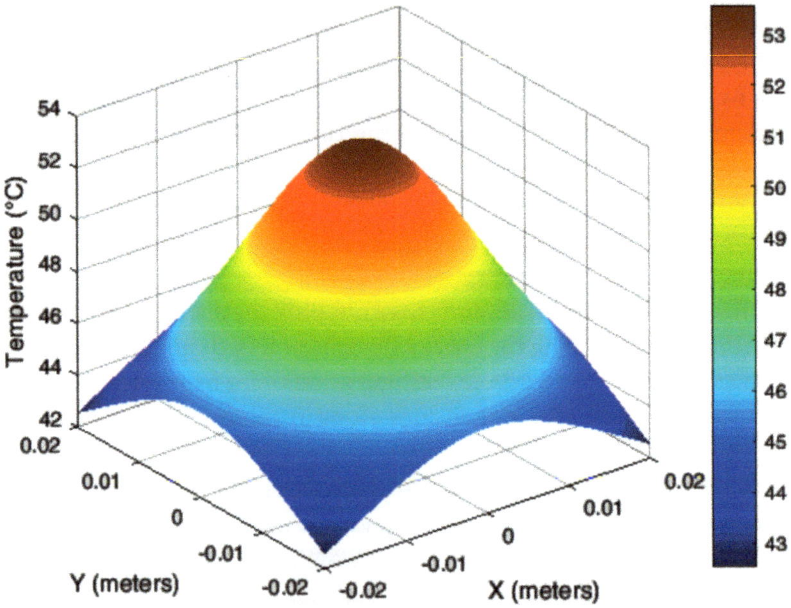

Figure 15.13 Temperature distribution in a tumor (2D cross-section).

```
% 3- Temperature Capping prevents unrealistic high temperatures
near the microwave source.
% 4- Energy per Gram ensures delivered energy per gram stays
within biological limits.

% Constants for Tissue Properties (Typical for soft tissue)
rho = 1000; % Tissue density in kg/m^3 (soft tissue)
C = 3600; % Specific heat capacity in J/kg°C (soft tissue)

% Parameters for Microwave Ablation
freq = 2.45e9; % Microwave frequency in Hz (2.45 GHz)
wavelength = 3e8 / freq; % Wavelength in meters

% SAR Calculation - Set in the typical range for microwave
ablation in soft tissue
SAR = 10; % Specific Absorption Rate in W/kg (typical value)

% Tumor Parameters
tumor_radius = 0.02; % Tumor radius in meters (2 cm)
antenna_power = 50; % Power delivered by antenna in Watts
treatment_duration = 10; % Treatment duration in minutes

% Simulation Grid
grid_size = 100; % Number of points for simulation grid
x = linspace(-tumor_radius, tumor_radius, grid_size);
y = linspace(-tumor_radius, tumor_radius, grid_size);
z = linspace(-tumor_radius, tumor_radius, grid_size);
[X, Y, Z] = meshgrid(x, y, z); % 3D grid for tumor volume

% Calculate Distance from Antenna (center of tumor)
distance = sqrt(X.^2 + Y.^2 + Z.^2);
distance(distance < 0.001) = 0.001; % Minimum threshold to
avoid singularities

% Initial Temperature Distribution
initial_temp = 37; % Initial temperature in °C (body temperature)

% Temperature Increase Calculation (ΔT)
temperature_increase = SAR * (antenna_power * treatment_duration
* 60) * (1 ./ (4 * pi * distance.^2)) / (rho * C);

% Apply Temperature Cap (Max Temperature Limit)
max_temp_increase = 100 - initial_temp; % Max temperature
increase to keep within biological range
temperature_increase = min(temperature_increase, max_temp_
increase); % Cap at max increase

% Final Temperature Distribution
```

```
final_temp = initial_temp + temperature_increase; % Final
temperature in °C

% Visualization of Temperature Distribution
figure;
surf(X(:,:,1), Y(:,:,1), final_temp(:,:,1)); % 2D Cross Section
(Z=0 plane)
title('Temperature Distribution in Tumor (2D Cross Section)');
xlabel('X (meters)');
ylabel('Y (meters)');
zlabel('Temperature (°C)');
colorbar;
colormap jet;
shading interp;

% Calculate Total Thermal Energy Delivered to the Tumor
tumor_volume = (4/3) * pi * tumor_radius^3; % Tumor volume in
cubic meters
mass_tumor = rho * tumor_volume; % Tumor mass in kilograms

% Total Thermal Energy Calculation (in Joules)
thermal_energy = SAR * antenna_power * treatment_duration * 60 *
mass_tumor; % Total energy in Joules

% Energy per Gram Calculation
energy_per_gram = thermal_energy / (mass_tumor * 1000); % Energy
per gram in J/g

% Display Results
disp(['Total Thermal Energy Delivered: ', num2str(thermal_
energy), ' Joules']);
disp(['Energy Delivered per Gram of Tumor: ', num2str(energy_per_
gram), ' Joules/gram']);

% Check if Energy per Gram is within Target Range (200-500 J/g)
if energy_per_gram > 500
  disp('Warning: Energy per gram is too high, consider reducing
SAR or power.');
elseif energy_per_gram < 200
  disp('Warning: Energy per gram is too low, consider increasing
SAR or power.');
else
  disp('Energy per gram is within the target range for
effective ablation.');
end
```

By understanding these principles and utilizing this code, clinicians and engineers can further optimize microwave ablation techniques, enhancing patient outcomes through more precise and effective tumor treatments. This approach integrates physics, engineering, and biology, illustrating the interdisciplinary nature of therapeutic microwave ablation.

15.3.4 Future Directions and Innovations

As antenna technology progresses, emerging trends focus on miniaturization, biodegradability, and smart substrates, enabling more sustainable, efficient, and adaptive healthcare systems. These directions are explored in the next section.

In the rapidly evolving field of antenna engineering, the development of metamaterial antennas and flexible, biodegradable antennas are two promising directions with significant implications for future technology. These innovations push the boundaries of traditional antenna applications, especially in areas requiring miniaturization, adaptability, and biocompatibility.

15.3.4.1 Metamaterial Antennas

Metamaterials are engineered structures with unique electromagnetic properties not commonly found in nature, such as negative refractive indices. By manipulating these properties, antenna designers can achieve improved performance characteristics, such as enhanced directivity, reduced size, and increased bandwidth. Negative permittivity and permeability values allow metamaterial antennas to concentrate electromagnetic waves more effectively, making them highly efficient and compact. These attributes are advantageous for applications in miniaturized devices, high-resolution imaging, and even stealth technology.

For example, a patch antenna utilizing metamaterial properties can achieve a significant reduction in size while maintaining effective radiation performance. By designing the substrate with a negative permittivity, the antenna's effective electrical size is altered, enabling it to resonate at lower frequencies than traditional materials.

A MATLAB code example of a metamaterial-inspired antenna is provided below, simulating a microstrip patch antenna with a metamaterial substrate. This example demonstrates how altering the substrate properties affects the antenna's behavior.

15.3.4.2 Simulating a Metamaterial-Inspired Patch Antenna

The following code simulates a microstrip patch antenna with a substrate engineered to exhibit negative permittivity. This effect emulates a metamaterial behavior in the antenna structure. Figure 15.14 calculates and compares the dimensions of a standard patch to a metamaterial one, while Figure 15.15 compares the bandwidth obtained by a standard patch, using FR4 dielectric, and a metamaterial patch.

MATLAB Program 45

```
% This code calculates and compares the dimensions and bandwidth
of two patch antennas:
% a standard patch antenna using FR4 substrate and a
metamaterial-inspired patch antenna.
% It displays the width, length, and area for each antenna and
plots their dimensions.
% It also estimates and compares their bandwidths using a
simplified quality factor approach.
```

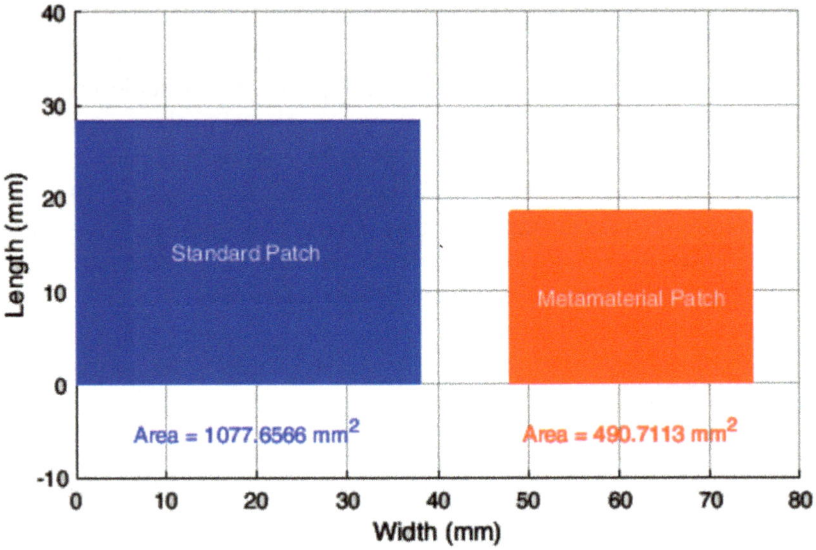

Figure 15.14 Patch antenna dimensions comparison.

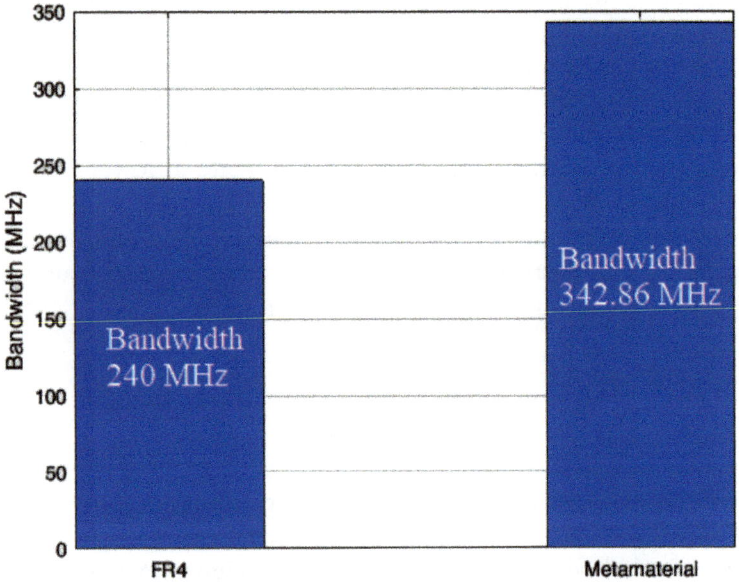

Figure 15.15 Bandwidth comparison between standard and metamaterial-inspired patch.

```
clc;
clear;
close all;

% Constants
c = 3e8; % Speed of light in vacuum (m/s)
f_0 = 2.4e9; % Resonant frequency (Hz)
h = 1.6e-3; % Substrate height (1.6 mm for both antennas)
```

```
% Material properties
epsilon_r_fr4 = 4.4; % Relative permittivity for FR4
(Standard Patch)
epsilon_r_meta = abs(-10); % Negative permittivity magnitude for
Metamaterial-inspired Patch

% Calculate dimensions for Standard Patch (FR4)
W_fr4 = c / (2 * f_0 * sqrt((epsilon_r_fr4 + 1) / 2)); % Width of
the patch
delta_L_fr4 = (0.412 * h * (epsilon_r_fr4 + 0.3) * (W_fr4 / h +
0.264)) / ((epsilon_r_fr4 - 0.258) * (W_fr4 / h + 0.8)); % Length
adjustment due to fringing
L_fr4 = c / (2 * f_0 * sqrt(epsilon_r_fr4)) - 2 * delta_L_fr4; %
Length of the patch

% Calculate dimensions for Metamaterial Patch
W_meta = c / (2 * f_0 * sqrt((epsilon_r_meta + 1) / 2)); % Width
of the patch
delta_L_meta = (0.412 * h * (epsilon_r_meta + 0.3) * (W_meta / h
+ 0.264)) / ((epsilon_r_meta - 0.258) * (W_meta / h + 0.8)); %
Length adjustment
L_meta = c / (2 * f_0 * sqrt(epsilon_r_meta)) - 2 * delta_L_meta;
% Length of the patch

% Convert dimensions to mm
W_fr4 = real(W_fr4) * 1e3; % Standard Patch Width in mm
L_fr4 = real(L_fr4) * 1e3; % Standard Patch Length in mm
W_meta = real(W_meta) * 1e3; % Metamaterial Patch Width in mm
L_meta = real(L_meta) * 1e3; % Metamaterial Patch Length in mm

% Calculate areas
Area_fr4 = W_fr4 * L_fr4; % Area in mm^2
Area_meta = W_meta * L_meta; % Area in mm^2

% Display Patch Dimensions
disp('Standard Patch Antenna (FR4) Dimensions:');
disp(['Width (W) = ' num2str(W_fr4) ' mm']);
disp(['Length (L) = ' num2str(L_fr4) ' mm']);
disp(['Area = ' num2str(Area_fr4) ' mm^2']);

disp('Metamaterial-inspired Patch Antenna Dimensions:');
disp(['Width (W) = ' num2str(W_meta) ' mm']);
disp(['Length (L) = ' num2str(L_meta) ' mm']);
disp(['Area = ' num2str(Area_meta) ' mm^2']);

% Plot the Patch Dimensions
figure;
hold on;

% Standard Patch
```

```
rectangle('Position', [0, 0, W_fr4, L_fr4], 'FaceColor', [0.3 0.3
1], 'EdgeColor', 'b');
text(W_fr4/2, L_fr4/2, 'Standard Patch', 'HorizontalAlignment',
'center', 'Color', 'w');
text(W_fr4/2, -5, ['Area = ' num2str(Area_fr4) ' mm^2'],
'HorizontalAlignment', 'center', 'Color', 'b');

% Metamaterial Patch
rectangle('Position', [W_fr4 + 10, 0, W_meta, L_meta],
'FaceColor', [1 0.3 0.3], 'EdgeColor', 'r');
text(W_fr4 + 10 + W_meta/2, L_meta/2, 'Metamaterial Patch',
'HorizontalAlignment', 'center', 'Color', 'w');
text(W_fr4 + 10 + W_meta/2, -5, ['Area = ' num2str(Area_meta) '
mm^2'], 'HorizontalAlignment', 'center', 'Color', 'r');

% Axis Labels and Title
xlabel('Width (mm)');
ylabel('Length (mm)');
title('Patch Antenna Dimensions Comparison');
legend('Standard Patch', 'Metamaterial Patch');
grid on;
axis equal;
hold off;

% Compare Bandwidth using a simple bar chart (Figure 2)
% Using simplified bandwidth estimation based on quality factor
(Q)
Q_fr4 = 10; % Estimated quality factor for FR4
Q_meta = 7; % Estimated quality factor for Metamaterial-
inspired antenna

% Bandwidth estimation formula: BW = f_0 / Q
BW_fr4 = f_0 / Q_fr4; % Bandwidth for FR4
BW_meta = f_0 / Q_meta; % Bandwidth for Metamaterial

% Plot Bandwidth Comparison with narrower bars
figure;
bar([real(BW_fr4) / 1e6, real(BW_meta) / 1e6], 'FaceColor', 'b',
'BarWidth', 0.4);  % Set BarWidth to make bars narrower

title('Bandwidth Comparison between Standard and Metamaterial-
inspired Patch');
set(gca, 'XTickLabel', {'FR4', 'Metamaterial'}); % Fix for
xticklabels function
ylabel('Bandwidth (MHz)');
grid on;

% Adding a legend to indicate what the color represents
legend('Bandwidth (MHz)', 'Location', 'northeast');

disp('End of Simulation');
```

15.3.4.3 Flexible and Biodegradable Antennas

In medical and environmental applications, there is a growing interest in antennas made from flexible, biodegradable materials. Using materials such as silk, magnesium, and cellulose, these antennas can be embedded temporarily in biological tissues or environmental sensors, designed to degrade naturally after a certain period. This property is especially advantageous for medical implants, where the antenna serves its purpose without needing surgical removal, minimizing long-term impacts on the body and reducing costs associated with permanent implants.

Flexible, biodegradable antennas are also used in wearable technologies and implantable devices for health monitoring, where adaptability to movement and environmental factors are critical. Future applications may include:

- *Medical diagnostics and therapeutics:* Temporary, dissolvable implants for monitoring and treatment.
- *Environmental sensing:* Disposable, environmentally friendly sensors that degrade after use, reducing electronic waste.
- *Wearable electronics:* Ultrathin, skin-conforming antennas for unobtrusive monitoring of vital signs and activity.

Combining metamaterials and biodegradable materials opens exciting possibilities for future antenna designs. For example, a flexible antenna utilizing metamaterials can achieve both high performance and adaptability, ideal for next-generation wearables and bio-integrated electronics.

15.4 Wave Propagation in Lossy Media

15.4.1 Introduction to Wave Propagation in Lossy Media

In a lossy medium, electromagnetic waves undergo attenuation due to the medium's inherent resistance, resulting in energy dissipation as the wave travels. Lossy media such as biological tissues, seawater, and various industrial materials are characterized by complex permittivity and permeability. Understanding wave behavior in these media is essential for applications in biomedical imaging, wireless underwater communication, and industrial material analysis, where signal attenuation and penetration depth play a significant role.

15.4.2 Mathematical Modeling of Lossy Media

To understand how waves behave in these materials, we begin by mathematically modeling the properties of lossy media using complex permittivity and permeability.

The propagation of electromagnetic waves in a lossy medium is governed by Maxwell's equations, modified to include the medium's complex permittivity and permeability.

15.4.2.1 Complex Permittivity and Permeability

The complex permittivity is expressed as:

$$\tilde{\varepsilon} = \varepsilon' - j\varepsilon'' \tag{15.9}$$

where ε' represents the medium's storage capability and ε'' represents the loss factor due to conductivity.

The complex permeability is expressed as:

$$\tilde{\mu} = \mu' - j\mu'' \tag{15.10}$$

where μ' is the real part and μ'' is the imaginary part that accounts for losses within the medium.

15.4.2.2 Complex Wave Number in Lossy Media

The complex wave number is expressed as:

$$\tilde{k} = \beta - j\alpha \tag{15.11}$$

where ω is the angular frequency of the wave, β is the phase constant, and α is the attenuation constant.

For a medium with complex permittivity, $\tilde{\varepsilon}$, and complex permeability, $\tilde{\mu}$, the wave number is given by:

$$\tilde{k} = \omega\sqrt{\tilde{\mu}\tilde{\varepsilon}} \tag{15.12}$$

15.4.2.3 Attenuation and Phase Constants

The attenuation constant α and phase constant β are derived as:

$$\alpha = \omega\sqrt{\frac{\mu\varepsilon}{2}}\left(\sqrt{1+\left(\frac{\sigma}{\omega\varepsilon}\right)^2} - 1\right)^{1/2} \tag{15.13}$$

$$\beta = \omega\sqrt{\frac{\mu\varepsilon}{2}}\left(\sqrt{1+\left(\frac{\sigma}{\omega\varepsilon}\right)^2} + 1\right)^{1/2} \tag{15.14}$$

The attenuation constant α determines how quickly the wave's amplitude decays with distance, typically measured in nepers per meter (Np/m). The phase constant β describes how the wave's phase progresses spatially and is measured in radians per meter. Together, these constants govern how electromagnetic waves behave in lossy media, affecting both signal strength and phase coherence over distance.

In these expressions, σ represents the conductivity of the medium. Higher conductivity results in greater attenuation and shorter penetration depth, which are critical parameters when designing systems for wave propagation through biological or industrial media.

15.4.2.4 Intrinsic Impedance in Lossy Media

The intrinsic impedance $\tilde{\eta}$ is defined as:

$$\tilde{\eta} = \sqrt{\frac{\tilde{\mu}}{\tilde{\varepsilon}}} = \eta' - j\eta'' \tag{15.15}$$

This complex intrinsic impedance influences both the amplitude and phase of reflected and transmitted waves.

15.4.3 Propagation Characteristics

These expressions establish the foundation for analyzing how energy attenuates and how phase shifts occur as electromagnetic waves propagate through resistive materials.

In practical scenarios, attenuation significantly impacts the penetration depth; the distance at which the wave amplitude decays to $1/e$ of its initial value. The penetration depth δ is inversely related to the attenuation constant α:

$$\delta = \frac{1}{\alpha} \tag{15.16}$$

The penetration depth δ quantifies the effective distance that a wave can propagate into a lossy medium before its amplitude drops to about 37% of its initial value. This corresponds to e^{-1}, a standard decay threshold used in electromagnetics. It plays a critical role in applications such as microwave imaging or subsurface sensing, where depth of signal penetration directly impacts system performance.

To visualize how wave attenuation evolves over distance, the following MATLAB simulation models electric field decay in a representative lossy medium such as biological tissue or seawater.

15.4.4 Simulation Using MATLAB

The following MATLAB simulation models the attenuation of an electromagnetic wave as it propagates through a lossy medium. It computes the complex wave number, intrinsic impedance, and attenuation constant and then simulates the electric field decay over distance.

MATLAB Program 46

```
% MATLAB Code for Simulating Wave Attenuation in Lossy Media
% This code models the attenuation of an electromagnetic wave
as it
% propagates through a lossy medium. It calculates the
complex wave
% number, intrinsic impedance, attenuation constant, and
then simulates
% the decay in electric field magnitude over distance due to
medium losses.

% Define physical constants and medium properties
f = 1e9;           % Frequency in Hz (1 GHz)
epsilon_0 = 8.854e-12; % Permittivity of free space (F/m)
```

```
mu_0 = 4 * pi * 1e-7; % Permeability of free space (H/m)
sigma = 0.01;         % Conductivity of the medium (S/m)
epsilon_r = 4;        % Relative permittivity
mu_r = 1;             % Relative permeability

% Derived parameters
omega = 2 * pi * f;                 % Angular frequency (rad/s)
epsilon = epsilon_0 * epsilon_r;    % Absolute permittivity
mu = mu_0 * mu_r;                   % Absolute permeability

% Complex permittivity and permeability
epsilon_c = epsilon - 1j * sigma / omega;
mu_c = mu;

% Calculate complex wave number and intrinsic impedance
k_c = omega * sqrt(mu_c * epsilon_c); % Complex wave number
alpha = real(k_c);              % Attenuation constant
beta = imag(k_c);               % Phase constant (not used
in plotting)
eta_c = sqrt(mu_c / epsilon_c); % Complex intrinsic impedance

% Define distance vector for propagation
d = linspace(0, 5, 500);    % Distance (m)
% Calculate the electric field as a function of distance
E0 = 1;                         % Initial field strength
E = E0 * exp(-alpha * d);          % Field decay due to attenuation
% Plot the results
figure;
plot(d, abs(E), 'k', 'LineWidth', 1.5);
xlabel('Distance (m)');
ylabel('Electric Field Magnitude');
title('Attenuation of Electric Field in Lossy Medium');
axis([0 0.2 0 1]);
grid on;
% Display results
fprintf('Attenuation constant, alpha = %.3f Np/m\n', alpha);
fprintf('Intrinsic impedance, eta = %.3f - %.3fj Ohm\n',
real(eta_c), imag(eta_c));
fprintf('Penetration depth = %.3f m\n', 1/alpha);

Display Results
Attenuation constant, alpha = 41.927 Np/m
Intrinsic impedance, eta = 188.225 - 4.227j Ohm
Penetration depth = 0.024 m
```

As shown in Figure 15.16, the electric field magnitude decays exponentially with distance due to attenuation in the lossy medium. The sharp drop within the first few centimeters highlights the impact of the attenuation constant and the resulting limited penetration depth. This MATLAB simulation models that behavior by computing key parameters such as the attenuation constant, intrinsic impedance, and penetration depth—factors essential for understanding how electromagnetic energy dissipates in resistive environments. Such insights are particularly valuable in applications such as biomedical imaging, underwater communication, and material analysis, where signal strength and field distribution must be carefully managed for effective system performance.

This modeling approach is directly applicable to the biomedical antennas discussed in Sections 15.3.2 and 15.3.3, where body tissues introduce similar lossy behavior. An accurate estimation of attenuation and penetration depth supports antenna design for safe and efficient operation within the human body or other complex media.

15.5 Solved Problems with MATLAB

In this section, we present a set of solved problems that apply MATLAB-based simulations to reinforce and extend the theoretical concepts discussed throughout this chapter. Each problem has been carefully designed to demonstrate key techniques in antenna design, intelligent surface modeling, and electromagnetic wave propagation. The solutions serve as practical illustrations of real-world scenarios in wireless communication systems, biomedical antenna applications, and lossy media environments.

The MATLAB codes are fully commented to ensure clarity and reproducibility. Readers are encouraged to experiment with the parameters to gain deeper insights into system behaviors and design trade-offs. By bridging analytical understanding with simulation-based validation, this section empowers learners, researchers, and engineers to advance from theoretical comprehension to practical implementation.

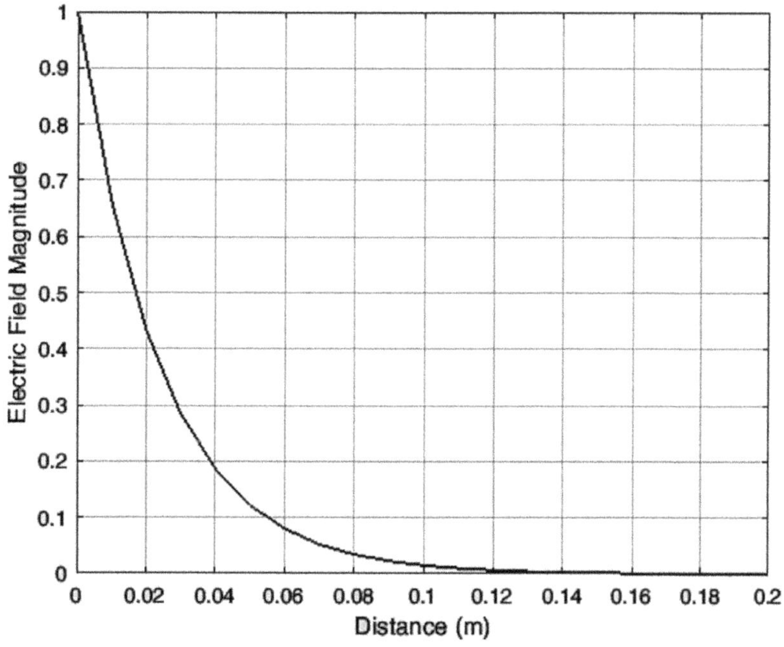

Figure 15.16 Attenuation of electric field in lossy medium.

The problems covered include:

- *Problem 1:* Simulation of RIS and beam steering performance.
- *Problem 2:* Comparative beamforming analysis of RIS and traditional reflect arrays.
- *Problem 3:* Design and performance evaluation of a patch antenna for biomedical applications.
- *Problem 4:* Wave propagation analysis in lossy environments such as seawater.

These problems exemplify the vital role of simulation in modern antenna engineering and radio wave propagation, providing a valuable foundation for further research and innovation in the field.

Problem 1: Simulation of RIS—Use MATLAB to simulate the reflection phase response of an 8 × 8 RIS array operating at 28 GHz.

(a) Model each RIS array element as a tunable phase-shifting surface. Let each element provide a phase shift between 0 and 2π in steps of $\pi/8$.

(b) Plot the far-field radiation pattern in polar for different incident angles (e.g., 30°, 45°), assuming uniform illumination from a fixed transmitter.

(c) Investigate and discuss the impact of element spacing (e.g., $\lambda/4$) on the beam steering capability of the RIS.

MATLAB Program 47 simulates and plots the far-field radiation pattern of an 8 × 8 RIS array with (spacing = 0.25λ, 30°, and 45°) and (spacing = 0.5λ, –30°, and –45°) at 28 GHz, as shown in Figure 15.17.

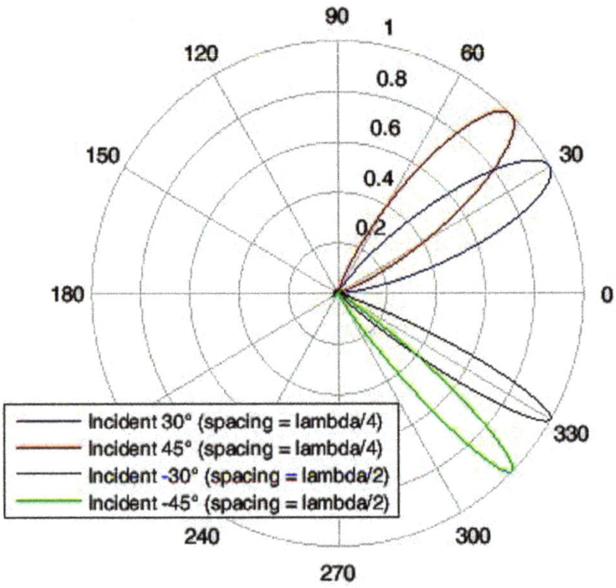

Figure 15.17 Far-field pattern of 8 × 8 RIS (spacing = 0.25λ, 30°, and 45°) and (spacing = 0.5λ, –30°, and –45°) at 28 GHz.

MATLAB Program 47

```
% MATLAB PROGRAM: RIS Far-Field Pattern Simulation at 28 GHz
% Purpose:
%  This code simulates the far-field radiation pattern of
a Reconfigurable
%  Intelligent Surface (RIS) array at 28 GHz, including a polar
plot with
%  correct main lobe and sidelobe orientation.
%
% Parameters
freq = 28e9;        % Frequency in Hz
c = 3e8;            % Speed of light in m/s
lambda = c / freq;  % Wavelength in meters

% Array configuration
N = 8;              % NxN RIS array
spacing = lambda / 4; % Element spacing

% Grid for RIS elements
[x_grid, y_grid] = meshgrid(0:N-1, 0:N-1);
x = x_grid * spacing;
y = y_grid * spacing;

% Incident angles (in degrees)
incident_angles = [30, 45];  % We'll simulate two cases
num_angles = length(incident_angles);

% Observation angles (for full-circle polar plot)
theta_deg = linspace(-180, 180, 2000);
theta_rad = deg2rad(theta_deg);

% Initialize array to store far-field patterns
E_all = zeros(num_angles, length(theta_rad));

% === Far-field computation ===
for a = 1:num_angles
  phi_inc = deg2rad(incident_angles(a));  % Incident angle
in radians
  steering_vector = zeros(size(theta_rad));

  for k = 1:length(theta_rad)
    phi_obs = theta_rad(k);
    sum_pattern = 0;

    for m = 1:N
      for n = 1:N
        delta_phase = 2 * pi / lambda * ...
          (x(m,n) * (cos(phi_obs) - cos(phi_inc)) + ...
           y(m,n) * (sin(phi_obs) - sin(phi_inc)));
```

```
            sum_pattern = sum_pattern + exp(1j * delta_phase);
         end
      end

      steering_vector(k) = abs(sum_pattern)^2;
   end

   % Normalize and store
   E_all(a,:) = steering_vector / max(steering_vector);
end

% === Polar Plot (Full-Angle, including sidelobes) ===
figure;
polar(theta_rad, E_all(1,:), 'b'); hold on;
polar(theta_rad, E_all(2,:), 'r');
legend('Incident 30°', 'Incident 45°', 'Location',
'southoutside');
title('Far-field Pattern of 8x8 RIS (spacing = 0.25, angle = 30°
and 45°) at 28 GHz');
```

Problem 2: Comparison of RIS and Reflect Arrays in Beamforming—Compare the performance of RIS and reflect arrays in beamforming applications using MATLAB.

(a) Model both an RIS and a conventional reflect array with identical physical dimensions and operating at 28 GHz. Use a 4 × 4 configuration for both.

(b) Simulate and plot their far-field radiation patterns for a fixed incident angle and a target reflected direction (e.g., normal to the surface).

(c) Evaluate and compare their beamforming gain, main lobe direction, and flexibility in dynamic beam steering.

MATLAB Program 48 compares and plots the beamforming performance or an RIS array and a fixed reflect array for a 4 × 4 configuration at 28 GHz with 45° incident wave, as shown in Figure 15.18.

MATLAB Program 48

```
% This code compares the beamforming performance of a
Reconfigurable Intelligent Surface (RIS) and a fixed
Reflect Array.
% Both systems are simulated for a 4x4 configuration at 28 GHz,
and their far-field patterns are analyzed for a 45°
incident wave.
% The goal is to assess dynamic steering capability and beam
gain differences.

% Define frequency, speed of light, and wavelength
freq = 28e9;
c = 3e8;
lambda = c / freq;
```

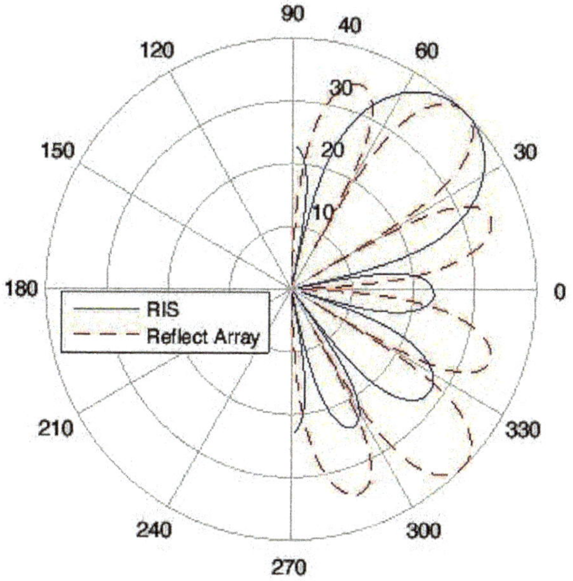

Figure 15.18 RIS versus reflect array beamforming (4 × 4 array) at 28 GHz with 45° incident wave.

```
% RIS/Reflect Array setup
N = 4;
spacing = lambda / 2;
[x, y] = meshgrid(0:N-1, 0:N-1);
x = x * spacing;
y = y * spacing;

% Observation angles from -90 to 90 degrees
theta = linspace(-90, 90, 1000);
theta_rad = deg2rad(theta);

% Fixed incident angle
phi_inc = deg2rad(45);

% Initialize far-field patterns
ris_pattern = zeros(size(theta));
reflect_array_pattern = zeros(size(theta));

% Compute far-field patterns
for k = 1:length(theta_rad)
  phi_obs = theta_rad(k);
  ris_sum = 0;
  reflect_sum = 0;

  for m = 1:N
    for n = 1:N
      r_phase = 2*pi/lambda * (x(m,n)*cos(phi_obs) +
y(m,n)*sin(phi_obs));
```

```
        delta_phase = 2*pi/lambda * ...
          (x(m,n)*(cos(phi_obs) - cos(phi_inc)) + ...
           y(m,n)*(sin(phi_obs) - sin(phi_inc)));

        ris_sum = ris_sum + exp(1j * delta_phase);
        reflect_sum = reflect_sum + exp(1j * r_phase);
      end
    end

    ris_pattern(k) = abs(ris_sum)^2;
    reflect_array_pattern(k) = abs(reflect_sum)^2;
  end

% Normalize and convert to dB
ris_pattern_dB = 10*log10(ris_pattern / max(ris_pattern)) + 40;
reflect_array_pattern_dB = 10*log10(reflect_array_pattern /
max(reflect_array_pattern)) + 40;

% Exclude negative values after adding 40 dB
valid_idx = (ris_pattern_dB >= 0) & (reflect_array_pattern_dB >=
0);

% Clip values to 40 dB max
ris_pattern_dB = min(ris_pattern_dB(valid_idx), 40);
reflect_array_pattern_dB = min(reflect_array_pattern_dB(valid_
idx), 40);
theta_polar = theta_rad(valid_idx);

% Polar plot
figure;
polar(theta_polar, ris_pattern_dB, 'b'); hold on;
polar(theta_polar, reflect_array_pattern_dB, 'r--');
legend('RIS', 'Reflect Array');
title('RIS vs. Reflect Array Beamforming (4x4 array) at 28 GHz
with 45° Incident Wave');
```

Problem 3: Antenna Design for Biomedical Applications—Simulate a small patch antenna for biomedical applications (e.g., implantable devices) operating at the 2.4-GHz ISM band using MATLAB.

(a) Design the antenna with appropriate dimensions (use a substrate with ε_r = 9.4 and thickness = 1.5 mm). Plot the geometry.

(b) Plot the return loss (S_{11}) versus frequency from 2.2 GHz to 2.6 GHz to evaluate impedance matching.

(c) Simulate the antenna's performance in a lossy medium that mimics human muscle tissue (ε_r = 52.7, conductivity = 1.73 S/m) and analyze the shift in resonant frequency and bandwidth.

MATLAB Program 49 estimates the physical dimensions of a microstrip patch antenna for biomedical use, at 2.4-GHz ISM band, based on dielectric properties, and evaluates and plots the return loss in both air and tissue, like environments, as demonstrated in Figure 15.19.

MATLAB Program 49

```
% This code designs a microstrip patch antenna for biomedical use
at the 2.4 GHz ISM band.
% It estimates physical dimensions based on dielectric
properties, and evaluates the return loss
% (S11) in both air and tissue-like environments using a
realistic Lorentzian model with proper resonance behavior.

% --------------------------
% Antenna Design Parameters
% --------------------------

freq = 2.4e9;  % Target operating frequency (2.4 GHz)
c = 3e8;       % Speed of light (m/s)
epsilon_r = 9.4; % Substrate relative permittivity
thickness = 1.5e-3; % Substrate thickness (m)

% Estimate patch width (W)
W = c / (2 * freq) * sqrt(2 / (epsilon_r + 1));

% Estimate effective permittivity
epsilon_eff = (epsilon_r + 1)/2 + (epsilon_r - 1)/2 * (1 + 12 *
thickness / W)^(-0.5);

% Estimate patch length (L)
```

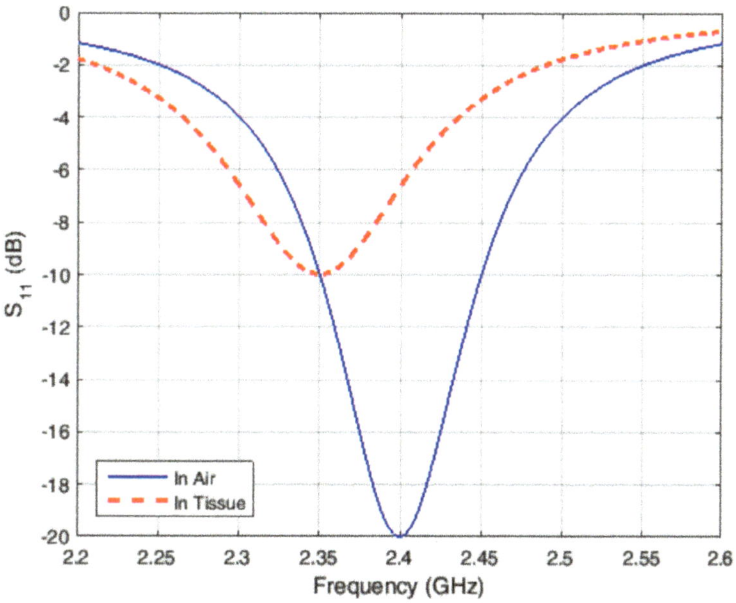

Figure 15.19 Return loss of patch antenna (realistic tissue effects).

```
L = c / (2 * freq * sqrt(epsilon_eff));

% Display calculated patch dimensions
fprintf('Patch dimensions: L = %.3f mm, W = %.3f mm\n', L * 1e3,
W * 1e3);

% --------------------------
% Return Loss Simulation
% --------------------------

% Frequency sweep
f = linspace(2.2e9, 2.6e9, 400);  % Frequency range from 2.2 GHz
to 2.6 GHz

% Resonant frequency and bandwidth
f0_air = 2.4e9;   % Resonance in air
BW_air = 0.05e9;    % Narrow bandwidth (high Q)
S11_min_air = -20;  % Deep notch in dB

f0_tissue = 2.35e9; % Shifted resonance in tissue
BW_tissue = 0.07e9;   % Broader bandwidth due to higher loss
S11_min_tissue = -10; % Shallower notch in dB

% Lorentzian model for realistic return loss
S11_air = S11_min_air ./ (1 + ((f - f0_air) / BW_air).^2);
S11_tissue = S11_min_tissue ./ (1 + ((f - f0_tissue) /
BW_tissue).^2);

% --------------------------
% Plotting
% --------------------------

figure;
plot(f / 1e9, S11_air, 'b', 'LineWidth', 1.5); hold on;
plot(f / 1e9, S11_tissue, 'r--', 'LineWidth', 1.5);

xlabel('Frequency (GHz)');
ylabel('S_{11} (dB)');
title('Return Loss of Patch Antenna (Realistic Tissue Effects)');
legend('In Air', 'In Tissue', 'Location', 'SouthWest');
grid on;
```

The return loss plot reveals that, at frequencies below 2.35 GHz, the antenna is better matched to the tissue medium, resulting in lower S_{11}. This indicates that more power is delivered to the tissue, where it is absorbed due to the medium's high permittivity and conductivity.

Above 2.35 GHz, the mismatch increases, causing higher reflection and less power absorption. The broader, shallower dip also signifies that the antenna's quality factor (Q) is lower in tissue due to increased losses.

Problem 4: Wave Propagation in Lossy Media—Write a MATLAB script to analyze electromagnetic wave propagation through a lossy medium (e.g., seawater) at 1 GHz.

(a) Use the complex permittivity and permeability of seawater ($\varepsilon_r = 81$, conductivity = 4 S/m, $\mu_r = 1$).
(b) Calculate and plot the attenuation constant (α) and phase constant (β) over the frequency range from 1 MHz to 1 GHz.
(c) Discuss the results in the context of underwater wireless communication, noting limitations and design considerations for antenna systems in high-loss environments.

MATLAB Program 50 computes and plots the attenuation constant and phase constant, using a complex permittivity mode of an electromagnetic wave propagating through a lossy medium like seawater, as shown in Figure 15.20.

MATLAB Program 50

```
% This code analyzes electromagnetic wave propagation through a
lossy medium (seawater) across a wide frequency range.
% It computes and plots the attenuation constant (alpha) and
phase constant (beta) using the complex permittivity model.
% Results help evaluate challenges in underwater communication
system design.

% Frequency range: 1 MHz to 1 GHz
f = logspace(6, 9, 500);
```

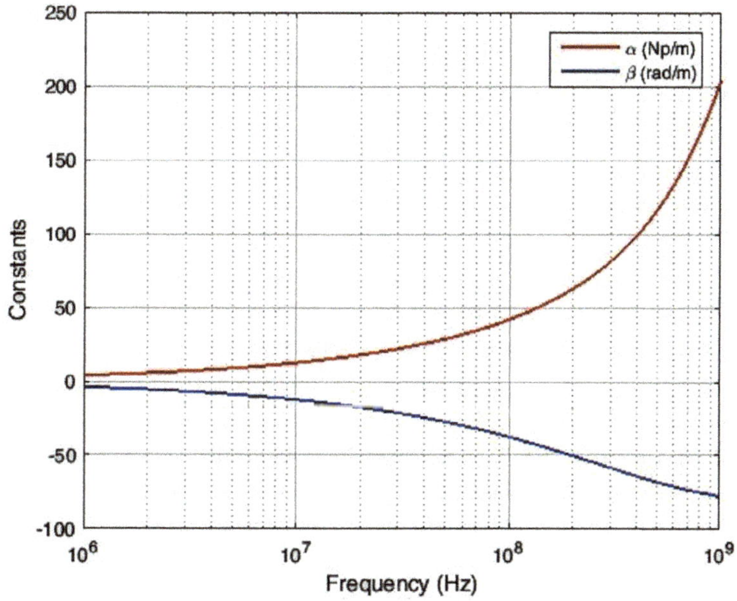

Figure 15.20 Wave propagation in seawater.

```
% Seawater properties
epsilon_r = 81; % Relative permittivity
sigma = 4;      % Conductivity (S/m)
mu_0 = 4*pi*1e-7; % Permeability of free space
epsilon_0 = 8.854e-12; % Permittivity of free space

% Angular frequency
omega = 2*pi*f;

% Complex permittivity
epsilon_c = epsilon_r*epsilon_0 - 1j*sigma./omega;

% Propagation constant
k = omega .* sqrt(mu_0 .* epsilon_c);

% Extract attenuation and phase constants
alpha = real(k); % Attenuation (Np/m)
beta = imag(k);  % Phase constant (rad/m)

% Plot results
semilogx(f, alpha, 'r', 'LineWidth', 1.5); hold on;
semilogx(f, beta, 'b', 'LineWidth', 1.5);xlabel('Frequency
(Hz)');
ylabel('Constants');
legend('\alpha (Np/m)', '\beta (rad/m)');
title('Wave Propagation in Seawater');
grid on;
```

Problems for Students

15.1 (a) Explain the concept of an RIS and its potential impact on wireless communication systems, particularly in the context of 5G and beyond.
(b) Using MATLAB, simulate the impact of RIS on the received signal strength for a wireless communication system operating at 28 GHz. Assume a fixed transmitting antenna and variable RIS configuration.

15.2 (a) Derive the expression for the reflection coefficient at a RIS element when the incident wave is at an angle θ and the RIS surface is reconfigurable.
(b) Using MATLAB, simulate the reflection coefficient for an RIS element as a function of incident angle θ (from 0° to 90°) and the reconfigurable parameters of the surface.

15.3 (a) Discuss the advantages of RIS in enhancing signal propagation in a multipath environment. How does it affect the signal-to-noise ratio (SNR) at the receiver?
(b) Using MATLAB, simulate a multipath environment with RIS and calculate the improvement in SNR compared to a nonreconfigurable scenario. Consider parameters such as path loss and RIS configuration.

15.4 (a) Define a reflect array and describe its working principle. How does it differ from traditional parabolic reflectors in terms of design and performance?

(b) Using MATLAB, simulate the far-field radiation pattern of a reflect array antenna. Assume that the reflect array consists of 10 elements operating at 10 GHz.

15.5 (a) Explain the concept of phase compensation in reflect arrays. How is it used to steer the beam in a desired direction?

(b) Using MATLAB, optimize the phase shifts of a 2×2 reflect array to direct the main lobe towards a target angle (e.g., 30°). Calculate the resulting radiation pattern and compare it to a nonoptimized array.

15.6 (a) Discuss the unique challenges in designing antennas for biomedical applications, particularly in-body communication and medical imaging.

(b) Using MATLAB, simulate the electromagnetic field distribution around a patch antenna operating at 2.4 GHz and placed in close proximity to human tissue. Consider the tissue properties such as permittivity and conductivity.

15.7 (a) Explain the role of antennas in wireless capsule endoscopy and the importance of miniaturization for such applications.

(b) Using MATLAB, simulate the radiation pattern of a small, miniaturized antenna designed for capsule endoscopy. Calculate the SAR in human tissue.

15.8 (a) Derive the expression for wave propagation in a lossy dielectric medium, accounting for both the attenuation constant and phase shift.

(b) Using MATLAB, simulate the propagation of a plane wave in a lossy medium at 2 GHz. Assume that the medium has a relative permittivity of 4 and a conductivity of 0.01 S/m. Calculate the attenuation and phase shift over a distance of 10m.

15.9 (a) Explain how waveguide modes are affected when the waveguide is filled with a lossy dielectric material.

(b) Using MATLAB, simulate the TE_{10} mode in a rectangular waveguide filled with a lossy dielectric. Calculate the attenuation constant and the cutoff frequency for a waveguide with dimensions 2 cm $\times$ 1 cm at 5 GHz.

15.10 (a) Discuss the effect of lossy media on antenna performance, specifically on radiation efficiency and bandwidth.

(b) Using MATLAB, simulate an antenna in a lossy medium and calculate the change in bandwidth and radiation efficiency compared to an ideal free-space scenario. Consider a monopole antenna operating at 1 GHz and a medium with a conductivity of 0.1 S/m.

Problems for Instructors

15.1 Design and Simulation of a Reflect Array: Design a planar reflect array using MATLAB for operation at 10 GHz.

(a) Calculate the required phase distribution for focusing the reflected wave at a point 10m away.

(b) Simulate the far-field radiation pattern and plot the gain in the principal plane.

(c) Analyze how element phase errors affect the beam efficiency.

15.2 Tissue-Specific Absorption in Biomedical Antennas: Using MATLAB, model the SAR for a patch antenna operating at 1.8 GHz in human tissue.
   (a) Simulate the field distribution inside a tissue phantom.
   (b) Calculate the SAR for varying input power levels (e.g., 1W, 2W).
   (c) Discuss how antenna design affects SAR compliance with safety standards.

15.3 Visualization of Wave Propagation in Lossy Media: Develop a MATLAB animation to visualize the propagation of a Gaussian-modulated sine wave through a lossy medium (e.g., fog or human tissue).
   (a) Include the effects of attenuation and dispersion in the animation.
   (b) Plot the amplitude and phase of the wave as functions of distance and time.
   (c) Analyze the challenges of designing antennas for lossy environments based on the results.

Selected Bibliography

Basar, E., et al., "Reconfigurable Intelligent Surfaces for 6G: Emerging Hardware Architectures, Applications, and Open Challenges," *IEEE Vehicular Technology Magazine*, 2024.

Basar, E., et al., "Wireless Communications Through Reconfigurable Intelligent Surfaces," *IEEE Access*, Vol. 7, 2019, pp. 116753–116773.

Basharat, S., et al., "Reconfigurable Intelligent Surfaces: Potentials, Applications, and Challenges for 6G Wireless Networks," *IEEE Wireless Communications*, Vol. 28, No. 6, 2021, pp. 184–191.

Benson, F. A., and T. M. Benson, *Fields, Waves, and Transmission Lines*, London: Chapman and Hall, 1991.

Gabriel, C., and S. Gabriel, *Compilation of the Dielectric Properties of Body Tissues at RF and Microwave Frequencies*, report by Brooks Air Force Technical Directorate, June 1996.

Hall, P. S., and Y. Hao, *Antennas and Propagation for Body-Centric Wireless Communications*, Norwood, MA: Artech House, 2012.

Hum, S. V., and J. Perruisseau-Carrier, "Reconfigurable Reflectarrays and Array Lenses for Dynamic Antenna Beam Control: A Review," *IEEE Transactions on Antennas and Propagation*, Vol. 62, No. 1, 2013, pp. 183–198.

Kiourti, A., and K. S. Nikita, "A Review of In-Body Biotelemetry Devices: Implantables, Ingestibles, and Injectables," *IEEE Transactions on Biomedical Engineering*, Vol. 64, No. 7, 2017, pp. 1422–1430.

Liu, Y., et al., "Reconfigurable Intelligent Surfaces: Principles and Opportunities," *IEEE Communications Surveys & Tutorials*, Vol. 23, No. 3, 2021, pp. 1546–1577.

Liu, F., D. H. Kwon, and S. Tretyakov, "Reflectarrays and Metasurface Reflectors as Diffraction Gratings: A Tutorial," *IEEE Antennas and Propagation Magazine*, Vol. 65, No. 3, 2023, pp. 21–32.

Malik, N. A., et al., "Implantable Antennas for Bio-Medical Applications," *IEEE Journal of Electromagnetics, RF and Microwaves in Medicine and Biology*, Vol. 5, No. 1, 2020, pp. 84–96.

Mohan, A., and N. Kumar, "Implantable Antennas for Biomedical Applications: A Systematic Review," *BioMedical Engineering OnLine*, Vol. 23, No. 1, 2024, p. 87.

APPENDIX

Derivation of the Power Transmission Coefficient at the Plane Boundary of Different Refractive Index Media

At the plane boundary between two homogeneous media of refractive index n_1 and n_2, as shown in Figure A.1, the tangential fields in the xz plane are such that the following electric and magnetic boundary equations hold:

$$E_{xi} + E_{xr} = E_{xt} \tag{A.1}$$

$$H_{yi} + H_{yr} = H_{yt} \tag{A.2}$$

where i, r, t are symbols representing the incident, reflected, and transmitted fields.

$$E_{xi} = Z_1 H_{yi} \cos\theta_i \tag{A.3}$$

$$E_{xr} = Z_1 H_{yr} \cos\theta_r = -Z_1 H_{yr} \cos\theta_i \tag{A.4}$$

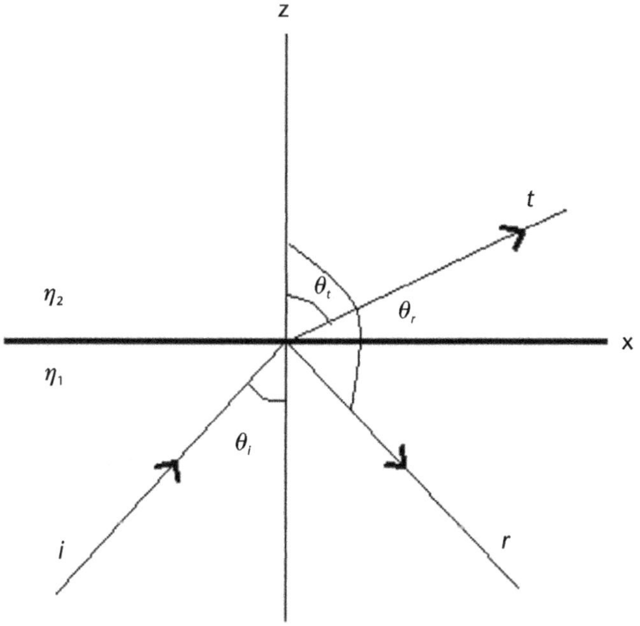

Figure A.1 Reflection and transmission of an incident wave at a boundary between two regions of different refractive indices.

$$E_{xt} = Z_2 H_{yt} \cos\theta_t \tag{A.5}$$

Substituting (A.2) to (A.5) into (A.1) gives

$$Z_1 H_{yi} \cos\theta_i - Z_1 \cos\theta_i \left(H_{yt} - H_{yi}\right) = Z_2 H_{yt} \cos\theta_t \tag{A.6}$$

Hence,

$$\frac{H_{yt}}{H_{yi}} = \frac{2Z_1 \cos\theta_i}{Z_2 \cos\theta_t + Z_1 \cos\theta_i} = \frac{2\cos\theta_i}{\dfrac{Z_2}{Z_1}\cos\theta_t + \cos\theta_i} \tag{A.7}$$

The power transmission coefficient, T_t, is defined as:

$$T_t = \frac{\text{Transmitted power}}{\text{Incident power}} \tag{A.8}$$

Hence,

$$T_t = \frac{\left|E_{xt} \times H_{yt}\right|}{\left|E_{xi} \times H_{yi}\right|} = \frac{E_{xt} H_{yt}}{E_{xi} H_{yi}} = \frac{Z_2}{Z_1}\left(\frac{H_{yt}}{H_{yi}}\right)^2 \tag{A.9}$$

$$T_t = \frac{Z_2}{Z_1} \frac{4\cos^2\theta_i}{\left(\dfrac{Z_2}{Z_1}\cos\theta_t + \cos\theta_i\right)^2} \tag{A.10}$$

Since

$$Z_1 = \sqrt{\frac{\mu_0}{\varepsilon_0 \varepsilon_{r1}}} = \frac{Z_0}{\sqrt{\varepsilon_{r1}}} = \frac{Z_0}{n_1} \tag{A.11}$$

$$Z_2 = \sqrt{\frac{\mu_0}{\varepsilon_0 \varepsilon_{r2}}} = \frac{Z_0}{\sqrt{\varepsilon_{r2}}} = \frac{Z_0}{n_2} \tag{A.12}$$

then

$$T_t = \frac{4\dfrac{n_1}{n_2}\cos^2\theta_i}{\left(\dfrac{n_1}{n_2}\cos\theta_t + \cos\theta_i\right)^2} = \frac{4n_1 n_2 \cos^2\theta_i}{\left(n_1\cos\theta_t + n_2\cos\theta_i\right)^2} \tag{A.13}$$

In a special case when $n_2 = 1$ and $\theta_i = \theta_t = 0$, (A.13) reduces to:

$$T_t = \frac{4n_1}{\left(n_1 + 1\right)^2} \tag{A.14}$$

About the Author

Osama W. Ata received his BSc (Hons) degree in electronic engineering from Hull University (United Kingdom) in 1980 and an MSc in electronic communications and PhD in microwave antenna systems from Cardiff University (United Kingdom), in 1981 and 1989, respectively. He was granted a postdoctoral fellowship at Bangor University (United Kingdom), a Fulbright Senior Fellowship at Stanford University (United States), and senior visiting fellowships at Villanova University (United States) and Victoria University of Technology (Australia). Pending his Fulbright fellowship, Dr. Ata qualified for a unique and prestigious Extraordinary Ability O1 visa in the United States (2000 to 2005), for his expertise in antenna research, and worked at the research facility of Sprint Telecommunications Corporate. Dr. Ata is a professor at Palestine Polytechnic University in Hebron, where he teaches courses in antennae and propagation, microwave communications, electromagnetics, transmission line theory, optical communications, radar systems, applied engineering mathematics, and research methodology. His research interests are in microwave horn antennae, microstrip antennae, radio frequency propagation, in-building penetration loss modeling and measurement, electromagnetic radiation, RF cellular planning, and mobile radio networks. He recently won the Abundance in Quality Scientific Research Production award and won the Best Paper award at the 2019 Jordan International Joint Conference Electrical Engineering and Information Technology (IEEE JEEIT).

Index

N

O

P

T

U

V

W

X

Y

Z

Recent Titles in the Artech House Antennas and Propagation Library

Christos Christodoulou, Series Editor
Vince Rodriguez, Series Editor

Adaptive Array Measurements in Communications, M. A. Halim

Antenna-on-Chip: Design, Challenges, and Opportunities, Hammad M. Cheema, Fatima Khalid, and Atif Shamim

Antenna Design for Cognitive Radio, Youssef Tawk, Joseph Costantine, and Christos Christodoulou

Antenna Design for CubeSats, Reyhan Baktur

Antenna Design with Fiber Optics, A. Kumar

Antenna Engineering and Radio Wave Propagation with MATLAB®, Osama W. Ata

Antenna Engineering Using Physical Optics: Practical CAD Techniques and Software, Leo Diaz and Thomas Milligan

Antenna Measurement Techniques, Gary E. Evans

Antennas for IoT, Prutha P. Kulkarni

Antennas and Propagation for Body-Centric Wireless Communications, Second Edition, Peter S. Hall and Yang Hao, editors

Antennas and Site Engineering for Mobile Radio Networks, Bruno Delorme

Antennas for Small Mobile Terminals, Kyohei Fujimoto and Koichi Ito, editors

Analysis, Design and Measurement of Small and Low-Profile Antennas, Kazuhiro Hirasawa

Analysis of Radome-Enclosed Antennas, Second Edition, Dennis J. Kozakoff

Applications of Geographic Information Systems for Wireless Network Planning, Francisco Saez de Adana, Josefa Gómez Pérez, Abdelhamid Tayebi Tayebi, and Juan Casado Ballesteros

The Art and Science of Ultrawideband Antennas, Second Edition, Hans G. Schantz

AWAS for Windows Version 2.0: Analysis of Wire Antennas and Scatterers, Antonije R. Djordjević, et al.

Broadband Microstrip Antennas, Girsh Kumar and K. P. Ray

Broadband Patch Antennas, Jean-François Zürcher and Fred E. Gardiol

Circularly Polarized Dielectric Resonator Antennas, Raghvendra Kumar Chaudhary, Rajkishor Kumar, and Rakesh Chowdhury

Design and Applications of Active Integrated Antennas, Mohammad S. Sharawi and Oualid Hammi

Design and Measurement of Antennas and Propagation in Mobile Cellular Systems, Hiroyuki Arai

Designer Notes for Microwave Antennas, Richard C. Johnson

Dielectric Resonator Antenna Handbook, Aldo Petosa

Dielectric Resonator Antennas, Biswajeet Mukherjee and Monika Chauhan

Electromagnetics and Antenna Technology, Alan J. Fenn

Electromagnetics, Microwave Circuit and Antenna Design for Communications Engineering, Peter Russer

Engineering Applications of the Modulated Scatterer Technique, Jean-Charles Bolomey and Fred E. Gardiol

Fixed & Mobile Terminal Antennas, Akhileswar Kumar

Foundations of Antenna Engineering: A Unified Approach for Line-of-Sight and Multipath, Per-Simon Kildal

Four-Arm Spiral Antennas, Joseph A. Mosko

Frequency-Agile Antennas for Wireless Communications, Aldo Petosa

Fresnel Zones in Wireless Links, Zone Plate Lenses and Antennas, Hristo D. Hristov

Handbook of Antennas for EMC, Second Edition, Thereza Macnamara with John McAuley

Handbook of Reflector Antennas and Feed Systems, Volume I: Theory and Design of Reflectors, Satish Sharma, Sudhakar Rao, and Lotfollah Shafai, editors

Handbook of Reflector Antennas and Feed Systems, Volume II: Feed Systems, Lotfollah Shafai, Satish Sharma, and Sudhakar Rao, editors

Handbook of Reflector Antennas and Feed Systems, Volume III: Applications of Reflectors, Sudhakar Rao, Lotfollah Shafai, and Satish Sharma, editors

Interference Suppression Techniques for Microwave Antennas and Transmitters, Ernest Freeman

Introduction to Active Phased Arrays, Thomas Sikina

Introduction to Antenna Analysis Using EM Simulators, Hiroaki Kogure, Yoshie Kogure, and James C. Rautio

Introduction to Antennas and RF Propagation Analysis, Dean James Friesen

LONRS: Low-Noise Receiving Systems Performance and Analysis Toolkit, Charles T. Stelzried, Macgregor S. Reid, and Arthur J. Freiley

Mechanical and Manufacturing Design of Antennas, Richard L. Burge

Measurement of Mobile Antenna Systems, Second Edition, Hiroyuki Arai

Microstrip Lines and Slotlines, Third Edition, Ramesh Garg, Inder J. Bahl, and Maurizio Bozzi

Microstrip Antenna Design, K.C Gupta and Abdelaziz Benalla

Microwave and Millimeter-Wave Remote Sensing for Security Applications, Jeffrey A. Nanzer

Microwave Cavity Antennas, Akhileswar Kumar and Hristo D. Hristov

Millimeter Wave Microstrip and Printed Circuit Antennas, Prakash Bhartia

Mode-Inspired Wideband and Multiband Planar Antennas, Neng-Wu Liu, Xiao Zhang, and Lei Zhu

Mobile Antenna Systems Handbook, Third Edition, Kyohei Fujimoto, editor

Modern Automotive Antenna Measurements, Lars J. Foged and Manuel Sierra Castañer, editors

Moment Methods in Antennas and Scattering, Robert C. Hansen

Multiband Integrated Antennas for 4G Terminals, David A. Sánchez-Hernández, editor

Near-Field Antenna Measurements, Dan Slater

Noise Temperature Theory and Applications for Deep Space Communications Antenna Systems, Tom Y. Otoshi

Over-the-Air Measurement for Wireless Communication Systems, Yihong Qi and James L. Drewniak

Phased Array Antenna Handbook, Third Edition, Robert J. Mailloux

Phased Array Antennas, Arthur A. Oliner

Phased Array Antennas with Optimized Element Patterns, Sergei P. Skobelev

Plasma Antennas, Second Edition, Theodore Anderson

Polarization in Electromagnetic Systems, Second Edition, Warren Stutzman

Practical Antenna Design for Wireless Products, Henry Lau

Practical Microstrip and Printed Antenna Design, Anil Pandey

Practical Phased Array Antenna Systems, Eli Brookner

Practical Simulation of Radar Antennas and Radomes, Herbert L. Hirsch

Printed MIMO Antenna Engineering, Mohammad S. Sharawi

Radio Propagation in the Urban Scenario, Giorgio Franceschetti, Antonio Iodice, and Daniele Riccio

Radiowave Propagation and Antennas for Personal Communications, Third Edition, Kazimierz Siwiak

Reconfigurable Antenna Design and Analysis, Mohammod Ali

Reflectarray Antennas: Analysis, Design, Fabrication and Measurement, Jafar Shaker, Mohammad Reza Chaharmir, and Jonathan Ethier

Reflector Lens Antennas: Analysis and Design Using Personal Computers, Carlyle J. Sletten, editor

RF Coaxial Slot Radiators: Modeling, Measurements, and Applications, Kok Yeow You

Shipboard Antennas, Second Edition, Preston Law

Significant Phased Array Papers, Robert C. Hansen

Smart Antenna Engineering, Ahmed El Zooghby

Solid Dielectric Horn Antennas, Carlos Salema, Carlos Fernandes, and Rama Kant Jha

Sparse Phased Array Antennas: Theory and Applications, Ashutosh Kedar

Ultrawideband Antennas for Microwave Imaging Systems, Tayeb A. Denidni and Gijo Augustin

Ultrawideband Short-Pulse Radio Systems, V. I. Koshelev, Yu. I. Buyanov, and V. P. Belichenko

Waveguide Components for Antenna Feed Systems: Theory and CAD, Jaroslaw Uher, Jens Bornemann, and Uwe Rosenberg

Wearable Antennas and Electronics, Asimina Kiourti and John L. Volakis, editors

For further information on these and other Artech House titles, including previously considered out-of-print books now available through our In-Print-Forever® (IPF®) program, contact:

Artech House
685 Canton Street
Norwood, MA 02062
Phone: 781-769-9750
Fax: 781-769-6334
e-mail: artech@artechhouse.com

Artech House
16 Sussex Street
London SW1V HRW UK
Phone: +44 (0)20 7596-8750
Fax: +44 (0)20 7630 0166
e-mail: artech-uk@artechhouse.com

Find us on the World Wide Web at: www.artechhouse.com